WETLANDS

WETLANDS

Editor

Gernot Paulis

scitus
academics

Wetlands

Edited by **Gernot Paulis**

This Book is Printed on Acid-Free Paper. ♾

ISBN: 9781681179759

Library of Congress Control Number: 2017949182

SCITUS Academics LLC,

616, Corporate Way, Suite 2, 4766,

Valley Cottage, NY 10989

www.scitusacademics.com

Notice

Printed in United States of America.

Preface

There is an environmental concern of ecological protection and conservation that should be incorporated with the water resources management policies. Wetlands are vital for human survival. They are among the world's most productive environments; cradles of biological diversity that provide the water and productivity upon which countless species of plants and animals depend for survival. Naturally occurring wetlands form when rain and runoff from watersheds saturate the ground or there is a continual flow of groundwater to the surface. Flood waters and coastal waters that immerse low-lying land near waterways can also cause a natural wetland to form. The first experiments using wetland macrophytes for wastewater treatment were carried out in Germany in the early 1950s. Since then, the constructed wetlands have evolved into a reliable wastewater treatment technology for various types of wastewater. The classification of constructed wetlands is based on: the vegetation type (emergent, submerged, floating leaved, free-floating); hydrology (free water surface and subsurface flow); and subsurface flow wetlands can be further classified according to the flow direction (vertical or horizontal). In order to achieve better treatment performance, namely for nitrogen, various types of constructed wetlands could be combined into hybrid systems. Wetlands play an integral role in the ecology of the watershed. The combination of shallow water, high levels of nutrients, and primary productivity is ideal for the development of organisms that form the base of the food web and feed many species of fish, amphibians, shellfish, and insects. Considering some of the wetlands to be as protectorates should help in the water resources conservation and enhance the national water management policy.

This book 'Wetlands' covers research studies on the issues, methods and impacts of wetlands management. It covers on topics relevant to freshwater, brackish and marine coastal wetland ecosystems. It serves multi- and interdisciplinary discussion for key issues in wetlands science, management, policy and economics.

Table of Contents

Chapter 1 Study on Different Substrates in Stable Surface Flow Wetland 1

Chapter 2 The Significant Surface-Water Connectivity of "Geographically Isolated Wetlands" 11

Chapter 3 Icelandic Inland Wetlands: Characteristics and Extent of Draining 23

Chapter 4 Multilayer Substrate Configuration Enhances Removal Efficiency of Pollutants in Constructed Wetlands 47

Chapter 5 Sediment Accumulation in Prairie Wetlands under a Changing Climate: The Relative Roles of Landscape and Precipitation 69

Chapter 6 Establishment and Application of Wetlands Ecosystem Services and Sustainable Ecological Evaluation Indicators 93

Chapter 7 Wetland Evapotranspiration: Eddy Covariance Measurement in the Biebrza Valley, Poland 111

Chapter 8 A Novel Protection and Utilization Pattern for Urban Wetland—As an

Example of Hemu Wetland in Hangzhou, China 137

Chapter 9 Applying Process-Based Models for Subsurface Flow Treatment Wetlands: Recent Developments and Challenges 155

Chapter 10 Main Driving Factors for Seacoast Vegetation in the Southern and Eastern Baltic 181

Chapter 11 Early Snowmelt Enhances the Carbon Sequestration of Hummock-Forming Sphagnum Mosses on Boreal Wetlands 203

Chapter 12 Design and Hydrologic Performance of a Tile Drainage Treatment Wetland in Minnesota, USA 219

Chapter 13 Public Support for Wetland Restoration: What is the Link With Ecosystem Service Values? 247

Chapter 14 Restoration of Ailing Wetlands 277

Chapter 15 Structural and Functional Loss in Restored Wetland Ecosystems 283

Chapter 16 Relationships Between Salinity and Short-Term Soil Carbon Accumulation Rates from Marsh Types Across a Landscape in the Mississippi River Delta 305

Chapter 17 Carbon and Hydrogen Isotopes as Tracers of Methane Dynamic in Wetlands 329

Chapter 18 **Nitrogen Removal in a Horizontal Subsurface Flow Constructed Wetland Estimated Using the First-Order Kinetic Model** **343**

Chapter 19 **Large Constructed Wetlands for Phosphorus Control: A Review** **363**

Index **417**

CHAPTER 1

Study on Different Substrates in Stable Surface Flow Wetland

Chengduan Wang[1] and Jun Zhang[2*]

[1]*Sichuan University of Science and Engineering, Zi gong, Sichuan, 643000, China*

[2]*School of Environment and Resources Southwest University of Science and Technology, Mian yang, Sichuan 621010, China*

ABSTRACT

This paper discussed the feasibility and advantage of the removal efficiency of low concentration sewage using three Stable Surface Flow Wetlands (SSFW) with bamboo splint and palm silk as substrates. Under the same hydraulic loading, the result of experiment indicated that the removal rate of COD,TP, TN were 73.97%, 61.42%, 28.98% respectively in SSFW with a bamboo splint substrate, and the corresponding removal rate of SSFW with a palm silk substrate were 78.37%, 69.42%, 24.4%; while substrate without a bamboo splint and a palm silk were 66.61%, 58.71%, 22.23%. The above information demonstrates that the removal efficiency of TN with a bamboo splint is better while the removal efficiency of COD and TP with palm silk is better. Bamboo splint and palm silk have their own advantages when they are applied to SSFW as substrate, which provide theoretical basis for the selection and matching of substrate.

Keywords: Stable surface flow wetland (SSFW); Bamboo splint; Palm silk; Substrate; Constructed wetland

1. INTRODUCTION

The Stable Surface Flow Wetland (SSFW), which provides the potential of benefits such as the stable operation, the small system clogging, and the superior quality effluents [1], is a constructed wetland technology on the basis of stabilization ponds and constructed wetlands. SSFW extends the flow path and increases the hydraulic retention time.

Pollutants can fully contact with substrates, and plants also fully contact with microbes, which make the physical, chemical and biological reactions more fully in the wetland. Meanwhile by adding the baffle to make the direction of flow in wetland bed run up and down alternately, the fluid state was more uniform, and the dead area was reduced to appear [2].

Substrate can not only provide carriers for the growth of plants and microbes, but it also removes pollutants directly by its sedimentation, filtration, and adsorption and so on. A large number of researches from home and abroad explored the purification of the constructed wetland by choosing of substrates and plant species and the planting conditions such as the size of the wetland bed, the breadth proportion, water operating conditions and temperature and so on [3-5].

This paper studied the removal effect of the SSFW to pollutants with different substrates with the purpose of finding a scientific and efficient substrate. An appropriate constructed wetland substrate plays a vital role in optimizing the sewage treatment technology of the constructed wetland.

At present there are only a few researches using the bamboo splint and palm silk as the constructed wetland substrates. As an excellent biological carrier, the surface of the bamboo splint contains mass of hydroxyl, making it hangs the membrane rapidly under the conditions of oligotrophic, low temperature and so on [6].

While the palm silk has a large area of surface, contains no sugar but more than 70% of cellulose, and are not liable to rotten due to insect bites. And they have strong water resistance, great intensity, and will not become fragile. As natural substrates, bamboo splint and palm silk have many good potential features as potential wetland substrates because of the advantages such as being produced simply, no toxic effects on microorganism, low cost, planting in a wide range and easy to access in the south.

2. TECHNOLOGICAL DESIGNS

2.1 The raw wastewater source and characteristics

The experimental sewage was obtained from the effluent coming from the fine grid in the Sewage Treatment Plant, which process 10,000 tons sewage a day in Southwest University of Science and technology. The scope of the sewage quality indicators was shown in **TABLE 1**.

Table 1: Inflow water quality of constructed wetlands

Water quality indicators	COD(mg/L)	TN(mg/L)	TP(mg/L)	pH
Range	300-1500	30-70	4-10	7-8

2.2 Technological process

Raw water can convert the large molecules of organic compounds into small ones when enter into the anaerobic tank, which can enhance the biodegradability of sewage. The water in the anaerobic tank flows into the reservoir automatically through the overflowing, and the dissolved oxygen in water can increase over 1mg/L by water-dropping aeration. Then it came into SSFW and purified the water by the substrate adsorption and filtration, plant absorption and microorganism's role of the degeneration and so on. The experiment uses three parallel SSFW treating devices and numbers are A, B and C separately.

A SSFW with the bamboo splint substrate was placed into the pond A, and a SSFW with the palm silk substrate was placed into the pond B; the pond C contains a SSFW without the bamboo splint substrate or palm silk substrate (**FIGURE 1**).

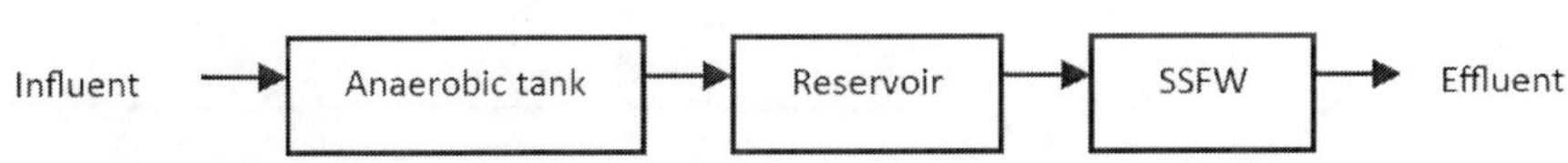

Figure 1: Flow chart of domestic sewage treatment process.

2.3 The main building and design parameters

The anaerobic tank was made of semi-buried reinforces concrete with dimensions of 1.0m × 1.0m × 1.8m (length × width × height) , the bottom elevation of -0.5m, the top elevation of 1.3m, the extra high of 0.3m. Because of Flowing into the reservoir through the overflow, anaerobic tank and the reservoir is designed to share the same wall height 1m. The above was covered with the glass plate, HRT=2d. The sludge in the anaerobic tank should be cleaned by hand regularly. The reservoir was made of reinforced concrete with dimensions of 0.5m × 1.0m ×1.3m (length × width × height), the extra high of 0.3m. SSFW was made of reinforces concrete with dimensions of 2.0m × 0.5m ×1.0m (length × width × height), the bottom slope of 1% and the extra high of 0.3m. In SSFW there are two glasses fiber reinforces plastic plates. It is about 0.7m and divides the SSFW into three functional zones. The lengths of them are 0.6m, 0.8m and 0.6m separately. The total thickness of substrate laid in the first zone is 25cm. From bottom to top are 15cm gravel and 10cm sand. Water hyacinth was planted in the first zone at a density of 25 plants·m^{-2}.The substrate in the second zone is the same as the third one. The total thickness is 40cm. The substrates from top to bottom are as follows: pond A was filled with gravel for 15cm, cinder 5cm, ceramist 5cm, sand 10cm and bamboo splint 10cm; pond B was filled with gravel for 15cm, cinder 5cm, ceramist 5cm, sand 5cm and palm silk 10cm; pond C was filled with gravel for 15cm, cinder 5cm, ceramist 5cm and sand 5cm. The *Cyprus alternifolius and Rumer pationtia* L were planted in the second zone and third zone respectively at a density of 15 plant·m^{-2}. And some uniform holes should be dig on the pipe of influent and effluent in order to make the distribution and catchments uniform (**FIGURE 2**).

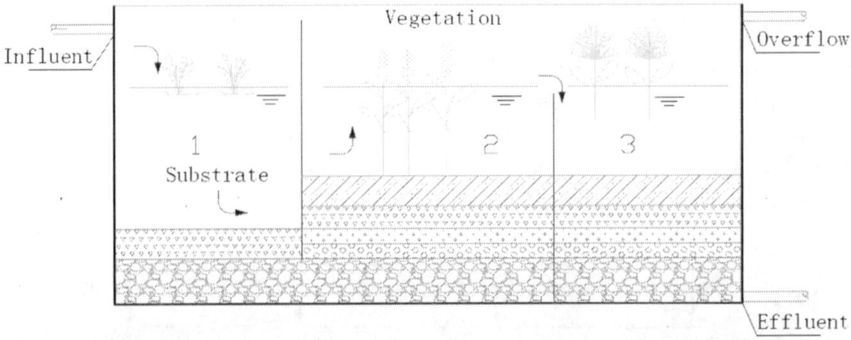

1-the first zone; 2-the second zone; 3-the third zone

Figure 2: Schematic diagram of SSFW installation.

2.3 Experimental operation

The SSFW system began to operate after the substrate was laid and vegetation was planted in late March 2011. System used intermittent water and sludge of anaerobic in the wastewater treatment plant was added to the anaerobic tank to enhance the system's resilience. After a week of acclimation, SSFW substrate table attached biofilm and the system enter the normal operation phase. The system was controlled by the PLC controller. Wastewater was fed from 9:00 to 18:30 every day. The total inflow is $0.65m^3$ every day and the Hydraulic Loading Rate (HLR) were $0.4m^3/(m^2 \cdot d)$. Three parallel processing units were sampled every four days. Then the water quality of influent and effluent should be monitored at the same time so that we can compare the treatment effect with each other. The temperature range was 3~17°C during the experiments.

2.4 Analytical methods and data analysis

Samples including Chemical Oxygen Demand (COD), Total Phosphorus (TP) and Total Nitrogen (TN) were analyzed in accordance with Standard Methods of State Environmental Protection of China [8] The TN was determined by alkaline potassium per sulfate digestion- UV spectrophotometer method. The COD was measured by potassium dichromate oxidation. The TP was measured by ammonium molybdate spectrophotometer method. Temperature and the pH were measured with a HI223 pH-meter (Hanna Co. Italy). The data were entered into a Microsoft Excel spreadsheet. Figures were analyzed by OriginPro8.

3. RESULTS AND DISCUSSION

The experiment joins the blank contrast system. **FIGURE 3-5** show that SSFWs have a good elimination effect to the pollutant when the bamboo splint and the palm silk substrate are in the experimental operation's process.

3.1 Cod removal efficiency and analysis

The average influent concentration was 477.88 mg/L by analyzing the COD removal rate of the A, B and C pond in **FIGURE 3**. The quality of B pool with the palm silk substrate achieved remarkable results for COD removal in the sewage. The average COD effluent concentration was 106.4 mg/L. While the average effluent COD concentration of A pool with bamboo silk and blank C pool were 125.46 mg/L and 161.82 mg/L respectively. The effects were not better than the above. The corresponding removal rate was 78.37%, 73.97% and 66.61%. Palm silk had some advantages for the treatment of low concentration of COD in the effluent because it contains 70% of cellulose and cellulose which can degrade in the course of soluble carbon sugar that provides carbon source for bacterial metabolism. The early experiment had good removal efficiency on COD, but after a period of operation, organic matter concentrated in the substrate surface, leading to the treatment effect in the latter part of the experiment had some fluctuations in COD. Sewage had been pre-treatment in anaerobic tank before it entered into SSFW, making the majority of organic molecules into small molecules and enhancing the biodegradability of sewage, which provides a guarantee for the follow-up COD removal.

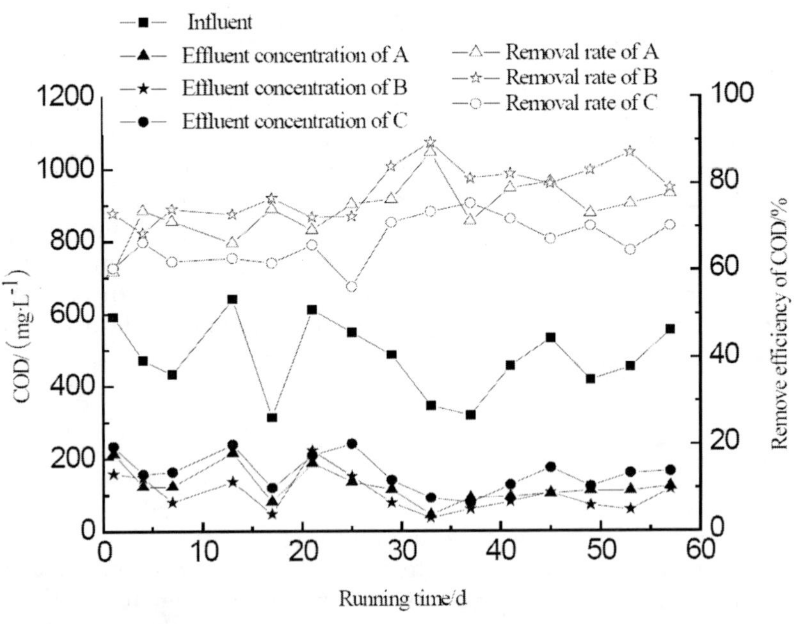

Figure 3: COD removal efficiency in A pond, B pond and C pond.

3.2 Tp removal efficiency and analysis

It can be seen from **FIGURE 4** that the TP removal rate of A, B and C pools are gradually rising. In the early of experiment, the rising curve is steep. It's because the gradually establishment of biofilm in SSFW system and the enhancement of microbial degradation. In the latter of experiment, gentle curves and eventually tend to equilibrium. Researchers have shown that 70% ~ 87% of the phosphorus in constructed wetland are removed by adsorption and precipitation of substrate [8]. Substrate composition and its physical and chemical properties have a significant effect on the purification of phosphorus. The bamboo splint and the palm silk used in the experiment are natural materials. **FIGURE 4** shows that pool B of SSFW with palm silk substrate has higher TP adsorption capacity than that of pool A of SSFW with bamboo splint substrate and C pool of SSFW with blank. The average effluent of TP concentration were 2.12mg/L, 2.7mg/L and 2.89mg/L, and the corresponding average removal rate of 69.42%, 61.42% and 58.71% separately. The palm silk substrate has obvious advantages for the adsorption of TP compared to the bamboo splint substrate. The palm silk substrate is more slender than the bamboo splint substrate. And on the same volume conditions, the palm silk have much larger surface area than the bamboo splint, making the palm silk on the sewage with a good adsorption performance. That is one of the important reasons why the palm silk performance better than the bamboo splint and the blank on the removal of phosphorus. Alkaline environment is more suitable for the growth of phosphorus than that of the acidic environment. By monitoring three SSFW systems, the pH was about 7.1~7.8, which is also conducive to the removal of phosphorus.

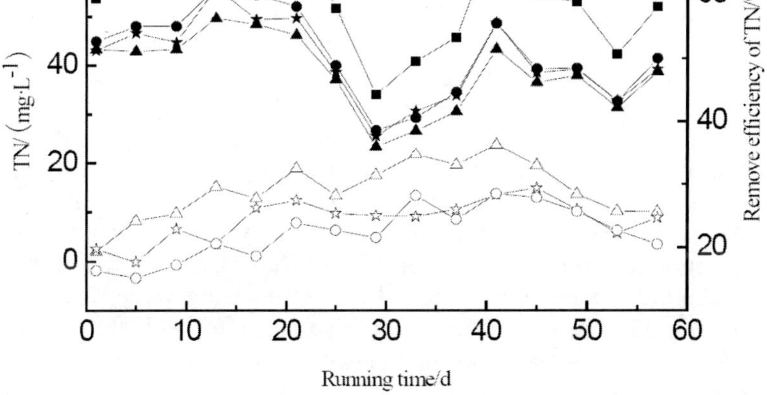

Figure 4: TP removal efficiency in A pond, B pond and C pond.

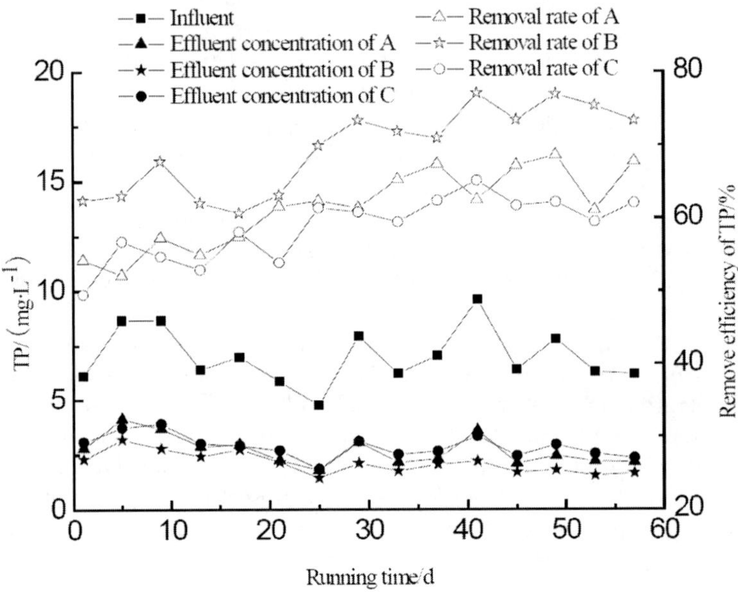

Figure 5: TN removal efficiency in A pond, B pond and C pond.

3.3 Tn removal efficiency and analysis

It can be seen from **FIGURE 5** that the pool A, B and C have some superiority on TN compared to the common constructed wetland. The way in which the sewage enters the SSFWs is uniform distribution and the water inlet is higher than the water surface, which has the function of water-dropping aeration and enhanced the TN elimination effect of SSFWs. People have studied that the ammonia volatility of purification way can be neglected when pH value of the sewage in the constructed wetland is less than 8 [9]. In this experiment, the pH value of SSFWs interior environment was between 7~8. Most of the other purification ways of TN were the nitration and the denitrification except the substrate absorption and the plant absorption of TN. The nitrification carried on under the good oxygen condition. And it runs smoothly because of the oxygenizement of the plant root system and atmospheric reaeration of free surface in the SSFWs. The bamboo splint substrate and the palm silk substrate were the disorderly stack in the SSFWs. The different crevice of them can intercept and fix suspension in the sewage so as to form a large number of suspended substances. These can fix nitrifying bacteria and enhance the nitration effect. The denitrification was one of the efficient ways to discharge the nitrogen. The concentrations of TN in the effluent were relatively low because the denitrification carried on under the anaerobic condition and the anaerobic environment space is relatively big and the denitrification is quite complete at the depth 0.7m of SSFW. This also shows that the denitrification mainly occurs at bottom of the SSFW.

FIGURE 5 also shows that the average effluent concentration of TN in A, B and C pools was 38.6mg/L, 41.07mg/L and 42.33mg/L respectively, and the corresponding average removal rate was 28.98%, 24.4% and 22.23%. The removal rate of TN in pool A was higher than that of B and C. The bamboo splint substrate surface area is rougher than the palm silk substrate. And it had developed porosity internal and better ventilation. It not only made oxygen in the atmosphere enter into the SSFWs, but it also promoted the formation of anaerobic environment to remove nitrogen for denitrification because of the high moisture content. The substrate with bigger porosity can provide microbial with more surface area to grow for the SSFW system, so microbial biomass was larger and community structure was more stable. Thus purification would be more efficiency and stable. In the later experiment, TN removal rate declined and the concentration of the effluent TN was high. There are two reasons for it. On one hand, denitrification was not fully and the productions accumulated continuously for the strong nitrification; On the other hand, when it came to mid-May, the *Rumer pationtia* L plants began to wither and leaves in the sewage released a certain amount of nitrogen.

4. CONCLUSIONS

(1) Both the bamboo splint and palm silk substrates are natural materials. They are widely planted and easily obtained in the Sichuan province. They are cheap and do no harm to the environment. They can easily make up for the secondary pollution caused by the other substrates shortages as SSFW substrates.

(2) The advantages of bamboo splint substrate are porosity larger and higher moisture content. While the palm silk have a large area of surface, contains no sugar but more than 70% of cellulose, and are not liable to insect. And they have strong water resistance, great intensity, and will not become fragile.

(3) Bamboo splint and palm silk have their own advantages when they are applied to the treatment of low strength domestic sewage in SSFW as substrates. The removal efficiency of TN with a bamboo splint is better while the removal efficiency of COD and TP with palm silk is better.

ACKNOWLEDGEMENTS

This paper was funded by the key Science Project (2011G009) of Zigong, Sichuan Province of PRC. The authors would like to thank the Project.

REFERENCES

1. Chengduan Wang (2008) Low-cost wastewater treatment course [M]. Beijing: Chemical Industry Press.
2. Jing-xi Tie, Zhi-hong Zheng, Zheng Zheng, Luo Xingzhang (2009) Study on the characteristics of three sub-surface flow constructed wetlands with different internal structures. Industrial Water Treatment 29: 51-53.
3. Gearheat RA (1992) Use of constructed wetland to treat domestic wastewater, city of Arcat California. Water Science & Technology 26: 1625-1637.
4. Zhang HC (2004)Study progress in constructed wetland ecosystems for sewage purification. Techniques and Equipment for Environmental Pollution Control 5: 11-14.
5. Kong LY, Ni JR (2007) Characteristics of the generalized structure of pollutant removal models for constructed wetlands. Acta Ecologica Sinica 27: 1428-1433.
6. Liu Ying, Lingling Wang (2003-2007) A way of carrier substrate and preparation [P].
7. State Environmental Protection of China (2002) Testing and Analysis Method for Water and Wastewater (4th Edition) [M]. Beijing: China Environmental Science Press : 252-354.
8. Reddy KR, Connor GA, Gale PM (1998) Phosphorus sorption capacities of wetland soils and stream sediments impacted by dairy effluent. Journal of Environmental Quality 27: 438-447.
9. Koottatep T (1997) Role of plant uptake on nitrogen removal in constructed wetlands located in the tropics. Water Sci Technol 36: 1-8.

CHAPTER 2

The Significant Surface-Water Connectivity of "Geographically Isolated Wetlands"

Aram J. K. Calhoun[1], David M. Mushet[2], Laurie C. Alexander[3], Edward S. DeKeyser[4], Laurie Fowler[5], Charles R. Lane[6], Megan W. Lang[7], Mark C. Rains[8], Stephen C. Richter[9], Susan C. Walls[10]

[1]*Department of Wildlife, Fisheries, and Conservation Biology, University of Maine, Orono, ME 04469, USA*

[2]*Northern Prairie Wildlife Research Center, US Geological Survey, 8711 37th Street SE, Jamestown, ND 58401, USA*

[3]*Office of Research and Development, US Environmental Protection Agency, 1200 Pennsylvania Ave NW (8623-P), Washington, DC 20460, USA*

[4]*School of Natural Resource Sciences, North Dakota State University, Fargo, ND 58108, USA*

[5]*School of Law, University of Georgia, Athens, GA 30602, USA*

[6]*Office of Research and Development, US Environmental Protection Agency, Cincinnati, OH 45268, USA*

[7]*National Wetland Inventory, US Fish and Wildlife Service, Falls Church, VA 22041, USA*

[8]*School of Geosciences, University of South Florida, Tampa, FL 33620, USA*

[9]*Department of Biological Sciences and Division of Natural Areas, Eastern Kentucky University, Richmond, KY 40475, USA*

[10]*Wetland and Aquatic Research Center, US Geological Survey, 7920 NW 71st Street, Gainesville, FL 32653, USA*

ABSTRACT

We evaluated the current literature, coupled with our collective research expertise, on surface-water connectivity of wetlands considered to be "geographically isolated" (*sensu* Tiner Wetlands 23:494–516, 2003a) to critically assess the scientific foundation of grouping wetlands based on the singular condition of being surrounded by uplands. The most recent research on wetlands considered to be "geographically isolated" shows the difficulties in grouping an ecological resource that does not reliably indicate lack of surface water connectivity in order to meet legal, regulatory, or scientific needs. Additionally, the practice of identifying "geographically isolated wetlands" based on distance from a stream can result in gross overestimates of the number of wetlands lacking ecologically important surface-water connections. Our findings do not support use of the overly simplistic label of "geographically isolated wetlands". Wetlands surrounded by uplands vary in function and surface-water connections based on wetland landscape setting, context, climate, and geographic region and should be evaluated as such. We found that the "geographically isolated" grouping does not reflect our understanding of the hydrologic variability of these wetlands and hence does not benefit conservation of the Nation's diverse wetland resources. Therefore, we strongly discourage use of categorizations that provide overly simplistic views of surface-water connectivity of wetlands fully embedded in upland landscapes.

Keywords: Clean Water Act, Connectivity, Geographic isolation, Hydrology, Streams, Upland, embedded, wetlands, Waters of the U.S.

1. INTRODUCTION

Throughout the world, small wetlands with seasonal hydrology are at great risk of loss or degradation and effective approaches to conserving their functions lag behind the increase in threats (Calhoun et al. 2016). For this reason, researchers and managers need to improve the understanding of vulnerable wetland functions and this includes both continuing research and clarifying regulations that do exist while considering alternative approaches. In this paper, we give one example of addressing this issue that has relevance to wetland managers globally. In a recent issue of *WETLANDS*, we published an essay entitled "Geographically Isolated Wetlands: Rethinking a Misnomer" (Mushet et al. 2015). In our paper, we described the declining relevance and confusing nature of the "geographically isolated wetlands" (GIWs) categorization as currently used in wetland science and policy in the United States. Leibowitz (2015) published a thoughtful response to our review in which he defended the use of the categorization and argued that there are important scientific, legal, and regulatory needs for identifying wetlands that are completely surrounded by uplands (i.e., GIWs, sensu Tiner 2003a). We have found that scientific responses to the legacies of the last decade's court actions and policy needs for wetland regulation under the Clean Water Act (CWA; 33 U.S.C. Chapter 26) have improved our understanding of the complexity of wetland hydrology, functions, and nexuses that transcend simple assessments of degree of upland embeddedness. Grouping wetlands by whether or not they are surrounded by uplands does not indicate a lack of a "significant nexus," and therefore does not provide a useful separation for meeting legal and regulatory information needs (Cohen et al. 2016).

We provide a brief review of key scientific findings to instantiate our thesis that having a static category based on upland embeddedness is no longer beneficial and, in fact, may be detrimental to conservation of these wetland resources and their influence on downgradient systems. The GIW term, or any term that implies that wetlands surrounded by uplands are in fact functionally isolated, is difficult to justify scientifically, difficult to apply pragmatically, subject to misuse and misinterpretation, and maps poorly onto the regulatory landscape. In this paper, we use the term upland-embedded wetland to describe a geospatial setting with no assumptions about connectivity or lack thereof and with no intent to replace the GIW term with "upland-embedded wetland". We focus on surface-water connections, as the GIW categorization has not been promoted as providing meaningful insights into other forms of connectivity (e.g., groundwater, biogeochemical, biotic) that clearly transcend degrees of upland embeddedness. We define surface water connectivity as flow of surface water (episodic, seasonal, or semi-permanent) between two unique landscape elements that may or may not be linked by an aquatic feature with a bed and bank (i.e., channel or other indicators of flow permanence).

2. DYNAMIC SURFACE-WATER CONNECTIONS

Upland-embedded wetlands occur along continuous spatial and temporal gradients, from highly connected to highly disconnected (Cohen et al. 2016). Research on upland-embedded wetlands demonstrates that many have surface-water connections to other aquatic landscape components (e.g., rivers, streams, lakes, other wetlands; Vanderhoof et al. 2016). A brief synthesis of key findings in the literature follows.

A conceptual model for thinking about how upland-embedded wetlands function at broader ecosystem scales is provided by Rains et al. (2016). They describe upland embedded wetlands as nodes in hydrological networks and state that these wetlands are "...integrally connected to uplands, other wetlands, and downgradient waters." The authors further describe complex lag, sink, and source functions of these wetlands and their resultant influences on surface-water and shallow-groundwater flows to downgradient waters (also see Golden et al. 2016). Rains et al. (2016) describe a wide range of surface-water connectivity displayed by wetlands, with wetlands identified as GIWs ranging from "infrequent/absent surface connectivity" (i.e., isolated) to "intermittent surface connectivity" (i.e., clearly not isolated). Likewise, Leibowitz (2015) describes GIWs that range from a wetland connected to a river by surface flow through a non-channelized swale to a geographically isolated wetland that is hydrologically isolated from a river. The key feature of the continuous range of surface-water connectivity described by both Leibowitz (2015) and Rains et al. (2016) is magnitude and timing, not the degree to which a wetland is surrounded by upland. While the "isolated" term has been used to describe the surface connections of all upland-embedded wetlands, the term describes only a subset of GIWs.

Consider work in the Prairie Pothole Region (PPR) of the Midwestern USA. Wetlands in this region have long been iconic examples of "geographically isolated" wetlands (Tiner 2003a) yet current research has documented high levels of hydrologic, biologic, and biogeochemical connectivity (Marton et al. 2015; Mushet et al. 2015; Cohen et al. 2016; Leibowitz et al. 2016; McLean et al. 2016). For example, Leibowitz et al. (2016) described the complex spill-and-fill and spill-and-merge surface-water connectivity of eight prairie pothole wetlands over a 26-year period (1979–2015). Their findings suggest that research exploring the effects of surface-water connections needs to address the specific types of connections and not broader categories. Further, in a detailed analysis of wetland hydrology, Hayashi et al. (2016) demonstrated how Midwestern USA prairie pothole wetlands and their upland catchments function as integrated units whose existence depends on the lateral movement of both subsurface and surface runoff water. Furthermore, they found that differences in surface-water connectivity among individual wetlands controlled ponded-water permanence, leading to a diversity of wetland functional types within wetland complexes.

The importance of surface-water connections to many wetlands considered to be "geographically isolated" is also supported by research that has documented high levels of hydrologic, biologic, and biogeochemical connectivity among vernal pools in California, USA (Golden et al. 2016; Rains et al. 2016). Western vernal pools are small depressional wetlands commonly connected by swales to one another and downgradient waters. The climate of this landscape is Mediterranean with pronounced wet and dry seasons. In the dry season, the variable source area from which streamflow is derived contracts and vernal pools may present as upland or upland- embedded wetlands. However, in the wet season, these vernal pools and swales become part of the river network system. These surface-water connections are not speculative or insubstantial, with measured surface-water connections for as many as 150–200 days being reported (Rains et al. 2006; Rains et al. 2008).

The condition of being wholly embedded within an upland matrix does not reliably indicate lack of surface water connectivity to other aquatic ecosystems. In short, there is a continuum of connectivity that applies to an individual wetland, complexes of wetlands, and wetlands within an ecoregion.

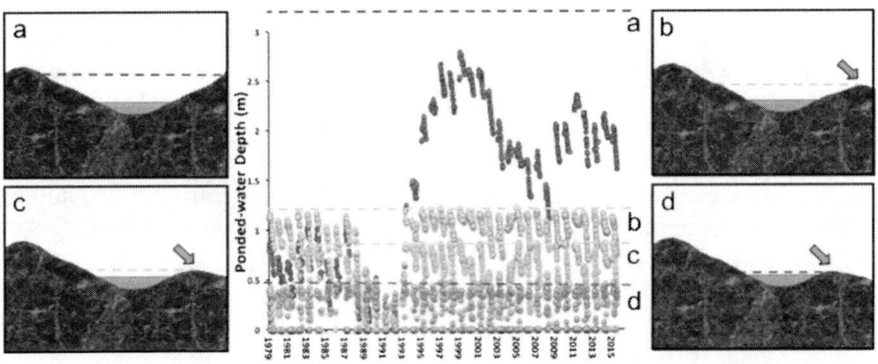

Figure . 1 Little knowledge about magnitude and timing of surface-water connectivity is gained by knowing that a wetland is surrounded by upland, i.e., is "geographically isolated." The above hydrograph displays water levels of four "geographically isolated" prairie-pothole wetlands (labeled a–d) at the Cottonwood Lake Study Area in Stutsman County, North Dakota, over a 36-year period (1979–2015). The drawings on the left and right of the hydrograph characterize the upland-embedded basins of the wetlands. External spill points (arrows), as defined by Leibowitz et al. (2016), set limits (color-coded dashed lines) to water storage and thus the magnitude of water losses from these wetland basins. Wetland P1 (**a**) is situated within a deep basin that does not have a realized external spill-point and thus does not contribute (i.e., spill) to down-gradient surface-water flows. By contrast, wetlands P8 (**b**), P3 (**c**), and T6 (**d**) each, to varying degrees, contribute to down-gradient flows when water levels reach an external spill point. The magnitude and timing of these surface-water flows vary greatly with similarly variable hydrological, geochemical and ecological effects

Furthermore, abiotic factors including soil type, precipitation patterns and geomorphology are often significant factors influencing degrees and nature of surface water connections, but these factors are not accounted for by the label "geographically isolated" (Fig. 1).

3. DISTANCE AS A SURROGATE FOR ISOLATION

A reexamination of the commonly used practice of identifying upland-embedded wetlands based on distance from a stream or large water body reveals that this methodology may result in a gross overestimate of the number of wetlands lacking significant surface flows to downstream waters (i.e., the condition that the GIW designation is assumed to identify). Vanderhoof et al. (2016) found notable variation among ecoregions in empirically measured distances at which wetlands connected via surface water to mapped streams, making it problematic to identify surface-water connected wetlands based on distance alone. For example, in the Des Moines Lobe ecoregion of the PPR, the authors found that 78% of surface-water connected wetlands were located within 400 m of a mapped stream. However, in the Drift Plains ecoregion of the PPR, only 52% of the connected wetlands were located within that same stream-buffer distance. Relative to these findings, most buffer distances previously used to identify upland-embedded wetlands (e.g., 76 m, Levin et al. 2002; 20–40 m, Tiner et al. 2002 and Tiner 2003b; 10 m, Frohn et al. 2009; 10 m, Reif et al. 2009; 20 or 40 m buffer for small streams and 300 m for large streams, Vance 2009; 10 m, Lane et al. 2012; 10 m, Lane and D'Amico 2016) are likely insufficient to judge surface water connectivity within some landscapes. As a result, numerous surface-water connected wetlands located beyond the threshold buffer distance are being grouped with wetlands lacking such connections. Not surprisingly, Lane and D'Amico (2016) found that increasing their 10-m buffer distance to 300 m resulted in a significant decrease in the number of putative GIWs in multiple ecoregions across the US. Further, Golden et al. (2016) found in their modeling assessment of the influence of GIWs on downgradient streamflow in the lower Neuse River Basin, North Carolina, USA, that the farther upland-embedded wetlands were located from downgradient streams, the greater their potential contributions to streamflow across long time scales (i.e., seasonally and annually). With the inclusion of all wetlands in the analyses, this effect disappeared. Therefore, many quantifications of upland-embedded wetlands likely have overestimated occurrence of non-connected wetlands since the multitude of connected wetlands outside the buffer distance are identified as isolated. Furthermore, Lang et al. (2012) found that commonly available stream vector datasets (e.g., the US Geological Survey National Hydrography Dataset [NHD]) used to quantify wetland–stream connections underestimate stream length, at least in relatively wet regions like the eastern US. This is partially explained by the fact that the NHD dataset was not designed to map ephemeral streams or streams < 1.6 km in length. Lang et al. (2012) concluded that these factors would lead many wetlands to be incorrectly considered to be disconnected from the stream network. This is counter to arguments that quantifications derived using buffers are conservative estimates of the numbers of upland-embedded wetlands (Leibowitz 2015).

Direct pre-identification of upland-embedded wetlands will continue to lessen the guesswork currently employed in establishing regulatory connectivity. For example, Wu and Lane (2016) developed a new approach to identifying wetland depressions in the PPR that accounts for dynamic filling, spilling and merging hydrological processes not considered in previous algorithms designed to identify such depressions (Leibowitz et al. 2016). Even low-tech methods involving using local knowledge and ground-truthing involving citizen-scientists can produce important information on current pools and past occurrences of connectivity. Levesque et al. (2016) describe a vernal pool conservation initiative in New England, USA, that recognizes the landscape-scale functions of vernal pools and encourages conservation of "poolscapes" in partnership with land trusts and other conservation groups who recognize the value of conserving ecosystem connections—work all driven by community based collaboration.

Proximity to mapped streams and other drainage features have been used as proxies for surface water connectivity (see above) because of the difficulty inherent in quantifying surface-water connectivity (Lane and D'Amico 2016). More advanced technologies and approaches provide promising solutions to better characterize connectivity. For example, other methods that could be examined include direct monitoring of inundation patterns using lidar intensity, multispectral and synthetic aperture radar data, predicting flow based on slope derived from lidar-based digital elevation models, and using process-based hydrologic models parameterized using geospatial data. Methodologies that move away from a categorical definition of geographically isolated wetlands and more closely approximate the adjacent versus non-adjacent definition will be better aligned with current legal/regulatory needs.

4. LEGAL/REGULATORY CONSIDERATIONS

In the years immediately following the U.S. Supreme Court's 2001 decision in *SWANCC* (*Solid Waste Agency of Northern Cook County* [*SWANCC*] vs. *US Army Corps of Engineers, 531 US 159*), there was a great deal of confusion regarding the concept of an "isolated" wetland. In scientific literature, this term was commonly used to describe various types of depressional wetlands (e.g., Damman and French 1987; Semlitsch and Bodie 1998; Bailey 1999). Following scientific usage, the Corps of Engineers promulgated a regulatory definition of "isolated wetland" for administration of their Nationwide Permit Program (NWP) 26 (33 CFR 330.2(e)). Prior to *SWANCC*, neither usage was relevant to Clean Water Act (CWA) jurisdiction (Downing et al. 2003). Following the *SWANCC* and later *Rapanos* (*Rapanos* vs. *United States, 547 U.S. 715,2006* decisions), existing definitions were muddied by case law that misinterpreted scientific and regulatory concepts of "isolation" and "adjacency" as end-members of waterbody functional connectivity. At its inception, the term "geographically isolated wetland" was meant to correct this misinterpretation and avoid further error (e.g., Tiner et al. 2002; Leibowitz 2003; Tiner 2003a). Unfortunately, the clarification presented in those seminal publications warning that geographic isolation should not be used to infer functional isolation did not communicate well to other communities of practice (e.g., evolution, ecology, genetics) in which "geographic isolation" has a very specific functional definition (Mushet et al. 2015). The science now shows that the degree of wetland surface-water connectivity cannot be assessed in a meaningful way by a simple determination of upland embeddedness (USEPA 2015; Rains et al. 2016; Cohen et al. 2016).

The recent Clean Water Rule (CWR, 80 FR 37054), which is currently stayed, does not use the GIW term, suggesting that federal agencies have moved beyond consideration of "geographic isolation" as a factor for determining CWA jurisdiction. Instead, the rule recognizes the best-available science by establishing five subcategories of wetlands (prairie potholes, Carolina and Delmarva bays, pocosins, western vernal pools in California, and Texas coastal prairie wetlands) that must be considered as "similarly situated" (that is, functioning as systems at the watershed scale) rather than as individual wetland basins, when determining their influence on navigable waters (CWR, 80 FR 37054). This consideration of the watershed-scale cumulative effects of wetlands and wetland complexes rather than individual basins is a large step forward from the localized, basin-scale assessments inherent in GIW categorization (Tiner 2003a; and Leibowitz 2015).

5. CONCLUSIONS

Recent research findings show that wetlands surrounded by uplands vary greatly in occurrence, type, as well as frequency, timing, and importance of surface-water connections to other aquatic systems (Rains et al. 2016; Cohen et al. 2016). Ambiguous generalizations about degrees of connectivity and isolation between upland-embedded wetlands and other wetlands and downstream waters are illogical (Mushet et al. 2015). The single condition of being surrounded by uplands currently used by wetland scientists to define "geographic isolation" does not provide a useful separation between wetlands that have a significant surface-water connection and those that do not. Upland embeddedness does not necessarily provide any indication that these wetlands are functionally "isolated".

Current research on connectivity of wetlands to downstream waters clearly shows that scientific needs are best met when gradients of surface-water connectivity are considered rather than through the use of a grouping defined by a threshold that does not reliably separate surface water connected/isolated wetlands, yet alone functionally connected/isolated wetlands. Embracing this knowledge requires a rethinking of our use of the "geographically isolated wetlands" misnomer and opens up advanced avenues to conserving wetland landscapes. Fully embracing the scientific knowledge gained since inception of the GIW grouping, knowledge that has identified the inherent connectedness of these "isolated" wetlands individually and as complexes, is needed to facilitate the long-term conservation of these important, and increasingly threatened, wetland resources. Conservation decisions cannot be made based on a broad category that, while created to help alleviate confusion over the term "isolated," has instead further muddied the waters. Recognizing the diverse functions supported by gradients of wetland connectivity will lead to better conservation of all wetland resources.

ACKNOWLEDGEMENTS

We thank Heather Golden and two anonymous reviewers for providing their critical reviews of an earlier draft of this manuscript. We also thank participants of a "Geographically Isolated Wetlands Research Workshop" convened and co-hosted by the U.S. Environmental Protection Agency Office of Research and Development and the Joseph W. Jones Ecological Research Center and held in Newton, GA, November 18–21, 2013, for contributing to the ideas and discussions within which this and our previous manuscript originated. We also thank Scott Leibowitz for providing the impetus needed to encourage our further explorations into the use and misuse of the GIW terminology. Any use of trade, firm, or product names is for descriptive purposes only and does not imply endorsement by the U.S. Government. USGS Cottonwood Lake Study Area water-level data used in Fig. 1 is available at: http://dx.doi.org/10.5066/F7707ZJ6. The findings and conclusions in this article are those of the authors and USGS and do not necessarily represent the views of the U.S. Fish and Wildlife Service.

This is Maine Agricultural and Forest Experiment Station Publication # 3517.

REFERENCES

1. Bailey MA (1999) Small isolated wetlands: vital to diversity. Alabama Wildlife Federation, Montgomery, AL. Alabama Wildlife Magazine Archives. (www.alawild.org/wetland.htm)
2. Calhoun AJK, Mushet DM, Bell KP, Boix D, Fizsimons JA, Isselin-Nondedeu F (2016) Temporary wetlands: challenges and solutions for protecting a "disappearing" ecosystem. Biological Conservation. doi:10.1016/j.biocon.2016.11.024
3. Cohen MJ, Creed IF, Alexander L, Basu NB, Calhoun AJK, Craft C, D'Amico E, DeKeyser E, Fowler L, Golden HE, Jawitz JW, Kalla P, Kirkman LK, Lane CR, Lang M, Leibowitz SG, Lewis DB, Marton J, McLaughlin DL, Mushet DM, Raanan-Kiperwas H, Rains MC, Smith L, Walls SC (2016) Do geographically isolated wetlands influence landscape functions? Proceedings of the National Academy of Sciences 113:1978–1986
4. Damman AWH, French TW (1987) The ecology of peat bogs of the glaciated northeastern United States: a community profile. U.S. Fish and Wildlife Service, Washington, DC, USA. Biological Report 85(7.16)
5. Downing DM, Winer C, Wodd LD (2003) Navigating through Clean Water Act jurisdiction: a legal review. Wetlands 23:475–493
6. Frohn RC, Reif M, Lane C, Autrey B (2009) Satellite remote sensing of isolated wetlands using object-oriented classification of Landsat-7 data. Wetlands 29:931–941
7. Golden HE, Sander HA, Lane CR, Zhao C, Price K, D'Amico E, Christensen JR (2016) Relative effects of geographically isolated wetlands on streamflow: a watershed-scale analysis. Ecohydrology 9:21–38
8. Hayashi M, van der Kamp G, Rosenberry DO (2016) Hydrology of prairie wetlands: understanding the integrated surface-water and groundwater processes. Wetlands. doi:10.1007/s13157-016-0797-9
9. Lane CR, D'Amico E (2016) Identification of putative geographically isolated wetlands of the conterminous United States. Journal of the American Water Research Association 52:701–722
10. Lane CR, D'Amico E, Autrey B (2012) Isolated wetlands of the southeastern United States: abundance and expected condition. Wetlands 32:753–767
11. Lang M, McDonough O, McCarty G, Oesterling R, Wilen B (2012) Enhanced detection of wetland-stream connectivity using LiDAR. Wetlands 32:461–473
12. Leibowitz SG (2003) Isolated wetlands and their functions: an ecological perspective. Wetlands 23:517–531
13. Leibowitz SG (2015) Geographically isolated wetlands: why we should keep the term. Wetlands 35:997–1003
14. Leibowitz SG, Mushet DM, Newton WE (2016) Intermittent surface water connectivity: fill and spill vs. fill and merge dynamics. Wetlands. doi:10.1007/s13157-016-0830-z
15. Levin GA, Suloway L, Plocher AE, Hutto FR, Miner JJ, Phillips CA, Agarwal J, Lin Y (2002) Status and function of isolated wetlands in Illinois. Illinois Natural History Survey, Champaign, Special Publication 23
16. Levesque V, Calhoun AJK, Bell KP (2016) Turning contention into collaboration: engaging power, trust, and learning in collaborative networks. Society and Natural Resources. doi:10.1080/08941920.2016.1180726

17. Marton JM, Creed IF, Lewis DB, Lane CR, Basu NB, Cohen MJ, Craft CB (2015) Geographically isolated wetlands are important biogeochemical reactors on the landscape. BioScience 65:408–418
18. McLean K, Mushet DM, Stockwell CA (2016) From "duck factory" to "fish factory": climate induced changes in vertebrate communities of prairie pothole wetlands and small lakes. Wetlands. doi:10.1007/s13157-016-0766-3
19. Mushet DM, Calhoun AJK, Alexander LC, Cohen MJ, DeKeyser ES, Fowler L, Lane CR, Lang MW, Rains MC, Walls SC (2015) Geographically isolated wetlands: rethinking a misnomer. Wetlands 35:423–431. doi:10.1007/s13157-015-0631-9
20. Rains MC, Dahlgren RA, Williamson RJ, Fogg GE, Harter T (2008) Geological control of physical and chemical hydrology in vernal pools, Central Valley, California. Wetlands 28:347–362
21. Rains MC, Leibowitz SG, Cohen MJ, Creed IF, Golden HE, Jawitz JW, Kalla P, Lane CR, Lang MW, McLaughlin DL (2016) Geographically isolated wetlands are part of the hydrological landscape. Hydrological Processes 30:153–160
22. Rains MC, Fogg GE, Harter T, Dahlgren RA, Williamson RJ (2006) The role of perched aquifers in hydrological connectivity and biogeochemical processes in vernal pool landscapes, Central Valley, California. Hydrological Processes 20:1157–1175
23. Reif M, Frohn RC, Lane CR, Autrey B (2009) Mapping isolated wetlands in a karst landscape: GIS and remote sensing methods. GIScience & Remote Sensing 46:187–211
24. Semlitsch RD, Bodie JR (1998) Are small, isolated wetlands expendable? Conservation Biology 12:1129–1133
25. Tiner RW (2003a) Geographically isolated wetlands of the United States. Wetlands 23:494–516
26. Tiner RW (2003b) Estimated extent of geographically isolated wetlands in selected areas of the United States. Wetlands 23:636–652
27. Tiner RW, Bergquist HC, DeAlessio GP, Starr MJ (2002) Geographically isolated wetlands: A preliminary assessment of their characteristics and status in selected areas of the United States. U.S. Department of the Interior, Fish and Wildlife Service, Northeast Region, Hadley
28. U.S. Environmental Protection Agency (USEPA) (2015) Connectivity of Streams & Wetlands to Downstream Waters: A Review & Synthesis of the Scientific Evidence. EPA/600/R-14/475F. U.S. Environmental Protection Agency, Washington, DC
29. Vance LK (2009) Geographically isolated wetlands and intermittent/ephemeral streams in Montana: extent, distribution, and function. Report to the Montana Department of Environmental Quality and the U.S. EPA. Montana Heritage Program, Helena
30. Vanderhoof MK, Christensen JR, Alexander LC (2016) Patterns and drivers for wetland consolidation in the Prairie Pothole Region, United States. Wetlands Ecology and Management. doi:10.1007/s11273-016-9516-9
31. Wu Q, Lane CR (2016) Delineation and quantification of wetland depressions in the Prairie Pothole Region of North Dakota. Wetlands 36:215–227

CHAPTER 3

Icelandic Inland Wetlands: Characteristics and Extent of Draining

Olafur Arnalds[1,2], Jon Gudmundsson[1], Hlynur Oskarsson[1], Sigmundur H. Brink[1], Fanney O. Gisladottir[1]

1 Agricultural University of Iceland, Árleyni 22,112 Keldnaholt, Iceland

2 United Nations University Land Restoration Training Program,Árleyni 22, 112 Keldnaholt, Iceland

ABSTRACT

Iceland has inland wetland areas with soils exhibiting both Andosol and Histosol properties which are uncommon elsewhere on Earth. They are generally fertile, with higher bird-nest densities than in similar wetlands in the neighboring countries, with nutrients released by rapid weathering of aeolian materials of basaltic nature. Icelandic inland wetlands cover about 9000 km^2 constituting 19.4 % of the vegetated surfaces of the island. The wetland soils are often 1–3 m thick and store 33 to >100 kg C m^{-2}. They have been subjected to broad-scale subsidy-driven draining for agricultural purposes. About 47 % of Icelandic inland wetlands are impacted by drainage. The ditch network extends about 30,000 km, mainly in lowland areas, where about 70 % of the wetland areas are impacted. There are >1 million wetland patches, most of them <1 ha. Much of the wetlands impacted from drainage are not used for intensive agriculture such as hay-making, however some are used for grazing. There is a need to prioritize the protection of undrained wetlands and their restoration based on a broad range of factors.

Keywords: Wetland draining, Andosols, Wetland patch, size, Gleyic

1. INTRODUCTION

High latitude wetlands with an elevated organic content make up a large proportion of Earth's wetlands (e.g., Mitsch and Gosselink 2007). Iceland, an island in the North-Atlantic Ocean, has diverse wetlands ranging from the coast to the interior highland, including marshes, estuaries, ponds, lakes and rivers. The wetlands addressed in this paper are the inland palustrine wetlands, mostly fens with both organic or inorganic soils (see definitions in Cowardin et al. 1979), which we here also term 'inland fen wetlands' or simply 'inland wetlands'. They were termed "mires" ("mýrar" in Icelandic) in an overview by Gardarsson (1998). Sloping fens are a common feature of the fiords and valleys formed during the last glaciation which cut into the Tertiary basalts whereas the greatest extent of topogenous fens are found within the gently sloping lowlands of west and south Iceland. Alluvial fens are located along some of the larger rivers, particularly in north and east Iceland. Palsa wetlands are found in isolated areas in the central highlands. Intertidal flats and inland marshes are not covered in this paper (559 $km^2$and 387 km^2, respectively, numbers based on the CORINE cover for Iceland).

The Icelandic wetlands are strongly influenced by the volcanic nature of the island and by intense aeolian deposition ("volcanic loess"), which separate Icelandic wetlands from most other wetlands on the globe. The volcanic and aeolian (loess) influences make these inland wetlands fertile with a relatively high pH and favorable nutrient availability (Arnalds 2004, 2015 Ch 8; Oskarsson et al. 2004). The fertility is reflected by high breeding bird densities, which often are >10 times higher in Iceland than for the same species in the UK and >2 times that of The Netherlands (Johannesdottir et al. 2014; Gunnarsson et al. 2015). The inland fen wetlands are important habitat for many bird populations in Iceland (Gunnarsson et al. 2006) with some of them important internationally (Einarsson et al. 2002). There have been extensive studies on changes in species composition of plants following drainage impacts (Magnusson 1998). Thorhallsdottir et al. (1998) found that the occurrence of rare wetland plant species was higher in areas with limited drainage disturbance. The extensive draining of Icelandic wetlands (addressed below) has had significant influence on both the distribution and overall abundance of Icelandic birds (Thorleifsson 1998; Skarphedinsson and Thorleifsson 1998). The wetland draining, along with the introduction of the American mink (*Mustela vison*), are considered to be the main reasons for the local extinction of the water rail (*Rallus aquaticus*), Iceland's only rail species (Skarphedinsson and Thorleifsson 1998).

Icelandic inland fen wetlands were subjected to widespread drainage during the twentieth century for agricultural purposes, with a large proportion of the wetlands being disturbed. Drainage for various purposes is still a threat to the remaining undisturbed wetlands. The Agricultural University of Iceland has generated nationwide geographical databases for land cover, including wetlands, and has surveyed the extent and impact of wetland drainage. The estimate of the extent and impact of drainage has been carried out in relation to the Icelandic national reporting to the UN Framework Convention on Climate Change (Hellsing et al. 2016). The main goal of this paper is to provide a general overview of the soils of the inland wetlands in Iceland and present an estimate of their extent, disturbance by drainage and the patch size of the remaining undrained wetlands based on relatively high resolution data.

2. BACKGROUND AND GENERAL CHARACTERISTICS OF THE WETLANDS

Iceland, a 103,000 km^2 volcanic island, is located just south of the Arctic Circle, between 63° and 66° north latitude, and 13° to 24° west longitude. The climate is relatively mild in spite of the high geographic latitude, the reason being the oceanic climate and the effect of the Gulf Stream that brings warm waters to the shores of Iceland. It is worth noting that continental areas at the same latitudes, in Siberia and Canada for example, experience much colder climates than Iceland, particularly in winter.

Iceland has about 30 active volcanic systems, fed by a volcanic mantle plume under the island (hotspot), with eruptions occurring every 3–5 years (Thordarson and Höskuldsson 2008). The oldest rocks are from the late Tertiary, but much of the surface is covered by rocks <3 million years old, with widespread Holocene lava and tephra deposits. Glaciers cover about 11 % of the country and many active volcanoes are located under ice. The volcanism provides the parent materials of Icelandic soils through tephra deposition and aeolian redistribution of volcanic materials, resulting in the formation of Andosols. The volcanic materials are primarily poorly crystalline basalt grains with rapid weathering rates, which results in the formation of allophane, ferrihydrite and imogolite (Arnalds 2015 Ch. 9). Areas outside the active volcanic regions are primarily made up of Tertiary basalt piles, but the Quaternary glaciation has cut deep valleys into the Tertiary rock formation.

The porosity of the Tertiary rocks has been partially plugged by secondary minerals such as calcite, silica minerals, zeolites and clays by chemical weathering (e.g. Ehlmann et al. 2012), resulting in slow hydraulic conductivity and higher water tables with more frequent occurrence of wetlands within the Tertiary Formation than in the active volcanic regions (Arnalds and Oskarsson 2009).

For centuries, the fen wetlands were used for grazing by livestock and harvested for hay. The wetland sward also provided sod for house construction (Sigurdardottir 2007). Intermittently flooded wetlands were important sources of hay in many parts of the country and often shared within the community. The more organic soils provided peat for fuel (Thorsteinsson 1975).

Draining the wetlands primarily took place during the latter part of the twentieth century and was subsidized by the government. The draining was part of the effort to enhance agricultural production and "food safety" in the post-World-War II era, but also an effort to reduce the migration of people from rural sections to fast growing urban areas (see Helgadottir et al. 2013).

The drainage efforts were sustained for almost 50 years in spite of substantial overproduction of meat and dairy products. Dairy and lamb meat production constitute the major proportion of Icelandic agriculture (data from Statistics Iceland February 2013; www.statice.is) and characterize the rural landscape (Fig. 1a). Currently there are about 26,000 dairy cows and 75,000 head of cattle in all, 460,000 winterfed sheep (> 1.3 million grazing sheep in summer) and about 80,000 horses. Sheep and cattle rearing is highly dependent on hay making in the summer as the cold winter means that animals are kept indoors in winter (see Helgadottir et al. 2013). Approximately half of the hayfields are on drained wetland soils (Wöll et al. 2014).

Previous estimations of the extent of wetlands include the soil map prepared by Johannesson (1960; see explanations by Arnalds 2015 Ch. 7), yielding 9–10,000 km^2 of what he termed 'peatlands'. However, *wetlands of various organic content* are a more appropriate term as is explained in the section on soils below.

This mapping was a noteworthy attempt, yet with relatively high uncertainty as the map is highly generalized with a very coarse scale (1:750,000). Other complete overviews include a vegetation map with the scale of 1:500,000 by the Icelandic Museum of Natural History (Gudjonsson and Gislason 1998) and the CORINE coverage, which is in large part a simplified version of the Agricultural University of Iceland (AUI) data presented here.

Figure. 1 A Typical Icelandic agricultural landscape. Hay fields are located between the ditches (photo A.L. Aradottir): **B** Undisturbed wetland patch. Vegetation dominated by *Eriophorum angustifolium*. Reduced iron (*blue*) and bog iron (reddish brown) present in the water. Many of the wetlands have frequently occurring gradients from wet to dry. Salix is present at the edge of this patch (*upper left*) in damp wetland. Typical highland landscape with mosaic of vegetation and soil types, including barren areas (deserts, *upper right*)

3. MATERIALS AND METHODS

The geographical data applied for the spatial analyses were derived from three main sources: the Agricultural University of Iceland (AUI) Nytjaland and IGLUD databases and the AUI soil map and soil databases. The AUI hosts several national geographical databases which include a soil database and a map, a database for broadly defined vegetation classes (Icelandic Farmland Database, Nytjaland in Icelandic) and the IGLUD (Icelandic Geographical Land Use Database), which was constructed in response to Iceland's commitments to the UN Framework Convention on Climate Change (UN FCCC; Gudmundsson et al. 2013). The Nytjaland database was made from supervised classification of Landsat 7 and Spot 5 satellite images, covering all of Iceland. Most of the work entailed separating the land into 10 vegetation classes, in addition to ice/glaciers and water/sea, but for a proportion of the country, 6 vegetation classes were used. The database was described in detail by Gisladottir et al. (2014) in Icelandic, but also in part in English by Arnalds and Barkarson (2003). The IGLUD database land use map of Iceland is compiled annually from a large number of available map layers, including the Nytjaland classification. The ditch network was digitized at the AUI manually using aerial and satellite photographs. Data on cultivated land (hay fields) were also assembled by digitizing the areas. The IGLUD database and its preparation was described by Gudmundsson et al. (2013) and Wöll et al. (2014). The soil map was constructed by combining soil metadata housed by the AUI (Ymir database) and the Nytjaland land cover database (Arnalds and Oskarsson 2009). This is a rather broad scale map (1:250,000) for general overview of the soils of Iceland.

The extent of wetlands and drainage was estimated using the AUI Nytjaland and IGLUD databases. Two main classes of the Nytjaland classification encompass wetlands: "saturated wetlands" and "damp wetlands". The IGLUD map layer of areas affected by drainage represent areas with average drawdown of the water table of >15 cm, which is, however, subjected to seasonal and annual differences. The preparation of the map layer of drained soils is based on a maximum 200 m distance from ditches. Other vegetation classes closer to 200 m distance from ditches are excluded, as well as areas of more than 10°slope. Areas <200 m from a ditch but isolated from it by a body of water are excluded from affected areas. The 200 m distance is based on considerable efforts by AUI to categorize drainage impacts (mostly unpublished data, see also Oskarsson 1998; Gisladottir et al. 2007, 2010; Gisladottir and Oskarsson 2011). The preparation of the map layer for drained soils has been further described by Wöll et al. (2014). Map layers for croplands (mostly hay fields) on wetland soils were made separately based on surveys of croplands and time series of land use conversions as described by Wöll et al. (2014). All data layers were combined to prevent overlapping or double counting of areas.

The impacts range from subtle changes in water flow, habitat diversity and nutrient and greenhouse gas dynamics (lowest density) to substantial lowering of the water table with pronounced releases of greenhouse gasses and vegetation changes, but many of the areas are also subjected to plowing (hay-making) and intensive grazing that cause ecosystem changes on top of changes caused by the drainage.

Levels of disturbance were estimated by using a kernel density method to calculate accumulated ditch density for areas closer than 200 m from a ditch. Then the impacted area was grouped into the following density intervals: 0.1–5, 5–10, 10–15, 15–20, and >20 km km^{-2}. Areas of slopes >10°, poorly vegetated land and road beds were excluded in the process. In addition, the size distribution of the inland wetlands in the AUI databases was determined, using the following size categories: <1, 1–5, 5–50, 50–100, 100–500 and >500 ha.

4. THE INLAND WETLANDS: SOILS, VEGETATION AND BIRDS

4.1 Soils

Most circumpolar areas of similar latitudes, such as in Finland, Russia, Alaska and Canada, have extensive wetlands of high organic content (peat), often dominated by permafrost (Jones et al. 2010). The mixture of poorly crystalline basaltic volcanic materials and peat, which dominates Icelandic wetland soils, is uncommon in the world, judging from our literature search. They differ from the paramo Histic Andosols of Ecuador (Buytaert et al. 2005, 2007) and the Azores (Madeira et al. 2007) in having a lower content of metal-humus complexes and a higher proportion of vitric materials deposited by continuous re-distribution of volcanic ash materials.

The Icelandic wetland soils rarely exhibit hydric soil properties (>100 % water content at 15 bar tension) in contrast to more developed Hydric Andosols of warmer climates. Aquic Andosols have been reported in many areas such as Japan (e.g., Shoji and Ono 1978), however usually much more developed and lacking the frequent volcanic additions and continuous aeolian deposition of volcanic ash which give the Icelandic soils their distinctive young and vitric character.

The soils of Icelandic inland wetlands have a wide range of organic content in the surface horizons that determine the classification of the soils. The C content is most often <25 %, considerably lower than the 36–54%C averages reported by Loisel et al. (2014) for northern peatlands. The soils of palsa areas (permafrost mounds) also have low organic content in surface horizons in contrast to most other Arctic palsa areas (Saemundsson et al. 2012). Carbon accumulation in the Icelandic soils, based on data from Oskarsson et al. (2004) and averaging total carbon stocks over 8000 yrs. indicates accumulation from 4 g m^{-2} $yr.^{-1}$ (Gleyic Andosols) to 25 g m^{-2} $yr.^{-1}$ (Histosols). The accumulation in Histosols is of the same order as 23 g m^{-2} $yr.^{-1}$ reported for northern peatlands by Loisel et al. (2014). Gudmundsson (2008) studied reported carbon accumulation between tephra-layers of known age for a Histosol in NW Iceland. The accumulation ranged between 9 and 20 g m^{-2} $yr.^{-1}$ during the Holocene (4 periods) with the average rate of about 15 g m^{-2} $yr.^{-1}$.

An Icelandic soil classification system (Arnalds and Oskarsson 2009) separates the inland wetland soils based on carbon content in the top 30 cm into *Gleyic Andosols* (<12 % C), *Histic Andosols* (12–20 % C) and *Histosols* (>20 % C). Andosols, by nature, accumulate higher organic content than other mineral soils and are differentiated from Histosols according the US Soil Taxonomy (Soil Survey Staff 1999) and the WRB (IUSS Working Group WRB 2014) at the 25 % C level (not 12–16 % as for the difference between Histosols and other mineral soils). These soils differ from the highly peaty Arctic soils, as they have a higher pH (often >5) and C/N ratio of 13–25, noticeably lower than the 34–77 average range reported for northern and Arctic peatlands (Loisel et al. 2014). Considerable amounts of nitrogen become readily available upon drainage. Part of the organic matter in all the wetland soils is in the form of organo-mineral complexes, which are characteristic of volcanic soils in general (see Arnalds 2015 Ch. 8). The variability in carbon content of Icelandic wetland soils is in large part due to differences in aeolian and tephra deposition contributing mineral matter to the soils, with substantially lower organic content close to volcanoes and the major dust sources. Variability in organic content is also caused by differences in drainage, flooding regimes and landscape positions (Arnalds 2004; Arnalds and Oskarsson 2009). Gleyic Andosols (Fig. 2) with relatively low organic content (often 2–6 %) are dominant within the active volcanic zone, but there is a gradient with carbon content gradually becoming higher with distance from active dust sources and volcanoes. Highly organic soils are therefore mostly found far away from active dust sources such as in West, Northwest and North Iceland. The main characteristic of the organic matter in the soils is that it is poorly decomposed and the Histosols would classify as Fibrists (Borofibrists and Cryofibrists) under Soil Taxonomy. They show considerable shrinkage when they are completely dried in the laboratory (often less than ½ of the original volume when dry), but limited or very slow shrinkage in the field when drained, which is in part attributed to the volcanic ash materials in the matrix (Bartoli and Burtin 2007).

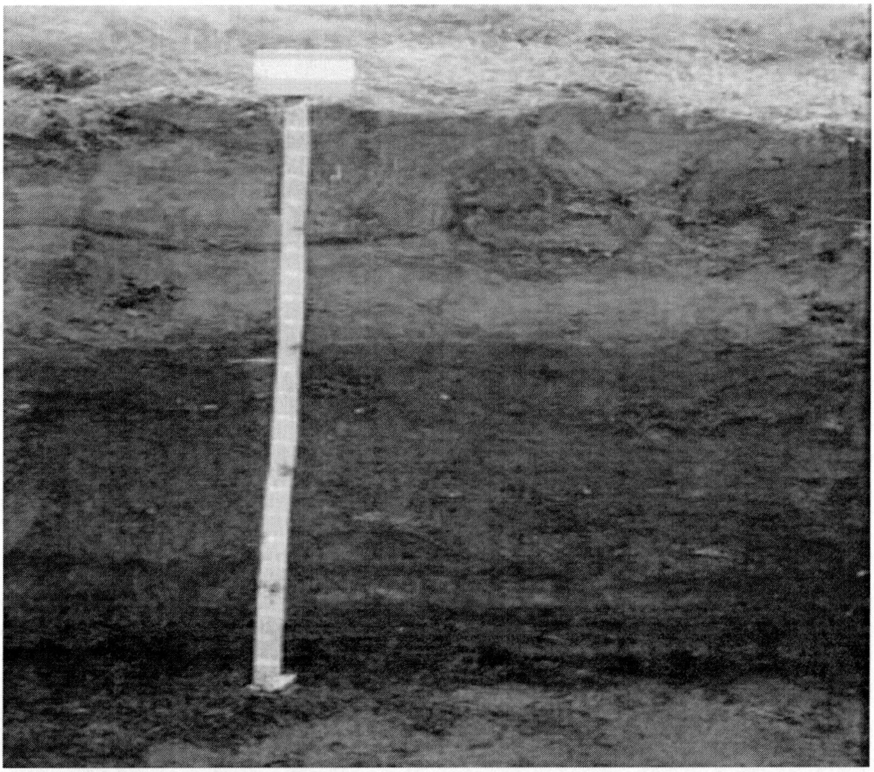

Figure. 2 Typical Gleyic Andosol, showing distinct tephra layers and signs of cryoturbation in the upper horizons. The boundary between the lighter colored upper part and darker colored lower part of the profile marks the settlement of Iceland about 1200 yrs. ago. Increased aeolian activity consisting in part of re-distribution of light colored tephra fragments influences the soil color. The lower part is more organic (>13–20%C) while the upper part has about 10%C; see Arnalds 2015 Ch. 8)

The Gleyic Andosols contain 10–20 % clays (allophane and ferrihydrite) on average, while the Histosols do not contain appreciable amounts of clays with Histic Andosols being intermediate. There is a gradient in pH from Gleyic Andosols (6–7) through Histic Andosols (5–6) to Histosols (4–5) attributable to decreasing aeolian input and increased organic content

(Arnalds 2015 Ch. 8). Bulk density ranges from 0.2 in organic horizons to about 0.7 g cm^{-3} in mineral horizons in Gleyic Andosols. The organic horizons have higher water retention than would be expected, but so do also the mineral horizons due to their andic soil properties (see Arnalds 2004, 2015 Ch. 7). The depth of the Icelandic inland wetland soils is quite variable, ranging from a few cm to several meters, but commonly 1–3 m.

4.2 Vegetation

The vegetation of Icelandic wetlands was described by Steindorsson (1964), but Magnusson (1998) specifically addressed the plants of drained wetlands. The overarching characteristic of Icelandic wetland vegetation is the predominance of vascular plants and virtual absence of *Sphagnum* moss dominated wetlands. This can most likely be ascribed to the general fertility of these systems arising both from the widespread aeolian deposition of mineral matter and the fact that there are almost no truly ombrotrophic inland wetlands in Iceland, most being minerotrophic or only partially ombrotrophic (Steindorsson 1964). Drainage changes the systems from minerotrophic to ombrotrophic in areas closest to the ditch.

The subsequent discussion takes note of the Agricultural University databases which separate the wetlands into "saturated wetlands" and "damp wetlands". Other divisions of Icelandic wetlands are discussed in Olafsson (1998). The plant species characteristic of the "saturated wetland" class include various *Carex* spp., such as *Carex bigelowii, C. lyngbyei, C. rostrata, C. chordorrhizia*, and *Equisetum* spp. Cotton grass (*Eriophorum angustifolium*) is quite common as are some heathland species such as dwarf-birch (*Betula nana*), blueberries (*Vaccinium uliginosum*), crowberries (*Empetrum nigrum*), common heather (*Calluna vulgaris*) and *Arctostaphylos uva-ursi*, but also willow species (*Salix phylicifolia, S. arctica* and *S. lanata*). Moss species of the genera *Hypnum, Philonotis, Mnium* and *Rachometrium* are widespread. Species characteristic of damp wetlands include many of the same species found in saturated wetlands, with lower dominance of *Carex* spp. and more frequent occurrence of woody species.

4.3 Birds

Icelandic wetlands in general are particularly important for birds with approximately 75 % of Icelandic bird species relying in part or fully on various types of wetlands as a habitat (Gudmundsson 1998). The Icelandic wetlands (including the inland fens) support about 20 internationally important bird species (Einarsson et al. 2002). They are of particular importance for waders (*Charadrii*) in the northern hemisphere

(Gunnarsson et al. 2006). Wetland birds include golden plover (*Pluvialis apricaria*), dunlin (*Calidris alpina*), snipe (*Gallinago gallinago*), whimbrel (*Numenius phaeopus*), black-tailed godwit (*Limosa limosa*), redshank (*Tringa totanus*), and meadow pipit (*Anthus pratensis*). Swans (*Cygnus cygnus*) and geese are also common, including the graylag goose (*Anser anser*) and pink-footed goose (*Anser brachyrhynchus*). Some of the highland wetlands are of particular importance, e.g., the Thjorsarver RAMSAR area, the world's largest breeding ground of pink-footed goose.

5. HISTORY OF THE DRAINAGE EFFORT

Draining wetlands facilitates oxidation of organic matter and release of nutrients, resulting in substantial increase in biomass production (Magnusson 1998), at least in the short-term. This prospect was of interest to the farming community during the early twentieth century. A law was passed in 1923 that initiated subsides for draining. Minor areas were drained annually during the first few decades as ditches were excavated by hand. After World War II, with the arrival of large and effective machinery for excavation, a broad-scale draining of the wetlands was initiated. The drainage activity gained momentum relatively rapidly, expedited by the government subsidies (Fig. 3).

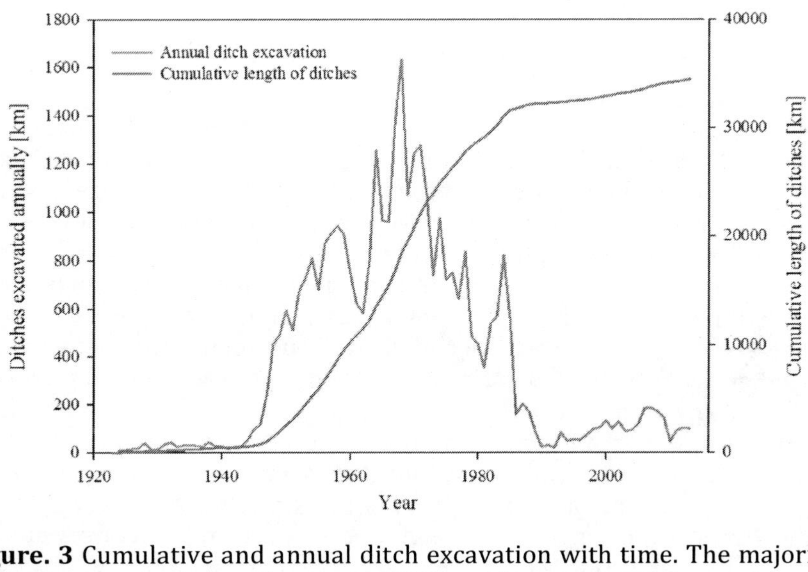

Figure. 3 Cumulative and annual ditch excavation with time. The majority of ditches were excavated prior to 1987, with annual activity peaking in 1968 (> 1600 km yr.$^{-1}$)

The draining peaked in 1968, when 1633 km of ditches were excavated. The effort decreased considerably after 1987 (Bjarnason 1982; Snaebjörnsson et al. 2010), when subsidies were cut dramatically, but it can also be postulated that much of the easily drained land near the farmsteads had been drained by that time. About 15 % of the drained wetlands are used for hay-making. The drained areas are also used for grazing to some extent. However, some of the drained land is not used for agriculture; the draining disturbance of those areas has no practical purpose.

6. RESULTS

6.1 The Extent of inland Wetlands

The wetlands, as represented in the AUI databases, consist of three classes: the saturated wetlands, damp wetlands and drained wetlands (partial to fully). The spatial distribution of these wetlands is presented in Fig. 4, and their altitudinal distribution is presented in Table 1.

The results of this GIS based analysis indicate that Icelandic inland wetlands cover about 9000 km^2. This area is based on vegetation classification using satellite images, with open water reflectance signals to augment the classification. It is likely that soil-based classification reflecting the presence of shallow ground water would increase this extent. However, data are not available at this time.

Wetlands occur in all geographic regions of Iceland and extend into highland elevations. However, about two-thirds of the wetlands are found below 200 m elevation, or 5795 km^2, which is equivalent to approximately 23 % of the lowland surface area and about 31 % of the vegetated land occurring below 200 m. About 20 % of wetlands are located between 200 and 400 m elevation, but their extent falls dramatically at higher elevations as does vegetation cover in general. Prominent inland fen wetlands occupy the southern lowlands (Fig. 4), with mostly inorganic Gleyic Andosols in the eastern part where aeolian and volcanic ash inputs are relatively high, together with Histic Andosols and higher carbon content towards the western part of the southern lowlands (Fig. 4). Prominent inland wetlands also occupy western Iceland, mainly Histosols and Histic Andosols. Inland wetlands are also common within the deep valleys of the Tertiary Formation in North and West Iceland, harboring mostly Histic Andosols and Histosols. Widespread wetlands are found in the northwestern and eastern highlands (Fig. 4).

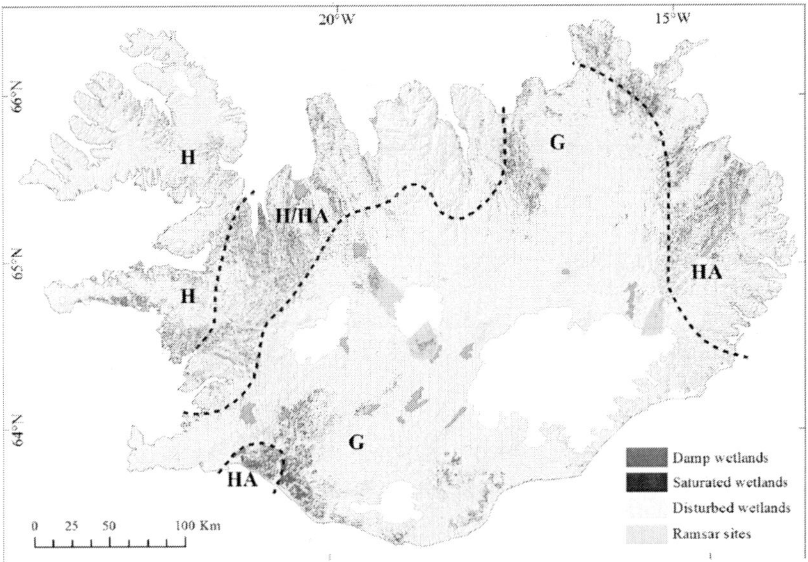

Figure. 4 Inland wetlands in Iceland. The most prominent areas are the southern and western lowlands together with the NW and NE highlands. Major wetland soil types dominating each region are indicated, enclosed by dotted lines. H: Histosols, HA: Histic Andosols, and G: Gleyic Andosols. Ramsar sites are indicated as green areas. The majority of the lowland wetlands have been impacted by drainage. Some impacted wetlands with altered vegetation cover and/or agricultural fields are not shown. Large water bodies are shown in blue. White areas on the map represent glaciers

Table 1 A geographical summary for Icelandic inland fen wetlands, sorted by elevation intervals. Wetlands are separated into saturated and damp wetlands. Total vegetation cover is shown in the top row for comparison. Data from the IGLUD database which is based in part on the Nytjaland database

	m.a.s.l.						
	0–200	200–400	400–600	600–800	800–1000	>1000	Total
	km^2						
Total area	25,257	18,106	22,293	18,519	8345	10,163	102,683
Vegetated land	18,665	12,758	10,550	3932	385	41	46,331
Undisturbed	1743	1611	1175	258	2	0	4789

wetlands$							
Saturated wetlands	*1112*	*1170*	*1010*	*236*	*2*	*0*	*3529*
Damp wetlands	*631*	*441*	*166*	*23*	*0*	*0*	*1260*
Drained wetlands	4052	136	3	0	0	0	4191
Total wetlands	5795	1747	1178	258	2	0	8980
	%						
Impacted wetlands	69.9	7.8	0.3	0	0	0	46.7

$*Saturated wetlands + damp wetlands* in the table

6.2 Drainage Disturbance

Our results show that 4195 km^2 or 47 % of all Icelandic inland wetlands have been impacted by drainage (Table 1), ranging from full drainage to subtle changes in water levels. The table shows only areas affected by draining (see Materials and Methods), but limited data are available for other wetland disturbances, except for hydro-electric reservoirs. The majority (70 %) of the lowland wetlands below 200 m elevation have been affected. Drainage of inland wetlands is rather limited above 200 m elevation, coinciding with the much limited agriculture at higher elevations.

6.3 Ditches and Ditch Density

The digitization of the entire ditch system revealed a total of about 29,700 km of ditches (see also Gisladottir et al. 2010). Preliminary results from a survey on the accuracy of the digitation showed that 91 % of ditches mapped were confirmed as ditches and that 5 % of ditches had gone undetected by the mapping effort (Wöll et al. 2014). This estimate of the total length of ditches is slightly lower than the previous estimate of Geirsson (1975), who estimated the total length of drainage ditches to be 32,700 km based on records of excavated volume obtained in relation to the subsidies rendered for making the ditches. The ditch system is further augmented by 61,600 km of sub-surface drainage made in 1961–1993, mostly in areas where ditches are in place (Geirsson 1975; Helgadottir et al. 2013).

The majority of the impacted area has low accumulated ditch density (0.1–0.5 km km^{-2}; Table 2) i.e. about 2300 km^2, or 67 %. The highest densities of >10 km km^{-2} make up 309 km^2 of the impacted area. The total extent of cultivated areas (hay-fields and other crops) in Iceland is about 1290 km^2, of which about half is on drained wetland soils (Wöll et al. 2014; see also Helgadottir et al. 2013).

Table 2 Wetland disturbance by drainage sorted by density of ditches

Density	Length of ditches	Impacted areas	Proportion of ditches	Proportion of total
km km^{-2}	km	km^2	%	
0.1–5	1818	2300	6.1	66.6
5–10	11,462	858	38.6	24.7
10–15	8479	231	28.6	6.7
15–20	5337	63	18.0	1.8
>20	2585	15	8.7	0.4
Total	29,680	3468	100	100

An example of an area with extensive wetland draining in South Iceland is provided in Fig. 5. The figure shows that almost the entire area is influenced by the draining, which is in accordance with the results for areas <200 m elevation in general. The hayfields and other cultivated land are within the areas of the highest drainage densities next to the farms. Many areas shown as intact in the figure are encircled by a ditch system and are likely to be somewhat impacted. Undisturbed areas are found on the riverine floodplains.

The impacts of low ditch densities are also affected by landscape characteristics such as slope and bedrock hydrology. It is clear that a large proportion of the drained land is not used for such agricultural production. An undefined, but substantial proportion of the remaining drained wetlands is used for periodic grazing (data not available), and some are used for afforestation.

6.4 Wetland Patch Size

The majority of the fragmented wetland patches are small or below one ha in size (Table 3), with over one million such patches registered in our GIS Nytjaland database and accounting for about 30 % of the wetland areas.

Damp wetland patches are smaller on average than the saturated wetland patches (data not shown). Only 16 % of the wetlands occur as patches >100 ha (1 km^2).

Larger patches occur as saturated wetlands on average compared to damp wetlands, which indicates that there is a higher probability of saturated surface conditions within the larger wetland patches. Damp wetlands are often found at the periphery of the saturated wetlands and are therefore relatively smaller than continuous saturated wetland patches.

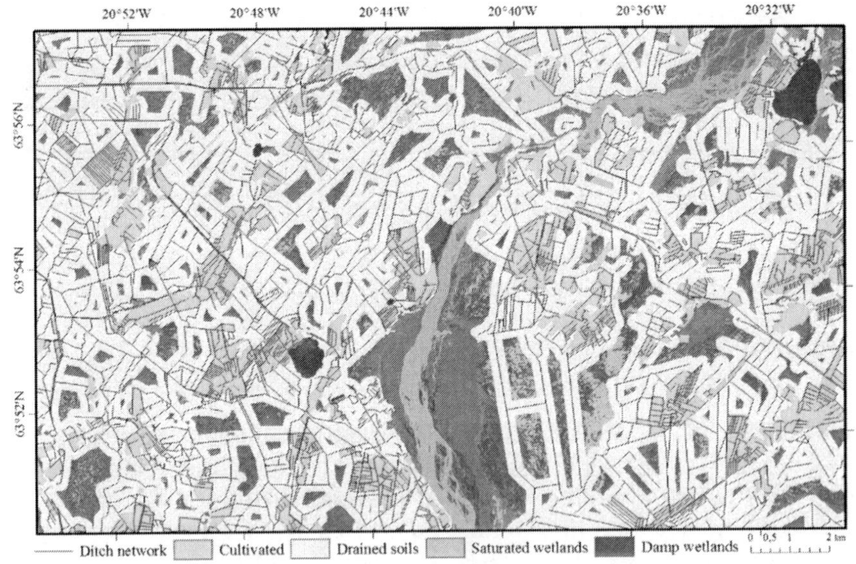

Figure. 5 Drained area in South Iceland. The river Thjórsá, partly in a gorge, runs through the center of the image. Undisturbed saturated and damp wetlands shown separately. The black lines display the ditch network. The map shows clearly that most of this area is less than 200 m from ditches, but a relatively small proportion of the area is cultivated (hay fields). Some areas marked undisturbed are surrounded by ditches but >200 from them. These areas are mostly likely somewhat impacted by the drainage, but the numbers for impacted areas are not included

Table 3 Number of wetland patches sorted by patch size for saturated wetlands and damp wetlands. Data from the Nytjaland AUI database and includes some disturbed/impacted wetlands

Patch size (ha)	Number of areas	Combined area (km^2)	%
<1	1,135,626	1631	29.6
1–5	61,859	1225	22.2
5–50	11,768	1446	26.2
50–100	486	334	6.1
100–500	318	588	10.7
>500	29	288	5.2

7. DISCUSSION

The soils of Icelandic wetlands are exceptional on a global scale for being a mixture of volcanic and organic soils, with extremely variable organic content. Their carbon content is usually much lower than in Arctic wetland soils (especially the Gleyic Andosols) and the C/N ratio is lower. They are more vitric in character than much of the Aquic and Hydric Andosols found in other volcanic areas. Their fertility is enhanced by a relatively favorable pH maintained by nutrient release from the rapid weathering of basaltic volcanic glass (see Arnalds 2008, 2015Ch. 9). Their uniqueness is highlighted by their dissimilarity with other northern circumpolar soils, which mostly constitute peat soils (Histosols) and/or Cryosols (permafrost) (Jones et al. 2010), while Icelandic wetlands are dominated by Andosols (mineral soils with variable organic content) with a relatively small proportion of Histosols, which also are influenced by aeolian deposition of volcanic materials.

The results show that a substantial part of the Icelandic inland wetlands have been disturbed by drainage. About 47 % of Icelandic wetlands have been drained, mostly below 400 m elevation. The results show that 70 % of the inland wetlands below 200 m elevation have been impacted to some degree, which is in good accordance with research on the southern and western lowlands (Thorhallsdottir et al. 1998; Oskarsson 1998). We suggest that the motivations for many of the ditches of the lowest densities (< 10 km km^2) were subsidy driven rather than by need for agricultural land, but such claims are difficult to corroborate.

The areas of intense drainage most often go through a vegetation composition shift from wetland to grassland plant communities, with higher cover of *graminoids* with various amounts of *salix* species and herbaceous plants, but the shift is also dependent on land use, such as grazing and fertilizer use, which tend to increase the abundance of grasses (see Magnusson 1998). We recognize that our estimate based on the 200 m distance from a ditch has its limitations. This distance is an estimated average, but will vary from one area to the other. Areas beyond the 200 m range enclosed by ditches, a common scenario in South Iceland, are also most likely impacted, with altered water and nutrient flows. This methodology gives a good first appraisal of drainage impacts in Iceland.

A minor proportion of the drained land (15 %) is used for hay production, mostly of high density drainage. Grazing by horses on both drained and undrained lowland wetlands has caused damage to the vegetative cover in many places (Magnusson 1998). The highland ecosystems are not affected by drainage efforts. They often feed important salmon rivers that have a significant impact on the rural economy. However, some highland wetlands have been disturbed by hydropower development, such as in the Blanda and Kárahnúkar areas in Northwest and East Iceland. There is widespread livestock grazing (mostly sheep) in the highlands, which has influence on wetlands, especially damp wetlands, which is likely to alter species composition, but research on these impacts is limited. Other disturbances have not been surveyed to date. It should also be noted that the nature of many of the inland wetlands, both highland and lowland, has been altered by increasing aeolian deposition rates after the Settlement of Iceland about 1200 years ago, with reduced carbon content per depth increment and often altered hydrology resulting from coarse sediments (silt and sand) being deposited on the areas.

The carbon dioxide emissions from drained wetlands outside cultivated areas are calculated as 5100–10,600 Gg CO_2 (data from Gudmundsson and Oskarsson 2014), or 7300 Gg CO_2 by applying the IPCC default emission factor for the area. The emissions of other greenhouse gasses (N_2O, CH_4) and off-site emissions are not included but add about 2300 Gg CO_2equivalents to the greenhouse gas impact of drainage (Hellsing et al. 2016). These emissions exceed the sum of all Icelandic non-LULUCF emissions (including smelters and transportation) reported for the year 2014 for Iceland (4597 Gg CO_2 equivalents) (Hellsing et al. 2016).

Icelandic landscapes are characterized by a great variety of ecotypes over short distances and this is reflected by the large number of small wetland patches. Small patches are important in terms of ecological functions and biodiversity, with a wide range of habitats provided over relatively short distances, while the small wetland patches also provide improved hydrological characteristics such as water regulation during snow melt on a landscape level. Small-scale mosaics of different habitat types can better fulfill different requirements for birds during the breeding season (e.g. nesting, feeding) than larger homogenous areas (Gunnarsson et al. 2006; Pickett and Siriwardena 2011; Benton et al. 2003). Our results show that much of the inland wetland patches are <5 ha with about 30 % of wetland areas consisting of patches smaller than 1 ha. Only a proportion of the small damp wetland patches were created by the drainage activities in areas below 200 m elevation. The importance of patch size is reflected in the current law, stating that wetland areas >3 ha in size should not disturbed. The 3 ha reference is based on much coarser mapping than presently available and potentially concealing the importance of smaller patches. The results presented here suggest that it is important to protect areas characterized by small wetland patches which give rise to the importance of the ecosystem mosaic rather than isolated habitat types and thus, small wetland patches are quite important. We argue that there should be adopted a smaller reference size than the current 3 ha (e.g. 0.5 ha) in light of the importance of the small wetland patches in the overall ecosystem mosaic. Further analysis of the 1–5 ha patch size shows that >40 % of wetland the area within this size class is between 1 and 2 ha. About 55 % of remaining wetlands are currently excluded from any protection.

8. CONCLUSIONS

Icelandic inland wetlands are among the most important ecosystems of the country because of their role in biodiversity, water regulation and carbon budgets and they have an international value by supporting large populations of migratory bird species. There is a need for increased research on the hydrology of Icelandic wetland soils and the impact of draining. Furthermore, there is a need to analyze the relationship between drainage levels and impacts on various ecosystem factors and processes, such as fragmentation, GHG budgets, nutrient cycling and release, bird life, plant composition and other biotic factors. Research on biodiversity in relation to wetlands also needs more attention in future studies.

The remaining undisturbed inland wetlands are still being threatened by such factors as agriculture, with improved crop growing conditions with the current warming of the climate (see Helgadottir et al. 2013), development and subdivision, land fragmentation (see Wald 2012), afforestation efforts and the generation of hydropower (large reservoirs). Some sectors, including the road authorities, have adopted strict policies to avoid wetland disturbance or restoring wetlands in place for wetland areas deemed necessary to disturb (Stefansdottir 2011). We conclude that it is important to increase the level of protection of all wetlands in Iceland. Lists have been made of areas in need of protection based on the importance to bird ecology (Ferdinand et al. 1973). The databases presented here have been shown to be of value for predicting wetland bird species and densities (Johannesdottir et al. 2014; Gunnarsson et al. 2015). Many wetlands are on a list proposed for nature protection ("Náttúruminjaskrá", Icelandic Natural Features Database, kept by the Environmental Agency of Iceland). However, there is a need to strengthen work to systematically prioritize the protection and restoration of Icelandic wetlands, which considers a broad range of factors such as size, continuity, ecology, hydrology, geomorphology, and the uniqueness of the wetlands on national, regional and global scales.

REFERENCES

1. Arnalds O (2004) Volcanic soils of Iceland. Catena 56:3–10
2. Arnalds O (2008) Soils of Iceland. Jökull - The Icelandic J Earth Sci 58:409–421
3. Arnalds O (2015) The soils of Iceland. World Soils Book Series. Springer, Dordrecht, the Netherlands
4. Arnalds O, Barkarson BH (2003) Soil erosion and land use policy in Iceland in relation to sheep grazing and government subsidies. Environ Sci Pol 6:105–113
5. Arnalds O, Oskarsson H (2009) Soil map of Iceland. Náttúrufræðingurinn 78:107–121 (In Icelandic, English summary, table and figure legends)
6. Bartoli F, Burtin G (2007) Organo-mineral clay and physical properties in COST-622 European volcanic soils. In: Arnalds O, Bartoli F, Buurman P, Oskarsson H, Stoops G, Garcia-Rodeja E (eds) Soils of volcanic regions in Europe. Springer, Heidelberg, Germany, pp. 469–491
7. Benton TG, Vickery JA, Wilson JD (2003) Farmland biodiversity: is habitat heterogeneity the key? Trends Ecol Evol 18:182–188
8. Bjarnason B (1982) Saga vatnsveita og framræslu (The story of water diversions and drainage). In: Snaebjornsson A (ed) Þættir um mýrajarðveg á Íslandi, Agricultural College Report 38:3–7. Agricultural College (now Agricultural University of Iceland), Hvanneyri, Iceland
9. Buytaert W, Wyseure G, De Bièvre B, Deckers J (2005) The effect of land-use changes on the hydrological behaviour of Histic andosols in South Ecuador. Hydrol Process 19:3985–3997
10. Buytaert W, Deckers J, Wyseure G (2007) Regional variability of volcanic ash soils in South Ecuador: the relation with parent material, climate and land use. Catena 70:143–154
11. Cowardin LM, Carter V, Golet FC, LaRoe ET (1979) Classification of wetlands and Deepwater habitats of the United States. U. S. Department of the Interior, Fish and Wildlife Service, Washington, D.C.
12. Ehlmann BL, Bish DL, Ruff SW, Mustard JF (2012) Mineralogy and chemistry of altered Icelandic basalts: Application to clay mineral detection and understanding aqueous environments on Mars. J Geophys Res 117:E00J16. doi:10.1029/2012JE004156
13. Einarsson Ó, Kristinsson H, Skarphéðinsson KH, Ottósson JG (2002) Verndun tegunda og svæða: tillögur Náttúrufræðistofnunar Íslands vegna Náttúruverndaráætlunar. The Icelandic Institute of Natural History. Reykjavik, Iceland (In Icelandic)

14. Ferdinand FHM, Thamdrup A, Haapanen H, Ormio CH, Ovesen A, Gardarson D (1973) Oversigt over vigtige våde fugleområder i Norden. Nordiske arbejdsgruppe vedrørende beskyttelse af våde fugleområder "wetlands". Copenhagen, Denmark, 366 pp
15. Gardarsson A (1998) Íslensk votlendi (Icelandic wetlands). In: Olafsson JS (ed) Icelandic Wetlands, Protection and Utilization (Íslensk votlendi, verndun og nýting). Háskólaútgáfan (University Publishing). University of Iceland, Reykjavik, pp. 13–35
16. Geirsson O (1975) Framræsla (Bog drainage for farming purposes). In: Garðarsson A (ed) Íslensk votlendi (Icelandic Wetlands). Landvernd Publ. no 4. Icelandic Environment Association, Reykjavik Iceland, pp. 143–154
17. Gisladottir IV, Oskarsson H (2011) The suitability of the WIV wetland index for Icelandic condtion (Athugun á hentugleika WIV-votlendisvísitölunnar við íslenskar aðstæður). Iceland Agricultural Congress (Fræðaþing landbúnaðarins) 2011:371–376
18. Gisladottir FO, Metusalemsson S, Oskarsson H (2007) Áhrifasvæði skurða, greining með fjarkönnunaraðferðum (drainage ditch impact areas, a remote sensing study). Iceland Agricultural Congress (Fræðaþing landbúnaðarins) 2007:371–376
19. Gisladottir FO, Gudmundsson J, Askelsdottir S (2010) Mapping and density analyses of drainage ditches in Iceland. Mapping and monitoring of Nordic vegetation and landscapes. Conference Proceeding. Norwegian Forest and Landscape Institute, As, Norway, pp. 43–46
20. Gisladottir FO, Brink SH, Arnalds O (2014) Nytjaland (The Icelandic Farmland Database). Agricultural University of Iceland Report No. 49. Reykjavik, Iceland.
21. Gudjonsson G, Gislason E (1998) Vegetation map of Iceland 1:500000. General overview map. Icelandic Institute of Natural History, Reykjavik, Iceland
22. Gudmundsson GA (1998) The importance of wetlands for birds. In: JS O (ed) Icelandic wetlands, protection and utilization (Íslensk votlendi, verndun og nýting). Háskólaútgáfan (University Publishing). University of Iceland, Reykjavik, pp. 121–129
23. Gudmundsson T (2008) Carbon sequestraton in soils (Kolefnisbinding í jarðvegi), vol IV. Agricultural Science Conference (Fræðaþing landbúnaðarins), Iceland, pp. 290–297Google Scholar
24. Gudmundsson J, Oskarsson H (2014). Carbon dioxide emission from drained organic soils in West-Iceland. Proceedings of the International conference: Soil carbon sequestration for climate food security and ecosystem services, European Commission Joint Research Centre and The Icelandic Soil Conservation Service, Ispra, Italy, pp. 155–159
25. Gudmundsson J, Brink SH, Gisladottir FO (2013) Preparation of a LULUCF land-use map for Iceland: development of the grassland layer and subcategories. Grassland Sci Europe 18:105–107
26. Gunnarsson TG, Gill JA, Appleton GF, Gíslason H, Gardarsson A, Watkinson AR, Sutherland WJ (2006) Large-scale habitat associations of birds in lowland Iceland: implications for conservation. Biol Conserv 128:265–275
27. Gunnarsson TG, Arnalds O, Appleton G, Mendez V, Gill JA (2015) Ecosystem recharge by volcanic dust drives broad-scale variation in bird abundance. Ecol Evol 5:2386–2396CrossRefPubMedPubMedCentral
28. Helgadottir A, Eythorsdottir E, Johannesson T (2013) Agriculture in Iceland – a grassland based production. Grassland Sci Europe 18:30–43
29. Hellsing VUL, Ragnarsdottir AS, Jonsson K, Andresson K, Johannsson T, Guðmundsson J, Snorrason A, Thorsson J, Einarsson S (2016) National Inventory Report 2016; submitted under the United Nations framework convention on climate change, emissions of greenhouse gases in Iceland from 1990 to 2014. Environmental Agency of Iceland, Reykjavik, Iceland
30. IUSS Working Group WRB (2014) World Reference Base for Soil Resources 2014. Intarnational soil classification system for naming soils and creating legends for soil maps. World Soil Resources Reports No. 106. FAO, Rome
31. Johannesdottir L, Arnalds O, Brink S, Gunnarsson TG (2014) Identifying important bird habitats in a sub-arctic area unergoing rapid land-use change. Bird Study 61:544–552
32. Johannesson B (1960) The Soils of Iceland. University Research Institute Dept. of Agriculture, Reports Series B – No 13. , University of Iceland, Reykjavik, Iceland

33. Jones A, Stolbovoy V, Tarnocai C, Broll G, Spaargaren O, Montanarella L (eds) (2010) Soil atlas of the northern circumpolar region. European Comission. Publications Office of the European Union, Luxembourg
34. Loisel J, Yu Z, Beilman DW, Camill P, Alm J, Amesbury D, Anderson D, Anderson S, Bochicchio C, Barber K, and 51 more authors (2014) A database and synthesis of northern peatland soil properties and Holocene carbon and nitrogen accumulation. The Holocene 24:1028–1042
35. Madeira M, Pinheiro J, Madruga J, Monteiro F (2007) Soils of volcanic systems in Portugal. In: Arnalds O, Bartoli F, Buurman P, Oskarsson H, Stoops G, Garcia-Rodeja E (eds) Soils of volcanic regions in Europe. Springer, Heidelberg, Germany, pp. 69–81CrossRefGoogle Scholar
36. Magnusson B (1998) Vegetation of drained mires in Iceland. In: Olafsson JS (ed) Icelandic Wetlands, Protection and Utilization (Íslensk votlendi, verndun og nýting). Háskólaútgáfan (University Publishing), University of Iceland, Reykjavik, pp 105–120.
37. Mitsch WJ, Gosselink JG (2007) Wetlands. John Wiley & Sons, New Jersey
38. Olafsson JS (ed) (1998) Icelandic wetlands, protection and utilization (Íslensk votlendi, verndun og nýting). Háskólaútgáfan (University Publishing), University of Iceland, Reykjavik
39. Oskarsson H (1998) Wetland draining in western Iceland. In: Olafsson JS (ed) Icelandic Wetlands, Protection and Utilization (Íslensk votlendi, verndun og nýting). Háskólaútgáfan (University Publishing), University of Iceland, Reykjavik, pp 121–129.
40. Oskarsson H, Arnalds O, Gudmundsson J, Gudbergsson G (2004) Organic carbon in Icelandic andosols: geographical variation and impact of erosion. Catena 56:225–238
41. Pickett SR, Siriwardena GM (2011) The relationship between multi-scale habitat heterogeneity and farmland bird abundance. Ecography 34:955–969
42. Saemundsson T, Arnalds O, Kneisel C, Jonsson HP, Decaulne A (2012) The Orravatnsrustir palsa site in Central Iceland – Palsas in an aeolian sedimentation environment. Geomorphology 167-168:13–20
43. Shoji S, Ono T (1978) Genesis of andosols at Ktakami, Iwate prefecture, Northeast Japan I Relationships between topography and soil formation. The Quat Res 16:247–254
44. Sigurdardottir S (2007) Torf til bygginga (turf for buildings). Smarit Byggdasafns Skagfirdinga VII, Sauðárkrókur, Iceland
45. Skarphedinsson KS, Thorleifsson E (1998) Keldusvín útdauður varpfugl á Íslandi (water rail, extinct bird in Iceland). Kvískerjabók. Sýslusafn Austur Skaftafellssýslu, Höfn í Hornafirði, Iceland, In, pp. 266–296
46. Snaebjörnsson A, Hjartardottir D, Blöndal E, Petursson JG, Eggertsson O, Halldorsson O (2010) Skýrsla nefndar um landnotkun. Athugun á notkun og varðveislu ræktanlegs lands (Committee report on land use. The use and preservation of arable land), vol 72. Ministry of Fisheries and Agriculture, Reykjavík, Iceland
47. Soil Survey Staff (1999) Soil taxonomy. A basic system of soil classification for making and interpreting soil surveys. 2nd edn. Agriculture Handbook No 436. USDA/NRCS, Washington
48. Stefansdottir M (2011) Ecological Restoration by the Road Authorities. In: AL A, Halldorsson G (eds) Restoration in Iceland, Agricultural University of Iceland and the Icelandic Soil Conservation Service, Reykjavik, pp. 56–59. isbn:ISBN:978–9979–881-08-7
49. Steindorsson S (1964) Gróður á Íslandi (Icelandic vegetation). Almenna bókafélagið, Reykjavík, Iceland
50. Thordarson T, Höskuldsson A (2008) Postglacial volcanism in Iceland. Jökull – The Icelandic J Earth Sci 58:197–228
51. Thorhallsdottir TE, Thorsson J, Sigurdardottir S, Svavarsdottir K, Jonsson MH (1998) Disturbance of wetlands in South Iceland. In: JS O (ed) Icelandic Wetlands, Protection and Utilization (Íslensk votlendi, verndun og nýting). Háskólaútgáfan (University Publishing). University of Iceland, Reykjavik, pp. 131–142
52. Thorleifsson E (1998) Effects of drainage on wetland birds in South Iceland. In: JS O (ed) Icelandic Wetlands, Protection and Utilization (Íslensk votlendi, verndun og nýting).

Háskólaútgáfan (University Publishing). University of Iceland, Reykjavik, Iceland, pp. 173–183
53. Thorsteinsson G (1975) Horfnir starfshættir og leiftur frá liðnum öldum. Örn og Örlygur, Reykjavík
54. Wald EC (2012) Land-use development in South Iceland 1900–2010. Faculty of Biology and Environmental Science, University of Iceland, Reykjavík, Iceland, MSc thesis
55. Wöll C, Hallsdottir BS, Guðmundsson J, Snorrason A, Thorsson J, Jonsson PVK, Andresson K, Einarsson S (2014) Emissions of greenhouse gases in Iceland from 1990 to 2012. National Inventory Report 2014. Submitted under the United Nations Framework Convention on Climate Change and the Kyoto Protocol, Environmental Agency of Iceland, Reykjavik, Iceland

CHAPTER 4

Multilayer Substrate Configuration Enhances Removal Efficiency of Pollutants in Constructed Wetlands

Shaoyuan Bai [1,2], Tao Lv [3], Yanli Ding [4,*], Xuefen Li [1], Shaohong You [1], Qinglin Xie [2] and Hans Brix [3,5,*]

[1]*Guangxi Key Laboratory of Environmental Pollution Control Theory and Technology, Guilin University of Technology, Guilin 541004, China*

[2]*Collaborative Innovation Center for Water Pollution Control and Water Safety in Karst Area, Guilin University of Technology, Guilin 541004, China*

[3]*Department of Bioscience, Aarhus University, 8000C Aarhus, Denmark*

[4]*College of Earth Sciences, Guilin University of Technology, Guilin 541004, China*

[5]*Sino-Danish Centre for Education and Research (SDC), 8000C Aarhus, Denmark*

ABSTRACT

This study aimed at optimizing horizontal subsurface flow constructed wetlands (CWs) to improve hydraulic performance and pollutant removal efficiency. A groundwater modeling package (MODFLOW) was used to optimize three design parameters (length-to-width ratio, inlet/outlet-to-length ratio, and substrate size configuration). Using the optimized parameters, three pilot-scale CWs were built to treat actual wastewater. For model validation, we used a tracer test to evaluate hydraulic performance, and investigated the pollutant spatial distributions and removal efficiencies. We conclude that MODFLOW is suitable for designing CWs, accurately predicting that increasing hydraulic conductivity from surface to bottom layers could improve performance. However, the effect of vegetation, which decreased the hydraulic conductivity of the surface layer, should be considered to improve simulation results. Multilayer substrate configuration, with increasing hydraulic conductivity from the surface to bottom layers, significantly increased pollutant removal compared with monolayer configuration. The spatial variation in pollutant transport and degradation through the filling substrate showed that the multilayer configuration was able to increase use of the available space and moderately reduced short-circuiting and dead zones. Thus, multilayer CWs had higher experimental retention times, effective volume fractions and hydraulic efficiencies, and lower short-circuiting compared with monolayer CWs operating under similar conditions.

Keywords: constructed wetlands; MODFLOW; hydraulic conductivity; hydraulic performance; flow distribution; substrate size configuration; tracer test; multilayer; pollutant spatial distributions

1. INTRODUCTION

As an economic, robust, and sustainable technology, horizontal subsurface flow constructed wetlands (hereafter simplified to CWs) have been widely used for the treatment of domestic, agricultural, and industrial wastewater [1,2]. In an ideally designed CW, water should flow horizontally from the inlet through the wetland bed to the outlet, with very little hydraulic dispersion. However, poor hydrodynamic behavior due to clogging is commonly found in full-scale CWs and have resulted in preferential flow pathways and dead zones [3]. Clogging may cause a large proportion of the water to pass over the bed as surface runoff, thus significantly reducing the effective volume of the system for removing waste [4]. The resulting ineffective wetland bed leads to incomplete biodegradation and reduces the performance of CWs [5,6].

In recent years, CW design has evolved to mitigate the adverse effects of clogging [4,7]. Previous studies aimed at increasing hydraulic retention time and maximizing contact opportunity between the substrate and wastewater to improve treatment performance [8]. Some important design parameters affecting retention time in CWs include length-to-width ratio, inlet/outlet-to-length ratio, and configuration of substrate size. According to García, et al. [9], the length-to-width ratio is one of the most important factors for enhancing the hydraulic and treatment performance of CWs. A relatively high length-to-width ratio can improve hydraulic behavior by reducing the internal dispersion of the water flow, thus forcing the water to pass through the whole cross-section of the wetland bed [10]. In recent years, the design of CWs with a low length-to-width ratio (<1) and large inlet cross-section was also designed to reduce the impact of total suspended solids (TSS) on the clogging in the first part of CWs' bed. However, spreading the influent over too wide an area presents a number of hydraulic challenges, especially for small systems, where the flow rate is close to zero [11]. The inlet/outlet-to-length ratio—defined as the ratio of the length of the inlet distribution zone or the outlet collection zone to the whole length of the CW—is also important to prevent the development of preferential paths and dead zones and, thus, affects treatment efficiency [12]. The substrate used in CWs is also an important factor affecting pollutant removal because most pollutants are removed through biodegradation by biofilm. Substrate configuration can be improved by placing finer media with low hydraulic conductivity on the top and coarse media with high hydraulic conductivity on the bottom, which should theoretically prevent water short-circuiting of the water flow by passing on the surface and decreasing the dead zones at the bottom. It would also make the water flow pattern more uniform. Here, we must note that previous studies have predominantly been based on random selection of parameters [7]. No studies have accurately quantified water flow distribution by integrating different design parameters in CWs to improve hydraulic and treatment performance.

The aims of the present study are to optimize the hydraulic performance of CWs through a computer model simulation and to validate the model by measuring the hydraulic performance and pollutant removal efficiencies of pilot-scale systems that were built to model specifications (Figure 1). The computer simulation study was conducted using the MODFLOW modeling package, which is a finite difference block-centered three-dimensional groundwater modeling package. The modeling package can simulate steady-state and transient groundwater flows for different hydrogeological systems. Based on the characteristics, the modeling package may be suitable for quantifying the flow patterns in CWs [13,14]. To optimize the flow distribution, we manipulated three main design parameters: length-to-width ratio, inlet/outlet-to-length ratios, and configuration of substrate size. The crucial determining factor for the flow uniformity was identified from the simulation results. Subsequently, additional models were built to optimize the original models, mainly aiming to select the best values for the determining factors. Based on the optimized model parameters, three pilot-scale CWs were established to treat actual domestic wastewater and validate the model by measuring the hydraulic performance, spatial distribution of the pollutants and removal efficiency.

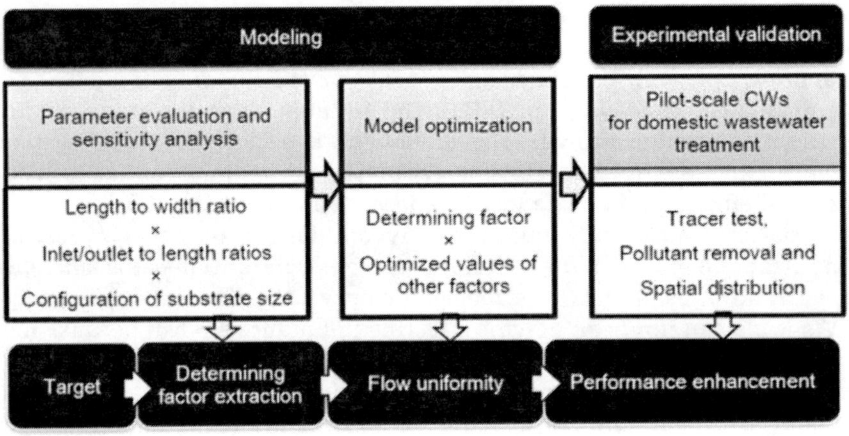

Figure 1. Flow chart of the study approach using simulation and constructed wetlands (CWs).

2. MATERIALS AND METHODS

2.1. Modeling

2.1.1. Model Domain Setup and Calculation

The software program Visual Modflow 4.0 (Waterloo Hydrogeologic, Kitchener, ON, Canada) was used to simulate the flow distributions in the CWs. A CW model domain was set that comprised an influent distribution zone, main substrate zone containing six equally spaced layers, and an effluent collection zone (Figure 2). The six layers can be combined or used separately. A regular grid of 10 × 10 cells was used throughout the model domain. A recharge boundary condition of 0.72 m^3/day was assigned to the distribution zone of CWs using the Visual Modflow Recharge package. A common wetland depth of 0.6 m was chosen for all the modeling. Also, a constant head boundary condition (0.6 m) was assigned to the effluent collection zone. The length of the inlet zone (influent distribution zone) and outlet zone (effluent collection zone) were assigned equal values. The hydraulic conductivity (K value) of the substrate in the influent distribution and effluent collection zones was always set to 500 m/day. The K values of the substrate in the main substrate zone were varied according to configuration of substrate size during the model simulation.

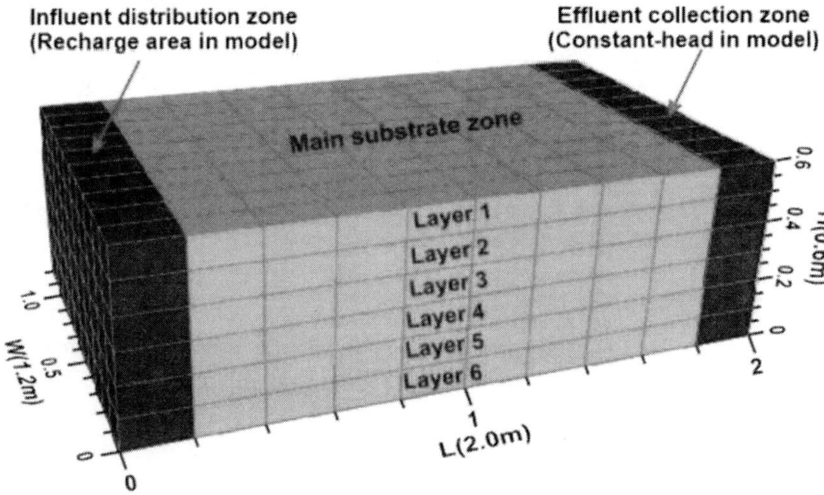

Figure 2. Three-dimensional view of the computational grid, boundary conditions and budget zones of the CW model domain.

The flow distributions in the CW model were calculated according to Equation (1) [15]:

$$\frac{\partial}{\partial x}\left(k_{xx}\frac{\partial h}{\partial x}\right)+\frac{\partial}{\partial y}\left(k_{yy}\frac{\partial h}{\partial y}\right)+\frac{\partial}{\partial z}\left(k_{zz}\frac{\partial h}{\partial z}\right)-w=S\frac{\partial h}{\partial t} \tag{1}$$

where k_{xx}, k_{yy}, k_{zz} are the hydraulic conductivities along the *x* (width), *y* (length), and *z* (depth) coordinate axes, which are assumed to be parallel to the major axes of hydraulic conductivity (m/day); *h* is the potentiometric head (m); *w* is a volumetric flux per unit volume (m^3/day water/m^3 substrate; day^{-1}) representing sources and/or sinks of water, with $w < 0$ for flow out of the CW system and $w > 0$ for flow into the system; *S* is the specific storage of the porous material (m^3 water/m^3 substrate/m hydraulic head; m^{-1}); and *t* is time (day).

2.1.2. Model Evaluation

The simulations were conducted for the CW models using three length-to-width ratios (1:1, 2:1, and 3:1). Each model was assigned one of the three inlet/outlet-to-length ratios (1:5, 1:10, and 1:20). Additionally, for each combination of length-to-width ratio and inlet/outlet-to-length ratio, three configurations of substrate size and layout were used: (1) a monolayer configuration (substrate with K value of 200 m/day); (2) an ascending three-layer configuration (equally thick substrate layers with K values of 50, 100, and 200 m/day, respectively from surface to bottom); and (3) a descending three-layer configuration (equally thick substrates with K values of 200, 100, and 50 m/day, respectively from surface to bottom). The combination of three different length-to-width ratios, inlet/outlet-to-length ratios, and configurations of substrate size resulted in a complete model design with a series of 27 test models (Table A1). The design factors affecting flow uniformity were evaluated according to the simulation results of this series of 27 test models.

2.1.3. Model Optimization

Subsequently, additional models were built to optimize the original models through the trial and error method, mainly aiming to determine the main factor and its interaction effects with the other two optimized factors. The flow distribution of each model was checked against the water budget, which was computed in the model by dividing the bed into different zones, including the distribution zone, main substrate zone and collection zone (Figure 2). The main substrate zone was divided into six layers of equal thickness, from layer 1 to layer 6. The water budget in the bed and fluxes from the distribution zone to the six layers were calculated by the Zone budget package. The water flow distribution was considered uniform when the fluxes in the six layers were equal to each other.

2.2. Experimental Validation

2.2.1. Experimental System Setup

Three parallel pilot-scale CWs were constructed according to the simulation results, with the dimensions of each system being 200 cm long, 120 cm wide, and 70 cm high (water depth of 60 cm) (Figure 3). The bed was composed of the distribution substrate, the main substrate and the collection substrate. The influent distribution zone and the effluent collection zone had the same dimensions of 20 cm × 120 cm × 70 cm (length × width × height) and were composed of gravel (30–80 mm in diameter) with a K value of 500 m/day.

According to the computer simulation results, the three CWs differed from each other in the configuration of their main filling substrate. The common materials used in the applied CW systems are sand or gravel, but not soil (particle diameter always below 0.1 mm). These materials with larger particle sizes can limit clogging problems and provide a high surface area for biofilm growth in CWs. Thus sand (for main CWs' bed) and gravel (for inlet and outlet collection zone) were chosen in the present study. CW1 was a monolayer substrate configuration, filled with the original mixed quartz sand (0–6 mm) with a K of 65 m/day. CW3 and CW6 were multilayer substrate configurations, composed of the quartz sand with different grain sizes arranged in layers with the hydraulic conductivity of each layer increasing with depth. CW3 comprised three layers of equal thickness (20 cm). The K values of these layers were 26, 36, and 64 m/day from the surface to the bottom layer, respectively, and the corresponding diameters were 0–0.4 mm, 0.4–0.6 mm, and 1–3 mm. CW6 was built up by six equal layers of 10 cm each. From the top to the bottom, the grain sizes were 0–0.4, 0.4–0.6, 0.6–0.9, 1–2, 2–4, and 4–6 mm, with K values of 26, 36, 43, 55, 75, and 176 m/day, respectively. For each CW, a total of 12 polyvinyl chloride (PVC) pipes were placed at a distance of 0.4, 0.8, 1.2, and 1.6 m from the inlet at depths of 0.1, 0.3, and 0.5 m (Figure 3). All of the PVC pipes were equipped with valves accessible outside each system for water sampling.

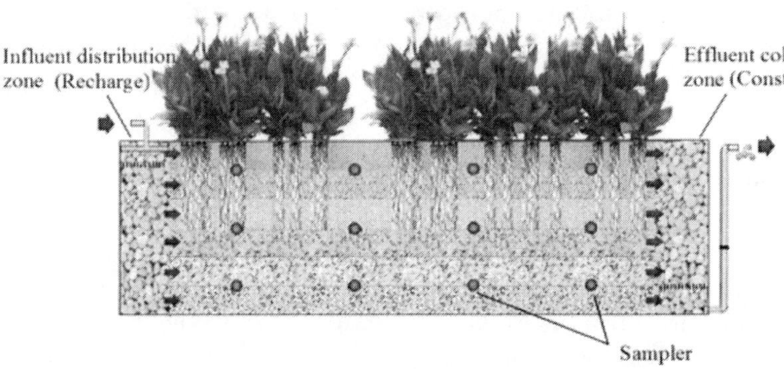

Figure 3. Longitudinal cross-section of each CW showing the sampling locations.

The CWs were established in December 2012 at Guilin University of Technology, in southwest China. All the wetlands were planted with the macrophyte species *Canna indica* at the density of 20 individuals per square meter. The laboratory facility was protected from rain by a glass roof but was exposed to natural temperature variations and light exposure. The research was carried out with actual domestic wastewater pretreated by a septic tank to reduce the suspended solids contents. The CWs were operated with a nominal hydraulic retention time (HRT) of 36 h for most of their life span with a continuous flow rate of 0.72 m^3/day. The experiment was conducted from January 2013 to September 2013, and the water temperatures ranged from 13 to 28 °C during the study period. The following ranges of factors were measured in the influent: the pH was 7.0–7.6, dissolved oxygen (DO) was 0.1–0.58 mg/L, oxidation-reduction potential (ORP) was −168 to −144 mV, concentration of chemical oxygen demand (COD) was 106–238 mg·L^{-1}, ammonia nitrogen (NH_4^+-N) was 5–56 mg·L^{-1}, nitrate nitrogen (NO_3^--N) was 0.5–1.8 mg·L^{-1}, total nitrogen (TN) was 20–68 mg·L^{-1}, and total phosphorus (TP) was 0.8–7.3 mg·L^{-1}.

2.2.2. Water Sampling and Measurement

A total of 14 sampling locations were placed within each CW to provide repeatable points for water quality sampling, and these locations included the inflow, the outflow, and sample valves (Figure 3). The water from the influent and effluent pipe was sampled once a week to evaluate the effects of substrate configuration on pollutant removal. The contaminant concentration profiles were determined using the water sample measurements taken monthly from the sampling valves, and the resulting mean values were used to create isogram maps. During water sampling, the ORP and temperature were measured in situ using a HANNA HI9828 multiparameter portable meter. The sample was then analyzed immediately in the laboratory for COD, NH_4^+-N, and TP by using standard methods (APHA, 1995).

2.2.3. Tracer Test

During the tracer tests, the wetlands were set a nominal hydraulic retention time (HRT) of 12 h. Tracer tests were conducted in each wetland for two sampling campaigns, January 2013 and September 2013 (absence and presence of growing vegetation, respectively), to analyze the hydraulic behavior of the CWs with different substrate configurations. The effects of plants on hydraulic performance were also evaluated by comparing the results of the two tracer tests. Each tracer test was conducted for 5 days by adding a single injection of sodium chloride solution (NaCl, 50 g/L) into the influent distribution zone (flow rate of 3 L/min, 4 min). Samples were taken from the effluent of each wetland and analyzed for chloride concentration every 1 h during the first 2 days and every 4 h during the following 3 days.

HRT was calculated according to the following ratio:

$$tn = V/Q \qquad (2)$$

where is Q the flow rate (m^3/day); and V is the effective volume (m^3).

Hydraulic efficiency (λ) is defined by the ratio:

$$\lambda = t_p/t_n \quad (3)$$

where t_p is the time taken for the tracer to reach a peak in the experimental residence time distribution (RTD) curve (h), and t_n is the nominal HRT (h). The hydraulic efficiency of a wetland can be considered high when $\lambda>0.75$, satisfactory when $0.5<\lambda<0.75$, and low when $\lambda<0.5$ [16].

2.3. Statistical Analyses

Multiway analysis of variance (ANOVA) was performed to detect the differences in water flow distribution among CWs with different design factors. One-way ANOVA was used to detect the effect of substrate configuration on CW performances. A p-level ≤ 0.05 was used to indicate significant differences. The analysis was performed with SPSS.21 (SPSS—Boston, MA, USA: International Business Machines Corp, 2012). The isogram maps of COD, NH_4^+-N, TP, and ORP data were created using MATLAB 7.0 (MathWorks, Natick, MA, USA) to identify data trends within the CWs.

3. RESULTS AND DISCUSSION

3.1. Effects of Design Parameters on Water Flow Distribution

The typical numerical simulation results are shown in Figure 4 for flux (water flow) distribution in each layer of CW models with different length-to-width ratios (3:1, 2:1, and 1:1), inlet/outlet-to-length ratios (1:5, 1:10, and 1:20), and substrate size configuration. The water flux in layer 1 (surface layer) was the highest among the six layers, and it decreased with increasing depth. Layer 6 (bottom layer) generally had the lowest flux compared with the other layers, particularly under monolayer conditions with a high K value (Figure 4c). This is consistent with previous research, which showed that preferential pathways exist in wetland beds, with water flowing mainly in the top layer and dead zones forming along the bottom of the bed, regardless of outlet position [17].

More evenly distributed water flow was correlated with higher length-to-width ratios in both mono- and multilayer CWs (Figure 4a). Higher length-to-width ratios increased the flux in the lower layers and reduced the flux in the upper layers, especially for layer 1. The length-to-width ratio of 3:1 was identified as the best of the three values tested for uniform flow distribution in both CW1 and CW3. Increased length-to-width ratios could shift the experimental retention time closer to the theoretical retention time, and result in higher plug flow and lower dead volume ratios due to the larger distance and the smaller cross-sectional area [17]. However, even under the best length-to-width ratio of 3:1, the flux in layer 1 was still significantly higher than that in layer 6 of CW1 ($p < 0.05$). In contrast, there were no significant differences among fluxes in the six layers ($p > 0.05$) in CW3 using the length to width ratio of 3:1. This indicates that substrate configuration was more important for achieving uniform flow distribution than the length to width ratio.

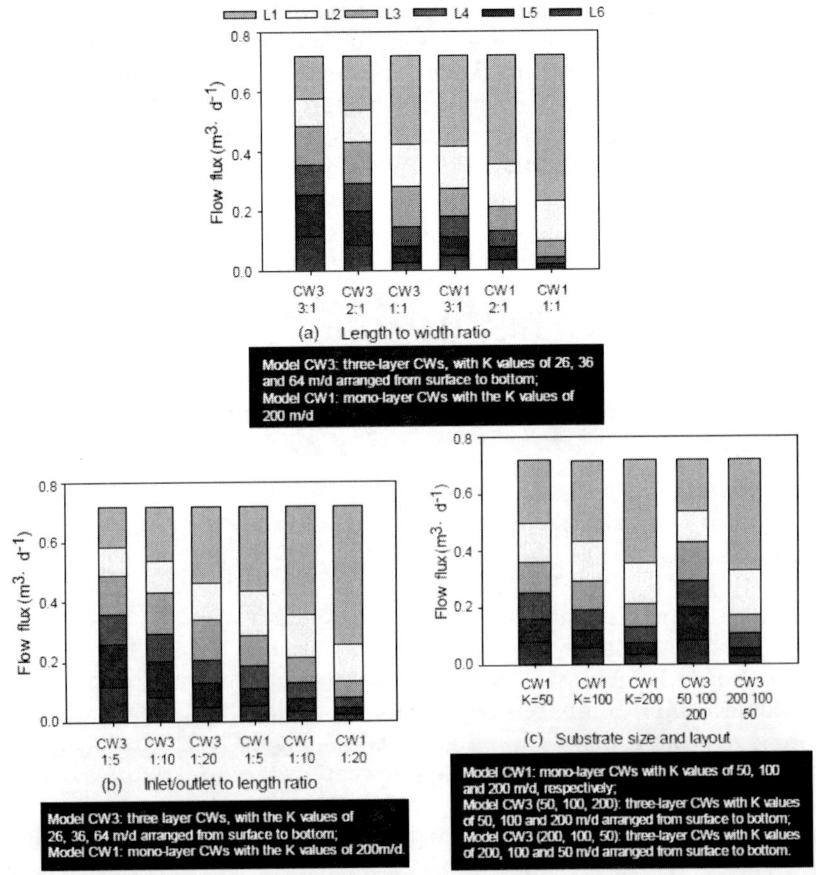

Figure 4. Numerical simulation results of water flux distribution in each layer of the CWs with (**a**) different length-to-width ratios (3:1, 2:1, and 1:1); (**b**) different inlet/outlet-to-length ratios (1:5, 1:10, and 1:20); and (**c**) different K values.

Increasing inlet/outlet-to-length ratio could also mitigate the non-uniformity of water flow in both mono- and multilayer CWs (Figure 4b). Lengthening the influent distribution and the effluent collection zones increased water flux in the bottom layers and reduced the flux in the surface layer, thereby allowing for more uniform water flow. The inlet or outlet length ratio of 1:5 was identified as the best of the three values tested for uniform flow distribution in both CW1 and CW3. However, the difference between layer 1 and layer 6 was still large in the monolayer structured CW1, even for the 1:5 ratio. In contrast, the differences among the six layers were not significant in CW3 under this same ratio. In view of this, even under the optimized inlet/outlet-to-length ratio, substrate configuration was more important for flow uniformity.

The effects of substrate size configuration on water flow distribution in the CWs are shown in Figure 4c. In the monolayer CWs (CW1), the differences among fluxes in the six layers increased with increasing values of substrate K. The increased flux in layer 1 from 0.22 m^3/day (K of 50 m/day) to 0.37 m^3/day (K of 200 m/day) corresponded to the decreased flux in layer 6 from 0.08 to 0.04 m^3/day.

Lower K values in monolayer CWs resulted in higher uniformity of flow distribution. Previous studies have indicated that finer porous media with lower K values could induce hydraulic behavior closer to an ideal plug flow system [9]. However, clogging has usually been found in CWs with substrates possessing K values below 20 m/day [18], thus, bigger substrate sizes with high K values are usually selected in applications to avoid clogging. Nonetheless, this increased substrate size may induce an uneven flow distribution, ultimately reducing the hydraulic performance and removal efficiency in monolayer CWs. In multilayer CWs, a structure using substrates with increasing K values from the surface to bottom layers could result in more uniform patterns of flow distribution (Figure 4a, CW 3, 50, 100, 200 m/day).

In contrast, a structure using substrates with decreasing K values from the surface to bottom layers (Figure 4a, CW 3, 200, 100, 50 m/day) could result in more uneven distribution of the flow field, which was even less uniform than in the monolayer CWs.

These results confirm that a rational arrangement of substrates according to the K values of porous media can reduce the formation of short-circuiting and dead zones [19]. Namely, composing the surface layer from low K value media can prevent water from traveling through the wetland bed on the fast-flow path, thus reducing short-circuiting in CWs. Also, filling the bottom layer with high K-value media (such as coarse porous media) will enhanced the water flux in the bottom layer, and offset the adverse effects of dead zones.

All of the three design parameters—length-to-width ratio, inlet/outlet-to-length ratios, and configuration of substrate size—had significant effects on the flow flux in each layer, and significant interactions were also observed among the three factors for the middle layers (2, 3, and 4). However, no two-way interaction was significant for the surface layer (layer 1). For the bottom layers (layer 5 and layer 6), significant interactions between configuration of substrate size and length to width ratios were observed (Table 1).

Table 1. Results (F-ratios) of analysis of variance (ANOVA) of design parameters on water flux: length to width ratios (L), inlet/outlet to length ratios (I), and configuration of substrate size (S). Layer 1 represents the surface layer and layer 6 is the bottom layer.

Variable	L	I	S	L × I	L × S	I × S
Layer 1	109 *	378 *	430 *	0.5	0.7	3
Layer 2	42 *	507 **	183 *	1 *	16 *	8 *
Layer 3	215 *	248 *	300 *	22 *	15 *	14 *
Layer 4	163 *	56 *	173 *	10 *	8 *	8 *
Layer 5	202 *	17 *	394 *	6 *	26 *	3
Layer 6	2526 *	29 *	3376 *	4	315 *	5 *

Notes: * Indicates statistically significant values ($p < 0.05$), ** $p < 0.01$.

3.2. Model Optimization

Using the selected length-to-width ratio of 3:1 and inlet/outlet-to-length ratio of 1:5, one monolayer CW and two multilayer CW models (3-layer and 6-layer) were built to optimize for configuration of substrate size through the trial and error method. Furthermore, a monolayer CW1 with the K value of 65 m/day was built simultaneously as a control to check for the validity of the simulation results.

The most optimized parameters for substrate size configuration in multilayer CWs to even out the distribution of water flow among the layers are shown in Figure 5. We identified two configurations of substrate size and layout that met the criteria of vertically uniform distribution of water flow: CW3 with three K values of porous media from surface to bottom (26, 36, 64 m/day) and CW6 with six K values from surface to bottom (38, 36, 43, 55, 75, and 176 m/day). The model results show that CW1 resulted in preferential flow along the surface, while the flow fluxes in CW3 and CW6 were vertically uniform.

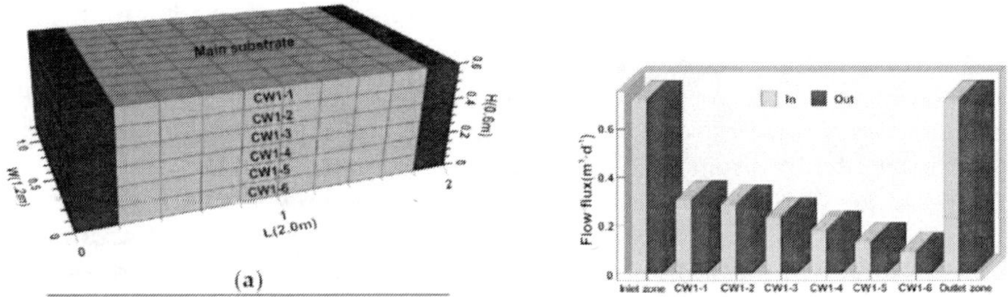

(a)

Zone	CW1-1 to CW1-6
K (m·d⁻¹)	65

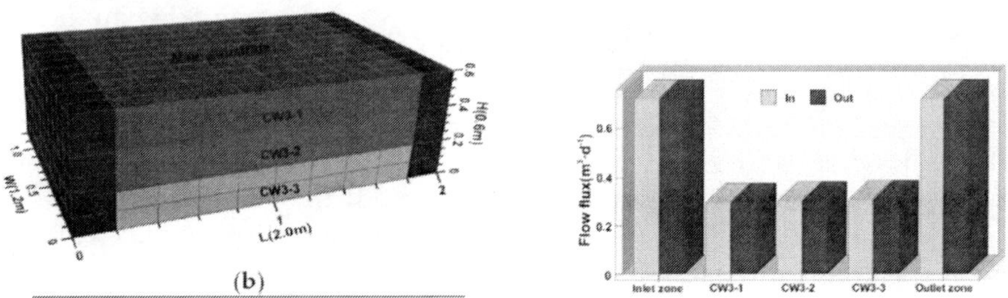

(b)

Zone	CW3-1	CW3-2	CW3-3
K (m·d⁻¹)	26	36	64

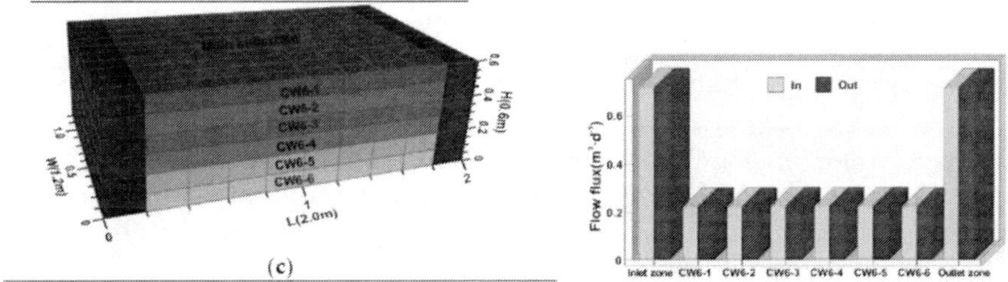

(c)

Zone	CW6-1	CW6-2	CW6-3	CW6-4	CW6-5	CW6-6
K (m·d⁻¹)	38	36	43	55	75	176

Figure 5. Setup of the in situ experiments, the assigned K values, and simulation results of CW1 (**a**); CW3 (**b**); and CW6 (**c**).

3.3. Experimental Validation

3.3.1. Hydraulic Performance

The residence time distribution (RTD) curves of the tracer (chloride), based on the tracer concentration at the outlets for CW1, CW3, and CW6, were affected by absence (Figure 6a, winter) and presence of vegetation growth (Figure 6b, summer). When vegetation was inactive, the peak tracer concentration times (t_p) in CW1, CW3, and CW6 were observed at 8.7, 9.5, and 10.5 h, which are shorter than the nominal hydraulic retention time (t_n of 12 h). The corresponding values of hydraulic efficiency (λ) were 0.73, 0.79, and 0.88. The values of λ in the multilayer CWs were much higher than that in the monolayer CW. Also, more precisely divided layers (6 layers), which increases the hydraulic conductivity more gradually through the layers, were favorable for improving λ. Moreover, the hydraulic performances of CW3 and CW6 can be considered to be high (λ>0.75) Even though the λ of CW1 was very close to 0.75, it should be noticed that the peak tracer concentration was clearly lower in CW1 than that in CW3 and CW6. Even after day 5 (120 h), the tracer could still be detected in the effluent of CW1. This observation suggests that in addition to the value of λ, the amount of tracer recovered could be important in evaluating the hydraulic performances of the three CWs. Overall tracer recoveries were 62%, 85%, and 86% for CW1, CW3, and CW6, respectively. Some loss of the tracer was probably caused by the presence of dead zones at the bottom of the bed. Tracer elements that reach the dead zone may stay for longer periods because short-circuiting water flow, with narrow and continuously fast flow paths and longitudinal dispersion, will not exchange fluid with slow-flowing regions [20].

When vegetation was active during the summer, the peak tracer concentration times(t_p) in CW1, CW3 and CW6 were observed at 9.3, 9.8 and 11.7 h, and the hydraulic efficiencies (λ) were 0.78, 0.82 and 0.98 for CW1, CW3 and CW6, respectively. The increased λ values when compared with the tests in winter indicate the positive effect of growing roots and rhizomes on the hydraulic performances of the three CWs. This has also been observed in previous research, where only the presence of plants seemed to help water to flow at least partially within the porous media; otherwise, water was shown to preferentially flow above the porous media, thus losing half the effective volume of the system [3,7,21]. The tracer recoveries of CW1, CW3, and CW6 were 72%, 82%, and 81%, respectively. For CW1, however, we observed 4 peaks and higher tracer recovery compared with winter values. This may have resulted from plant growth in CW1, which decrease K values of the surface layer substrate. The changes then promote smoother transport of the tracer ions from the bottom dead zones.

The experimental results support the simulation results, which indicated that a multilayer substrate structure with increasing K values from the surface to bottom could improve flow uniformity, in an indirect way. The direct measurement of the flow patterns inside the lab-scale pilot plants needed to be further investigated. The growth of roots and rhizomes could fill the spaces in the porous media in the surface layer, thus decreasing the K values in the surface layer, which would be beneficial for balancing the flow distribution in monolayer CWs. However, the effects of plants were not considered in the simulation models, and these effects should be investigated in future studies.

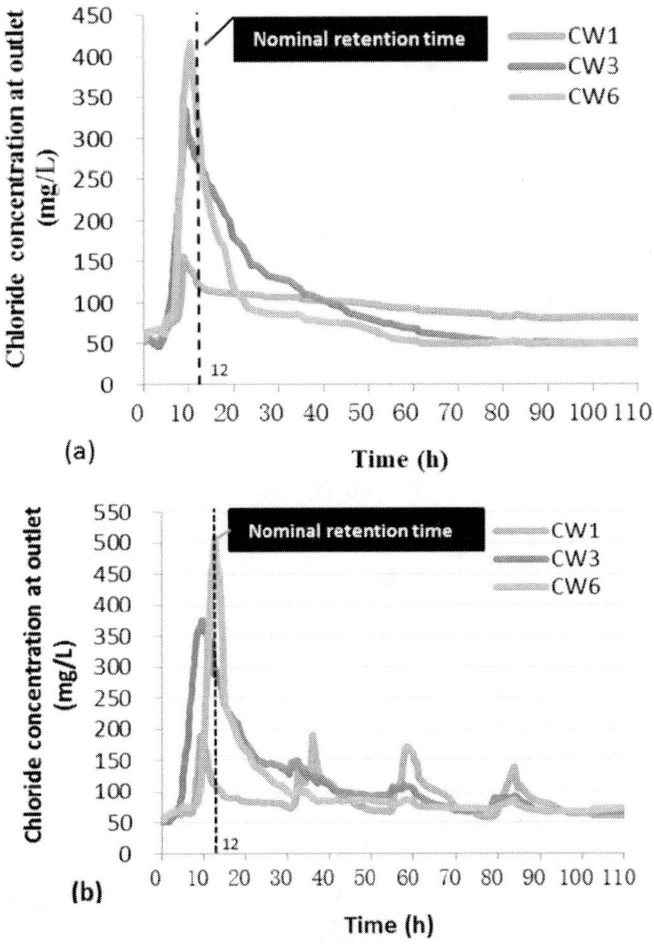

Figure 6. Residence time distribution (RTD) curves of the tracer (chloride) based on tracer concentration at the outlets of CW1, CW3, and CW6, (**a**) without growing plants and (**b**) with growing plants.

3.3.2. Overall Pollutant Removal Efficiencies

As shown in Figure 7, CW6 was significantly better at removing pollutants (72.9% for COD, 83.2% for TP, and 78.4% for NH_4^+-N) than CW1 and CW3; it was especially better than CW1 (38.2% for COD, 50.5% for TP, and 46.9% for NH_4^+-N). A significant effect of the substrate configuration on pollutant removal was observed ($p < 0.05$). The results indicate that multilayer configuration can achieve significantly higher removal rates of pollutants.

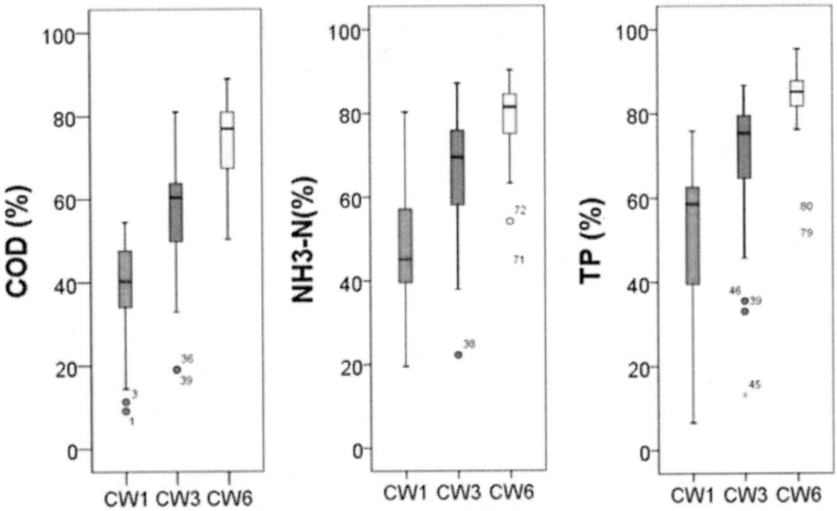

Figure 7. Correlation charts of removal efficiencies and substrate configuration for chemical oxygen demand (COD), total phosphorus (TP), and ammonium (NH_4^+-N).

3.3.3. Water Quality and Spatial Distribution of Pollutants

The spatial distribution of ORP in the CWs is shown in Figure 8. The overall ORP in the substrates of the three CWs indicated anaerobic conditions (less than −70 mV). The mean ORP in the influent water was around −159 mV, whereas the values in the effluents of CW1, CW3, and CW6 were −110, −101, and −96 mV, respectively. The proportion of the bed volume with ORP above −100 mV did not differ significantly between CW3 and CW6, but the proportion was significantly larger than that in CW1. This result indicates that multilayer CWs can improve the ORP in the wetland bed by improving hydraulic performance.

The distributions of COD, NH_4^+-N, and TP in the three CWs are shown in Figure 9. In CW1 (Figure 9a), the COD concentrations were generally lower in the upper than in the bottom layers. The higher COD concentration in the bottom could be attributed to the decline of ORP in dead zones. In CW3 (Figure 9b) and CW6 (Figure 9c), COD displayed a rapid and uniform horizontal decrease in concentration with the majority of removal occurring in the first half of the bed (0–1.0 m along the length). The minimum COD region (lower than 60 mg/L) was limited in CW1, and only appeared from 110 cm onwards to the outlet in the surface layer (from 0 to 20 cm depth), whereas the minimum COD region was more extensive in CW3 and CW6. A decrease of COD to lower than 60 mg/L occurred in nearly half of the whole cross-sectional area, especially on the surface of the bed near the outlet.

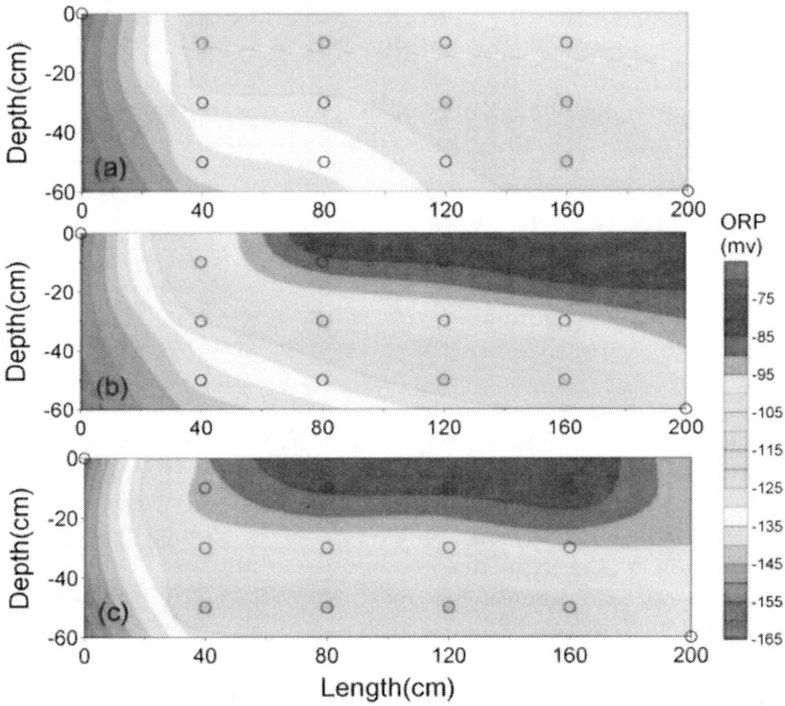

Figure 8. Profiles of oxidation-reduction potential (ORP) in CW-1 (**a**); CW-3 (**b**); and CW-6 (**c**).

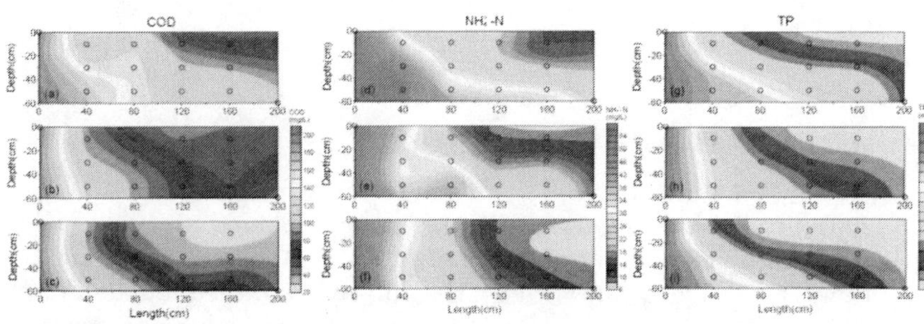

Figure 9. Spatial distribution of the pollutants COD (**a–c** for CW1, CW3, and CW6, respectively), NH4+-N (**d–f** for CW1, CW3, and CW6, respectively), and TP (**g–i** for CW1, CW3, and CW6, respectively) in the three experimental CWs.

For NH_4^+-N, the distributions showed similar trends to COD. The mean outlet NH_4^+-N concentrations were 27, 16, and 9 mg/L in CW1, CW3, and CW6, respectively. In CW1 (Figure 9d), there was an obvious decrease of NH_4^+-N in the upper layers. However, the degradation in the bottom layer was not significant. In CW3 (Figure 9e) and CW6 (Figure 9f), however, there was much faster removal of NH_4^+-N, especially in CW6. The NH_4^+-N removal rate in CW6 almost matched the trend found in horizontal plug flow. The minimum NH_4^+-N region in CW6 (lower than 10 mg/L) was significantly larger than those in the other two CWs.

The mean outlet TP concentrations were 0.7, 0.3, and 0.2 mg/L in CW1, CW3, and CW6, respectively. In CW1 (Figure 9g), higher levels of TP remained in the bottom, reaching 1.5 mg/L at 160 cm along the flow path, whereas TP in upper layers at the same horizontal position was only 0.5 mg/L. The minimum TP region, lower than 0.5 mg/L, was obviously more extensive in CW3 (Figure 9h) and CW6 (Figure 9i) than in CW1. Especially in CW6, the minimum region appeared from 30 to 200 cm along the surface layer, which amounts to nearly half of the whole cross-sectional area. The spatial distribution of TP in the three CWs indicates the positive impact of multilayer CWs on TP removal. However, it should be noted that phosphorus removal largely depends on the adsorption by the substrate in CWs [22,23]. When the phosphorus concentration of the effluents exceeds a certain level, which was not reached in our experiments, this indicates that most of the substrate in the CW systems has reached adsorption saturation levels for phosphorus [24,25].

The patterns of spatial distribution and removal of pollutants in the three CWs indicate that the multilayer filling configuration facilitates full use of the bed space available and moderately reduces short-circuiting and the extent of the dead zone. The higher removal efficiencies of CW6 compared to CW3 showed that a more fine-scale division of layers favors higher CW performance.

4. CONCLUSIONS

The configuration of substrate size and layout played a crucial role in both the hydraulic behavior and pollutant removal efficiency of the CWs. A multilayer configuration with increasing hydraulic conductivity from the surface to bottom can make better use of the available bed space and moderately reduce short-circuiting and the extent of dead zones, significantly improving removal performance compared with the monolayer configuration. The similarity between the simulation and experimental results suggests that MODFLOW software can be effective in designing constructed wetland systems. This software could be adopted to design CWs and achieve uniform distributions of wastewater over the processing bed. However, the effects of precipitation, evapotranspiration, and plant growth on hydraulics should also be considered in future studies. Also, the hydraulic model can be extended with various process-based models derived from the various processes occurring in wetlands, such as adsorption/desorption and microbial degradation and conversion.

APPENDIX A

Table A1. Modeling test series with the parameters of length-to-width ratios, inlet/outlet-to-length ratios, and three configurations of the substrate size.

Models	Length to Width Ratios	Inlet/Outlet to Length Ratios	Configurations of Substrate Size, K (m/Day) (Listed in Order of Surface to Bottom Layers)
Model 1	1:1	1:5	monolayer, K = 200
Model 2	1:1	1:5	three-layers, K = 50, 100 and 200
Model 3	1:1	1:5	three-layers, K = 200, 100 and 50
Model 4	1:1	1:10	monolayer, K = 200
Model 5	1:1	1:10	three-layers, K = 50, 100 and 200
Model 6	1:1	1:10	three-layers, K = 200, 100 and 50
Model 7	1:1	1:20	monolayer, K = 200
Model 8	1:1	1:20	three-layers, K = 50, 100 and 200
Model 9	1:1	1:20	three-layers, K = 200, 100 and 50
Model 10	2:1	1:5	monolayer, K = 200
Model 11	2:1	1:5	three-layers, K = 50, 100 and 200
Model 12	2:1	1:5	three-layers, K = 200, 100 and 50
Model 13	2:1	1:10	monolayer, K = 200
Model 14	2:1	1:10	three-layers, K = 50, 100 and 200
Model 15	2:1	1:10	three-layers, K = 200, 100 and 50
Model 16	2:1	1:20	monolayers, K = 200
Model 17	2:1	1:20	three-layers, K = 50, 100 and 200
Model 18	2:1	1:20	three-layers, K = 200, 100 and 50
Model 19	3:1	1:5	monolayer, K = 200
Model 20	3:1	1:5	three-layers, K = 50, 100 and 200
Model 21	3:1	1:5	three-layers, K = 200, 100 and 50
Model 22	3:1	1:10	monolayer, K = 200
Model 23	3:1	1:10	three-layers, K = 50, 100 and 200
Model 24	3:1	1:10	three-layers, K = 200, 100 and 50
Model 25	3:1	1:20	monolayer, K = 200
Model 26	3:1	1:20	three-layers, K = 50, 100, and 200
Model 27	3:1	1:20	three-layers, K = 200, 100 and 50

ACKNOWLEDGMENTS

This work was funded by the National Natural Science Foundation of China (No. 51408147, 41404116, 51638006) and the Natural Science

Foundation of Guangxi (No. 2014GXNSFBA118234, Guikehe1599005-2-2). It was further supported by the Guangxi Scientific Experiment Center of Mining, Metallurgy and Environment (KH2012ZD004) and the project of high level innovation team and standing scholar in Guangxi colleges and universities (002401013001). The language was edited by Esther Chang.

REFERENCES

1. Brix, H.; Schierup, H.H.; Arias, C.A. Twenty years experience with constructed wetland systems in Denmark—What did we learn? *Water Sci. Technol.* **2007**, *56*, 63–68.
2. Vymazal, J. The use of sub-surface constructed wetlands for wastewater treatment in the Czech Republic: 10 years experience. *Ecol. Eng.* **2002**, *18*, 633–646.
3. Ríosa, D.A.; Vélez, F.T.A.; Pena, M.R.; Parra, C.A.M. Changes of flow patterns in a horizontal subsurface flow constructed wetland treating domestic wastewater in tropical regions. *Ecol. Eng.* **2009**, *35*, 274–280.
4. Brovelli, A.; Carranza-Diaz, O.; Rossi, L.; Barry, D.A. Design methodology accounting for the effects of porous medium heterogeneity on hydraulic residence time and biodegradation in horizontal subsurface flow constructed wetlands. *Ecol. Eng.* **2011**, *37*, 758–770.
5. Knowles, P.R.; Griffin, P.; Davies, P.A. Complementary methods to investigate the development of clogging within a horizontal sub-surface flow tertiary treatment wetland. *Water Res.* **2010**, *44*, 320–330.
6. Knowles, P.; Dotro, G.; Nivala, J.; García, J. Clogging in subsurface-flow treatment wetlands: Occurrence and contributing factors. *Ecol. Eng.* **2011**, *37*, 99–112.
7. Pedescoll, A.; Sidrach-Cardona, R.; Sa'nchez, J.C.; Carretero, J.; Garfi, M.; Be'cares, E. Design configurations affecting flow pattern and solids accumulation in horizontal free water and subsurface flow constructed wetlands. *Water Res.***2013**, *47*, 1448–1458.
8. Morales, L.P.; Franco, M.; Garvi, D.; Lebrato, J. Influence of the stone organization to avoid clogging in horizontal subsurface-flow treatment wetlands. *Ecol. Eng.* **2013**, *54*, 136–144.
9. García, J.; Aguirre, P.; Barragán, J.; Mujeriego, R.; Matamoros, V.; Bayona, J.M. Effect of key design parameters on the efficiency of horizontal subsurface flow constructed wetlands. *Ecol. Eng.* **2005**, *25*, 405–418.
10. García, J.; Chiva, J.; Aguirre, P.; lvarez, E.A.; Sierra, J.P.; Mujeriego, R. Hydraulic behaviour of horizontal subsurface flow constructed wetlands with different aspect ratio and granular medium size. *Ecol. Eng.* **2004**, *23*, 177–187.
11. Kadlec, R.H.; Wallace, S. *Treatment Wetlands*; CRC Lewis Regional Relevance: Boca Raton, FL, USA, 2008.
12. Suliman, F.; Futsaether, C.; Oxaal, U.; Haugen, L.E.; Jenssen, P. Effect of the inlet–outlet positions on the hydraulic performance of horizontal subsurface-flow wetlands constructed with heterogeneous porous media. *J. Contam. Hydrol.* **2006**, *87*, 22–36.
13. Liolios, K.A.; Moutsopoulos, K.N.; Tsihrintzis, V.A. Modeling of flow and bod fate in horizontal subsurface flow constructed wetlands. *Chem. Eng. J.* **2012**, *200–202*, 681–693.
14. Ronkanen, A.K.; Kløve, B. Hydraulics and flow modelling of water treatment wetlands constructed on peatlands in Northern Finland. *Water Res.* **2008**, *42*, 3826–3836.
15. Harbaugh, A. Modflow-2005, the US geological survey modular ground-water model—The ground-water flow process. In *US Geological Survey Techniques of Water Resources Investigation Book*; U.S. Department of the Interior, U.S. Geological Survey: Washington, DC, USA, 2005.
16. Persson, J.; Somes, N.L.G.; Wong, T.H.F. Hydraulics efficiency of constructed wetlands and ponds. *Water Sci. Technol.***1999**, *40*, 291–300.

17. Alcocer, D.J.R.; Vallejos, G.G.; Champagne, P. Assessment of the plug flow and dead volume ratios in a sub-surface horizontal-flow packed-bed reactor as a representative model of a sub-surface horizontal constructed wetland. *Ecol. Eng.* **2012**, *40*, 18–26.
18. Pedescoll, A.; Uggetti, E.; Llorens, E.; Granés, F.; García, D.; García, J. Practical method based on saturated hydraulic conductivity used to assess clogging in subsurface flow constructed wetlands. *Ecol. Eng.* **2009**, *35*, 1216–1224.
19. Schincariol, R.A.; Schwartz, F.W. An experimental investigation of variable density flow mixing in homogeneous and heterogeneous media. *Water Resour. Res.* **1990**, *26*, 2317–2329.
20. Lightbody, A.F.; Nepf, H.M.; Baysb, J.S. Modeling the hydraulic effect of transverse deep zones on the performance of short-circuiting constructed treatment wetlands. *Ecol. Eng.* **2009**, *35*, 754–768.
21. Suliman, F.; French, H.K.; Haugen, L.E.; Søvik, A.K. Change in flow and transport patterns in horizontal subsurface flow constructed wetlands as a result of biological growth. *Ecol. Eng.* **2006**, *27*, 124–133.
22. Jiang, C.; Jia, L.; He, Y.; Zhang, B.; Kirumba, G.; Xie, J. Adsorptive removal of phosphorus from aqueous solution using sponge iron and zeolite. *J. Colloid Interface Sci.* **2013**, *402*, 246–252.
23. Tuszynska, A.; Kolecka, K.; Quant, B. The influence of phosphorus fractions in bottom sediments on phosphate removal in semi-natural systems as the 3rd stage of biological wastewater treatment. *Ecol. Eng.* **2013**, *53*, 321–328.
24. Vohla, C.; Kõiv, M.; Bavor, H.J.; Chazarenc, F.; Mander, Ü. Filter materials for phosphorus removal from wastewater in treatment wetlands—A review. *Ecol. Eng.* **2011**, *37*, 70–89.
25. Dunne, E.J.; Coveney, M.F.; Marzolf, E.R.; Hoge, V.R.; Conrow, R.; Naleway, R.; Lowe, E.F.; Battoe, L.E. Efficacy of a large-scale constructed wetland to remove phosphorus and suspended solids from Lake Apopka, Florida. *Ecol. Eng.***2012**, *42*, 90–100.

CHAPTER 5

Sediment Accumulation in Prairie Wetlands under a Changing Climate: the Relative Roles of Landscape and Precipitation

Susan K. Skagen[1], Lucy E. Burris[2], Diane A. Granfors[3]

[1]*U.S. Geological Survey, Fort Collins Science Center, 2150 Centre Avenue, Bldg. C, Fort Collins, CO 80526, USA*

[2]*Colorado State University, Fort Collins, CO 80523, USA*

[3]*U.S. Fish and Wildlife Service, 1011 East Tudor Rd, Anchorage, AK 99503, USA*

ABSTRACT

Sediment accumulation threatens the viability and hydrologic functioning of many naturally formed depressional wetlands across the interior regions of North America. These wetlands provide many ecosystem services and vital habitats for diverse plant and animal communities. Climate change may further impact sediment accumulation rates in the context of current land use patterns. We estimated sediment accretion in wetlands within a region renowned for its large populations of breeding waterfowl and migrant shorebirds and examined the relative roles of precipitation and land use context in the sedimentation process. We modeled rates of sediment accumulation from 1971 through 2100 using the Revised Universal Soil Loss Equation (RUSLE) with a sediment delivery ratio and the Unit Stream Power Erosion Deposition model (USPED). These models predicted that by 2100, 21–33 % of wetlands filled completely with sediment and 27–46 % filled by half with sediments; estimates are consistent with measured sediment accumulation rates in the region reported by empirical studies. Sediment accumulation rates were strongly influenced by size of the catchment, greater coverage of tilled landscape within the catchment, and steeper slopes. Conservation efforts that incorporate the relative risk of infilling of wetlands with sediments, thus emphasizing areas of high topographic relief and large watersheds, may benefit wetland-dependent biota.

Keywords: Climate change, Northern Great Plains, Revised Universal Soil Loss Equation, Sedimentation, Stream Power Erosion Deposition Model, Wetland

1. INTRODUCTION

Naturally formed depressional wetlands of midcontinental North America provide vital habitats that support biodiversity at many trophic levels, including terrestrial wildlife species (Tiner 2003). These wetlands also deliver numerous ecosystem services, including surface-water storage and flood-damage protection, ground-water recharge, and global cycling of nutrients (van der Kamp and Hayashi 1998; Tiner 2003; Smith et al. 2011). Many of these wetlands are completely surrounded by upland and are highly vulnerable to land conversion, habitat fragmentation, water pollution, ground-water withdrawal, and drought (Donald et al. 1999; Gleason et al. 2003; Tiner 2003). Sediment accumulation can also threaten the viability and hydrologic functioning of many wetlands embedded within the agricultural landscapes of the Great Plains (Smith et al. 2011; Burris and Skagen 2012; Daniel et al. 2014). For example, nearly 90 % of the shallow depressional wetlands, or playas, in the west-central Great Plains are predicted to completely fill with sediments by 2100 (Burris and Skagen 2012).

The Prairie Pothole Region (PPR) within the north-central United States (U.S., U.S. PPR) is renowned for its ecological value and has the greatest density of naturally formed wetlands in the continental U.S., with >2.6 million wetlands in North Dakota and South Dakota (Reynolds et al. 2006). These prairie wetlands provide breeding habitat for 50–80 % of the continent's waterfowl (Batt et al. 1989), including ca 4 million breeding pairs of dabbling ducks (Reynolds et al. 2006), and support >7 million en route migrant shorebirds during spring migration (Niemuth et al. 2006;

Skagen et al. 2008). Many wetlands of the U.S. PPR have been modified or eliminated though agricultural intensification over the past 200 years (Tiner 1984; Dahl 2011), and conversion of native landscapes to cropland continues today (Sylvester et al. 2013). In response to a recent rise in commodity prices and demand for biofuels, a pulse of grassland conversion to corn and soybean cropping has occurred in the last decade on lands with high erosion risk and in close proximity to wetlands (Johnston 2013; Wright and Wimberly 2013).

Sediment accumulation in wetlands is a natural process that is greatly accelerated by cultivation practices (Gleason and Euliss 1998; Smith et al. 2011; Burris and Skagen 2012). Sediments eroded from surrounding upland soils accumulate in wetland basins and decrease basin volume, leading to altered hydrologic function, shortened effective life-span, and diminished ecological utility of wetlands (Gleason and Euliss 1998; Tangen and Gleason 2008). As little as 0.5 cm of sediment is sufficient to significantly reduce emergence of both wetland seedlings and invertebrates, the primary food base of wetland-dependent vertebrates (Jurik et al. 1994; Gleason et al. 2003). Furthermore, when sediments diminish basin depth to <1 m, tall robust emergent perennials such as cattails (*Typha* spp.) easily outcompete annual species to eliminate open shoreline and open water habitats important to waterfowl and shorebirds (Gleason 2001:11).

Impending climate change may significantly impact rainfall runoff-driven sediment accumulation rates in the context of current land use patterns if precipitation patterns and amount differ substantially from contemporary climates. Increases in annual precipitation, changing seasonal precipitation patterns, and higher summer temperatures are predicted across the northern Great Plains by atmospheric-oceanic general circulation models (AOGCMs; Christensen et al. 2007; Ballard et al. 2014).

Here we examine the relative roles of increasing precipitation and land use context in the sedimentation process across the U.S. portion of the Prairie Pothole Region.

We modeled rates of sedimentation, sediment depths, and resultant depths of palustrine wetlands across the U.S. PPR from 1971 through 2100. We estimated erosion using two methods, (1) the Revised Universal Soil Loss Equation (RUSLE; Renard et al. 1997), a field-scale analysis approach applied at the regional scale, which included upland land use, general topography, erosivity (estimated from precipitation), soil texture, and conservation practice inputs; and (2) the Unit Stream Power Erosion Deposition model (USPED; Mitasova et al. 1996), an improved method that also accounted for topographic complexity of the landscape. Both methods assume overland flow of runoff water from rainfall as the sole mechanism for sediment transport. Conservation planning for these critical wetlands and their associated biota will be informed by the spatially explicit predictions of wetland sediment accumulation patterns.

2. METHODS

2.1 Study Area

Modern day pothole wetlands within the Prairie Pothole Region were formed as the ice sheet of the Wisconsin glaciations retreated 12,000 to 10,000 years ago (Kehew and Teller 1994), leaving depressions that filled with water to form wetlands of varying types: lakes and semipermanent wetlands that are hydrologically connected to groundwater, and shallower temporary and seasonal wetlands that fill with precipitation and runoff (Winter and Rosenberry 1995). Our 350,000 km^2 study area in the U.S. PPR extended across parts of North Dakota, South Dakota, Minnesota, and Iowa, USA, and spanned several major physiographic regions (Fig. 1). Elevation ranges from 200 to 750 masl (Gesch et al. 2002; Gesch 2007). The eastern lower elevation and relatively flat areas are known primarily as the drift prairies, although for simplicity, we have grouped this physiographic region with Glacial Lake Agassiz, which extends along the eastern border of North Dakota. Along the south-central and western sides of the study area rise the Prairie and Missouri coteaus, hilly moraines of glacial debris now covered with grassland and wetlands.

The rich prairie soils of the drift prairies have been extensively converted to agriculture, with <1 % remaining in native prairie in Minnesota and Iowa. Drainage of 50 % of wetland basins in Minnesota and 97 % in Iowa have accompanied this conversion (Dahl 1990; Dahl 2011). The landscape has been less altered in the Dakotas, with about 40 % of wetlands drained and 27 % remaining in native cover; the thinner, rockier soils of the coteaus are better suited to grass-based agriculture (i.e., grazing and haying) than production of crops through tillage. The climate is spatially variable and highly dynamic between years in temperature (ranging from −40 °C in winter to 40 °C in summer) and in annual precipitation (ranging from 300 mm $yr.^{-1}$ in the northwestern region to 900 mm yr^{-1} in the southeast; Millett et al. 2009).

2.2 Spatial Data

Wetland basin perimeters were developed by the U.S. Fish and Wildlife Service Habitat and Population Evaluation Team as defined in Cowardin (1982); Cowardin et al. (1995) and Johnson and Higgins (1997), where each basin is defined by the deepest water regime within contiguous wetlands delineated by the National Wetlands Inventory (NWI; U.S. Department of the Interior 2009. We included wetlands ranging from temporary (holding water for a few weeks at the start of the growing season), seasonal (holding water approximately one month into the growing season), semipermanent (holding water throughout the growing season) to permanent (or lakes, holding water throughout the year) (Cowardin et al. 1995). We excluded riverine wetlands and the "saturated" wetland category that was delineated only in Minnesota.

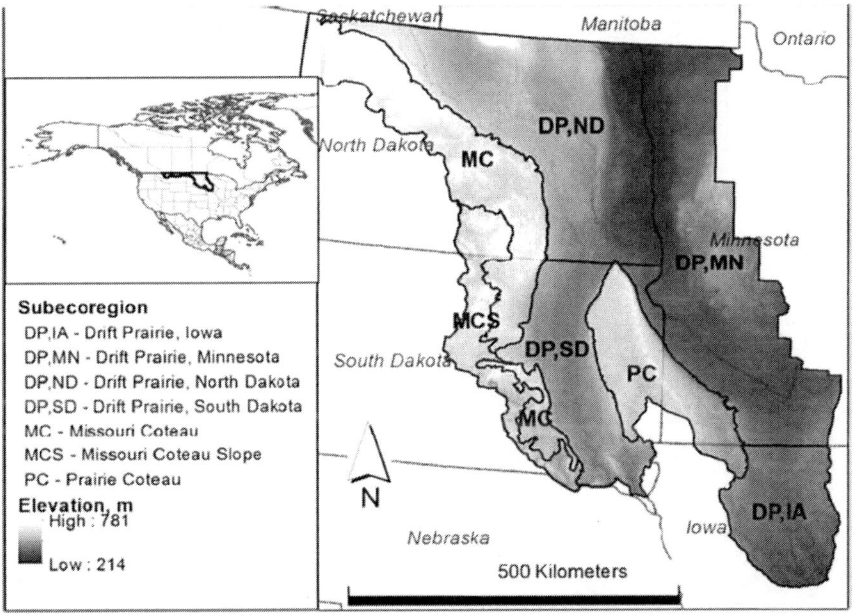

Figure. 1 Subecoregions and elevation profile within the study region portion of the U.S. Prairie Pothole Region (U.S. PPR). Inset map shows the full extent of U.S. PPR

We used the ESRI ArcGIS WATERSHED command (ESRI 1999-2008) with 1 arc-second elevation data (Gesch et al. 2002; Gesch 2007) to establish flow direction with wetland basins as pourpoints and to determine the surrounding catchment area that could contribute sediment into each wetland basin (Online Resource 1, Fig. S2). We estimated basin volume by assigning basins to subecoregions of the PPR and using equations developed by Gleason et al. (2007: Table 1) based on survey-based morphometry data.

Table 1 Predicted loss of palustrine wetland basins by regime by 2100 due to sediment accumulation. Revised Universal Soil Loss Equation ($RUSLE_{sdr}$) model with a sediment delivery ratio and the Unit Stream Power Erosion Deposition (USPED) model were applied to subecoregions of the U.S. Prairie Pothole Region using nine averaged scenario-model combinations of projected climate change models

Subecoregion	Method	Temporary		Seasonal		Semipermanent		All Palustrine Wetlands	
		Total Wetlands	Wetlands Filled (%)	Total Wetlands	Wetlands Filled (%)	Total Wetlands	Wetlands Filled (%)	Total Wetlands	Wetlands Filled (%)
Drift Prairie, ND	$RUSLE_{sdr}$	166,302	51,286 (31)	250,962	80,200 (32)	42,581	11,218 (26)	459,845	142,704 (31)
	USPED		40,439 (24)		45,880 (18)		4904 (12)		91,223 (20)
Drift Prairie, MN	$RUSLE_{sdr}$	44,934	18,747 (42)	104,209	53,555 (51)	38,357	16,652 (43)	187,500	88,955 (47)
	USPED		12,201 (27)		25,182 (24)		6546 (17)		43,929 (24)
Drift Prairie, IA	$RUSLE_{sdr}$	8656	4836 (56)	10,191	7179 (70)	2991	1665 (56)	21,838	13,681 (63)
	USPED		3009 (35)		4223 (41)		1076 (36)		8308 (38)
Drift Prairie, SD	$RUSLE_{sdr}$	145,660	41,478 (29)	94,286	23,869 (25)	23,688	5762 (24)	263,634	71,110 (27)
	USPED		36,083 (25)		16,810 (18)		3782 (16)		56,676 (22)
Prairie Coteau	$RUSLE_{sdr}$	42,589	17,683 (42)	57,900	20,764 (36)	30,720	8400 (27)	131,209	46,847 (36)
	USPED		12,912 (30)		12,666 (22)		4358 (14)		29,936 (23)
Missouri Coteau Slope	$RUSLE_{sdr}$	6013	2112 (35)	5910	2057 (35)	2054	836 (41)	13,977	5004 (36)
	USPED		1871 (31)		1424 (24)		620 (30)		3915 (29)
Missouri Coteau	$RUSLE_{sdr}$	138,743	43,746 (32)	285,325	83,155 (29)	66,146	16,082 (24)	490,214	142,982 (29)
	USPED		36,996 (27)		44,928 (16)		8220 (12)		90,144 (18)
All subecoregions	$RUSLE_{sdr}$	552,897	179,889 (33)	808,783	270,779 (34)	206,537	60,615 (30)	1,568,217	511,282 (33)
	USPED		143,510 (26)		151,115 (19)		29,506 (15)		324,131 (21)

All calculations were conducted on 30-m × 30-m grid cells. We estimated sediment accumulation rates for catchments larger than our minimum mapping unit (900 m^2) and for which all input values (slope, land use, soil erodibility, and rainfall erosivity) were available.

For computations of erosivity (R), we used historical precipitation data (1981–2010) for reference conditions and downscaled 8 km resolution precipitation data from three AOGCMs (the Australian Commonwealth Scientific and Industrial Research Organization model CSIRO-MK3.5, the Canadian Centre for Climate Modeling and Analysis model CGCM2.1MR, and the Japanese Centre for Climate System Research Model for Interdisciplinary Research on Climate model MIROC Version 3.2) and three IPCC 4 (United Nations Intergovernmental Panel on Climate Change Fourth Assessment) emissions scenarios: A1B, A2, and B1 (in broad terms, representing continuing current greenhouse gas emission rates, significant emissions increases, and emissions reduction, respectively) (Coulson et al. 2010; Pachauri and Reisinger 2007).

Erosivity (R) was computed based solely on total annual precipitation after Renard and Freimund (1994: Eqns. 11, 12):

$$R = 0.04830P^{1.610}, \text{when} P < 850 \text{mm yr}^{-1} \text{or} \quad (1)$$

$$R = 587.8\text{–}1.219\ P + 0.004105\ P^2, \text{when } P \quad (2)$$

At 850 mm yr^{-1}, values from the two equations were essentially the same. *R* was in units of MJ· mm·ha^{-1}·hr^{-1}·yr^{-1}. *R* was calculated for each 8-km grid cell for each year and summed for the 30-yr periods 1980–2010, 2011–2040, 2041–2070, and 2071–2100. Separate *R* grids were created for each climate model, emission scenario, and 30-yr period.

Soil erodibility (K) and oven dry weight (bulk density, ρ) were extracted from the Soil Survey Geographic (SSURGO) database from the National Resource Conservation Service (NRCS; Soil Survey Staff 2009). The cover factor (C) was determined from landcover data. We used state-level landcover data based on 2000–2001 Landsat Thematic Mapper satellite imagery that differentiated grass, hay, and cropland, the primary land uses in the U.S. PPR (Thogmartin et al. 2004). We compared land use and 1997 National Resources Inventory sub-county C-factor values (Natural Resources Conservation Service 2000) to assign values of 0.19, 0.02, and 0.01 to cropland, hayland, and grassland, respectively (Online Resource 1, Table S1). Here grasslands are considered native, and haylands are modified by grazing, mowing, irrigation, or seeding.

2.3 Model Application

We used RUSLE and USPED to estimate sediment loss from each catchment (drainage area surrounding each wetland basin) into its associated wetland basin. RUSLE was originally designed for use on short slopes without redeposition of sediments along the slope (hereafter, intermediate redeposition). RUSLE can be used with a sediment delivery ratio (*sdr*; USDA Soil Conservation Service 1983) to account for the unmodeled processes of intermediate redeposition on long and/or complex slopes ($RUSLE_{sdr}$). USPED was based upon RUSLE but uses a stream power model in a geographical information systems (GIS) framework to account for intermediate redeposition and differing rates of erosion on slopes (Mitasova et al. 2003). Because USPED accounts for intermediate redeposition of sediments, it provides a more conservative estimate of sediment accumulation in wetlands than does RUSLE without a sediment delivery ratio.

Due to its use of local topographic information and given sufficiently high quality of elevation information, USPED may provide a more accurate sedimentation estimate than RUSLE using a generalized sediment delivery ratio ($RUSLE_{sdr}$).

For each catchment, the RUSLE model (Renard et al. 1997) was

$$T_{up} = \mathrm{sdr}\Sigma n_i R_i K_i LS_i C_i P / \rho_i, \tag{3}$$

where T_{up} was annual soil loss ($m^3 \cdot yr^{-1}$) from the catchment (summation over n catchment unit areas) per unit time. R was rainfall erosivity ($MJ \cdot mm \cdot ha^{-1} \cdot hr^{-1} \cdot yr^{-1}$), and K was the soil erodibility factor ($t \cdot ha \cdot hr \cdot ha^{-1} \cdot MJ^{-1} \cdot mm^{-1}$; Foster et al. 1981). LS was the dimensionless topographic factor that accounted for slope length and steepness, given by $(\text{upland unit area length}/22.13)^{0.6}(\sin \text{slope})^{1.3}$ with the exponents selected for rill erosion (Moore and Wilson 1992). C was the cover management factor, dimensionless, determined from land cover. P was the conservation practice factor, dimensionless, and customarily set to 1 for large areas unless comparing specific field practices.

We calculated sediment delivery ratio, *sdr* (range 0, no sediment reached the slope bottom, to – 1, all sediment reached the slope bottom), using the guideline provided by the USDA Soil Conservation Service (1983: Fig. 6-2) based on catchment size. The bulk density, ρ ($g \cdot cm^{-3}$), was used to convert from metric tons to cubic meters of sediment. See Burris and Skagen (2012) for additional details of RUSLE methodology.

The USPED model (Online Resource 1, Fig. S1; Mitasova et al. 1996) was

$$T_i = R_iK_iC_iP(A_i/22.13)^{1.6}(\sin B_i/0.0896)^{1.3}/\rho_i, \tag{4}$$

where T was soil transport capacity; R, K, C, P, and ρ were as described for RUSLE. The topographic factor, $A^{1.6}$ · (sin $B^{1.3}$), was computed from elevation data, where A was upslope contributing area calculated using GIS raster methods as within cell flow accumulation · cell size (30 m), B was slope angle, and the exponents were selected for rill erosion (Moore and Wilson 1992). The constants, 22.13 and 0.896, made the topographic factor dimensionless. To account for within slope changes in transport capacity (i.e., some portions of the catchment may be deposition rather than erosion locations), T_{ed}, the effective erosion/deposition from the slope, was computed as

$$Ted = \sum_i^n d(T_i\cos a_i)/dx + d(T_i\sin a_i)/dy \tag{5}$$

where the summation was over n catchment unit areas, and α was the slope aspect (degrees, in the direction of steepest slope), and x and y were the length and width of each grid cell (30 m). (Mitasova et al. 1996, 2003; Mitas and Mitasova 1998; Mitasova and Mitas 1999; Warren et al. 2005).

For both models, we assumed that the volume of sediment displaced from the catchment was equal to that deposited in the wetland. For simplicity, we assumed wetlands were flat bottomed and filled evenly, thus sediment depth, $SDepth = T_{up}/A_{wetland}$ for RUSLE and $SDepth = T_{ed}/A_{wetland}$ for USPED, where $A_{wetland}$ was the area of the wetland, m^2. The volume of sediment was compared to the basin volume to determine the loss of wetlands due to complete filling.

2.4 Data Analyses

To determine future change in precipitation across the study area, we calculated the difference between future average annual precipitation and 1981–2010 average annual precipitation by climate model (CSIRO, CGCM, or MIROC), scenario (A1B, A2, or B2), and future period (2011–2040, 2041–2070, or 2071–2100) at 126 random locations. Locations were at least 30 km apart to minimize spatial correlation. We used analysis of covariance to test for significant differences in the dependent variable annual sediment accumulation rate based on the covariates time period and land use using SAS 9.3 (SAS Institute 2002–2010).

We used the average annual deposition rate from 1981 to 2010 to estimate historic sediment accumulation at each wetland basin. Using projected R-factors of the nine future climate model/scenario combinations, we estimated sediment accumulation for each future period averaging annual accumulation rates over the 30-year period. Wetland fill times were the number of years needed to fill the remaining wetland depth (based on volume loss). We assumed surface elevations corresponded to the vintage of the underlying Digital Elevation Model (DEM; USGS National Elevation Dataset Data Source Index, <http://ned.usgs.gov/). Most models were from 1970, so generally our estimated wetland depth loss due to sediment accumulation started in 1970. We counted wetlands filled by 2010, during each future period, and those that did not fill by 2100. We also determined the number of wetlands that lost 50 % of their volume by 2100 and those with average annual sediment rates greater than 0.5 cm yr^{-1}. We used the simplifying assumption that land use did not change from ca 1970 through 2100.

We compared our results against sediment accumulation rates empirically determined by two field studies in the region or nearby, Gleason (2001) and Preston et al. (2013). We selected wetlands comparable in basin size, catchment size, and geography in grassland and cropland contexts and using reported sample sizes, we created sampling distributions of mean (95 % confidence interval, CI) annual sediment accumulation rates based on 1000 random samples (e.g., for comparison with wetlands in cropland reported by Gleason [2001], we drew 1000 samples of 12 wetlands). We compared our results to the ^{210}Pb results reported by Gleason (2001).

Because RUSLE and USPED are both general models based on widespread applicability across the U.S. (Renard et al. 1997), we investigated the relative importance of model variables in influencing rates of sediment accumulation in wetlands in the PPR. For this evaluation, we randomly selected ten 1 % and one 10 % sample of wetlands stratified by regime. These smaller samples improved computational tractability and improved the likelihood of sample independence. We evaluated the relative importance of six variables, four variables in the $RUSLE_{sdr}$ model plus catchment and basin size, with annual sediment rate as the dependent variable, by contrasting *t*-values from an analysis of variance (Bring 1994). Specifically, we expressed a ratio of the *t*-value of the variable of interest divided by the maximum *t*-value of all considered variables. We saw no significant difference between the 10 small samples and the single large sample, so only the large (10 %) sample is reported here. To further examine the relative contribution of land use and precipitation to sediment accumulation determined by $RUSLE_{sdr}$, we compared total sediment accumulation between land uses: grassland and cropland and between two 30-yr. time periods: 1981–2010 and 2041–2070 using analysis of variance (ANOVA). To do so, from the above-mentioned 10 % sample, we selected a subset that included only wetlands that were not predicted to fill before 2070 and were situated in uplands of at least 75 % cropland or 75 % grassland.

3. RESULTS

We delineated 1,568,217 catchments and estimated sediment infilling in their associated wetland basins across seven subecoregions within the study area; see Fig. 1 for delineation of the subecoregions. Distribution and characteristics of the wetland basins and the surrounding landscape varied by subecoregion (Fig. 2). Temporary and seasonal wetlands predominated in absolute number in all subecoregions and were most numerous along the Missouri Coteau and in the drift prairies of North and South Dakota (Table 1, Fig. 2). Grassland predominated in the coteau regions and cropland in the drift prairies.

During the climate reference period (1981–2010), annual precipitation across the study area averaged 586 mm yr^{-1} (range 346–905 mm yr^{-1} across space and time) and increased from west to east (Fig. 3).

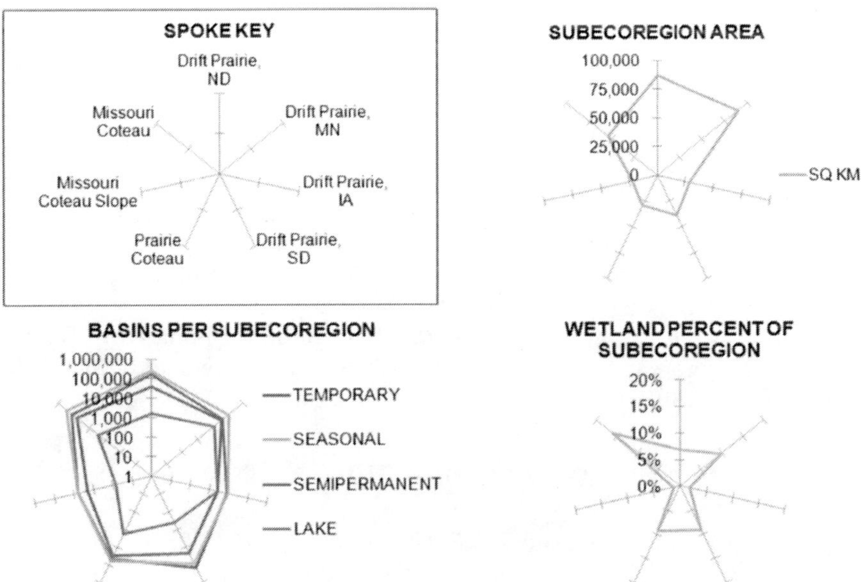

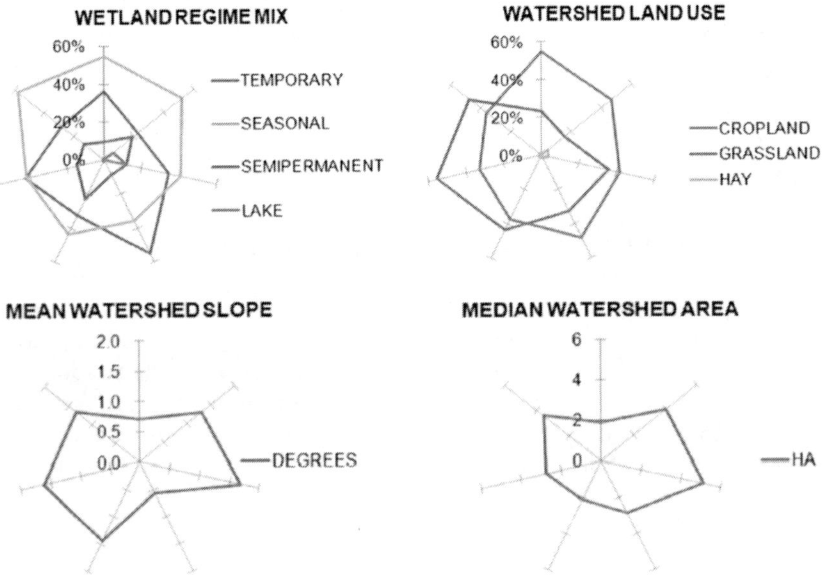

Figure. 2 Variation in wetland basin and landscape characteristics by subecoregions within the U.S. Prairie Pothole Region (U.S. PPR) portrayed in Fig. 1

During the reference period, there were distinct cycles of wet and dry periods, indicating the range in precipitation, with the wet period (1994–2001) averaging 17.6 mm yr^{-1}more than the reference period, and the dry period (1987–1992) averaging 75.9 mm yr^{-1} less than the reference period. On average across all climate models, emission scenarios, time periods, and sample locations, annual precipitation was projected to increase by 3 mm (± 1.47 se, n = 3402) above the 1981–2010 period (t = 2.04, df = 1, p = 0.042). The individual climate models varied in their predictions; MIROC predicted decreasing precipitation (mean across all emission scenarios, time periods, and sample locations (95 % confidence limit, CL): -35 mm, (−42,-27), whereas CGCM and CSIRO models predicted increasing precipitation (CGCM: 16 mm (8,24,); CSIRO: 28 mm (20,35).

3.1 Rates of Sediment Accumulation

Of the approaches we used (RUSLE, $RUSLE_{sdr}$, and USPED), the RUSLE model applied without a sediment delivery ratio gave the largest estimates of the number and percentages of wetlands completely filled by sediments by 2100 across all subecoregions and wetland types (Table 1; Online

Resource 1, Table S2). This method represented a theoretical maximum sediment accumulation. RUSLEsdr and USPED account for sediments settling out along the slope before reaching the wetland and therefore provided more conservative estimates than RUSLE. Across all categories, RUSLE estimated that 43 % of wetlands filled with sediments by 2100, whereas RUSLEsdr and USPED, generally in agreement, estimated that 33 % and 21 % of all wetlands filled with sediments, respectively.

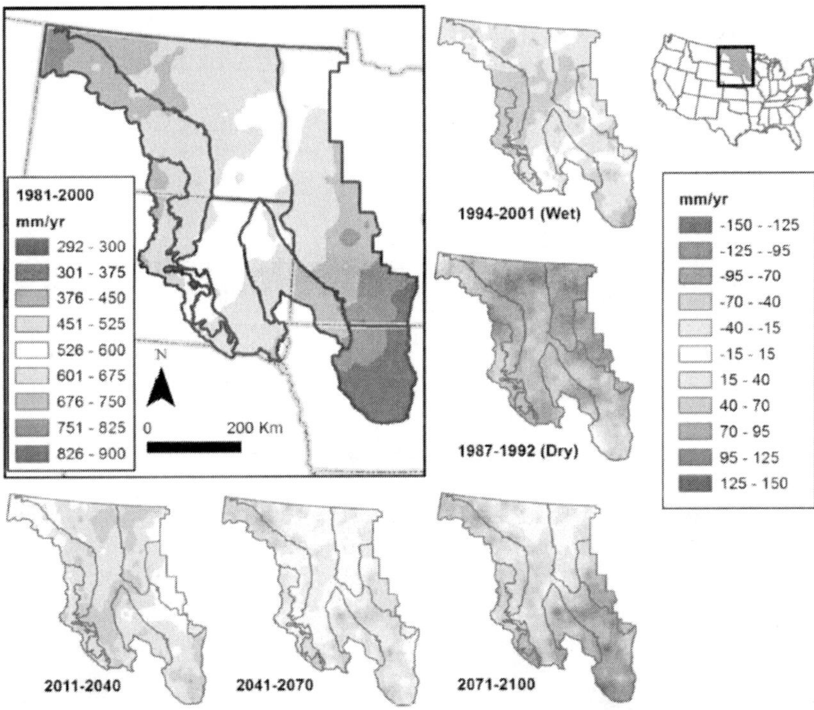

Figure. 3 Historic and predicted average annual precipitation (mm yr^{-1}) within the study region portion of the U.S. Prairie Pothole Region (U.S. PPR). Upper left: Average annual precipitation during a 30-yr reference period, 1981–2010; upper right: Average difference in annual precipitation relative to the reference period during a deluge (wet) period, 1994–2001; center right: Precipitation difference during a drought (dry) period, 1987–1992, relative to the reference period. Bottom row: Average annual precipitation relative to the reference period during three future periods: 2011–2040, 2041–2070, and 2071–2100 using the downscaled climate model CGCM2.1MR, A2 emissions scenario. Lines denote subecoregions portrayed in Fig. 1

The largest divergences between the latter two methods were in the drift prairies of Minnesota and Iowa where estimates varied by 20–29 %, depending on wetland regime (Table 1). Because RUSLEsdr and USPED provide more conservative and potentially more accurate values and were generally in agreement with one another, hereafter we report only projections of these two approaches.

Vulnerability to infilling did not vary markedly by wetland regime, although temporary wetlands and seasonal wetlands were slightly more likely to fill with sediments by 2100 (Fig. 4a; RUSLEsdr 33, 34 %, USPED 26, 19 %) than semipermanent wetlands (RUSLEsdr 30 %, USPED 15 %). The largest absolute numbers of wetlands expected to infill with sediment by 2100 occurred along the Missouri Coteau (RUSLEsdr 142,982, USPED 90,144) and in the drift prairies of North Dakota (RUSLEsdr 142,704, USPED 91,223). By percentage, wetlands in the Drift Prairie of Iowa and Minnesota were particularly vulnerable to infilling (RUSLEsdr 63 %, 47 %; USPED 38 %, 24 %, respectively).

We calculated two additional metrics that depict sediment accumulation rates: (1) the numbers (and percentages) of basins that will lose 50 % of their volume by 2100, and (2) the numbers (and percentages) of basins that will accumulate at least 0.5 cm of sediment annually through 2100 (Fig. 4b,c). Although estimates vary according to subecoregion, wetland regime, and method used, approximately 27–46 % ($RUSLE_{sdr}$ 46 %, USPED 27 %) of wetlands will lose half of their volume in the next century and 11–16 % ($RUSLE_{sdr}$ 16 %, USPED 11 %,) of basins will fill at a rate of >0.5 cm yr^{-1}. As with complete infilling, the largest numbers of vulnerable wetlands were in the drift prairies of North Dakota and along the Missouri Coteau, and by percentage, wetlands in the Drift Prairie of Iowa and Minnesota were particularly vulnerable (Fig. 4).

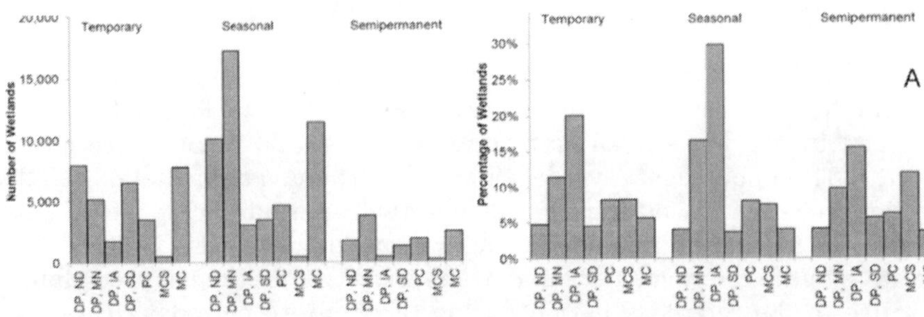

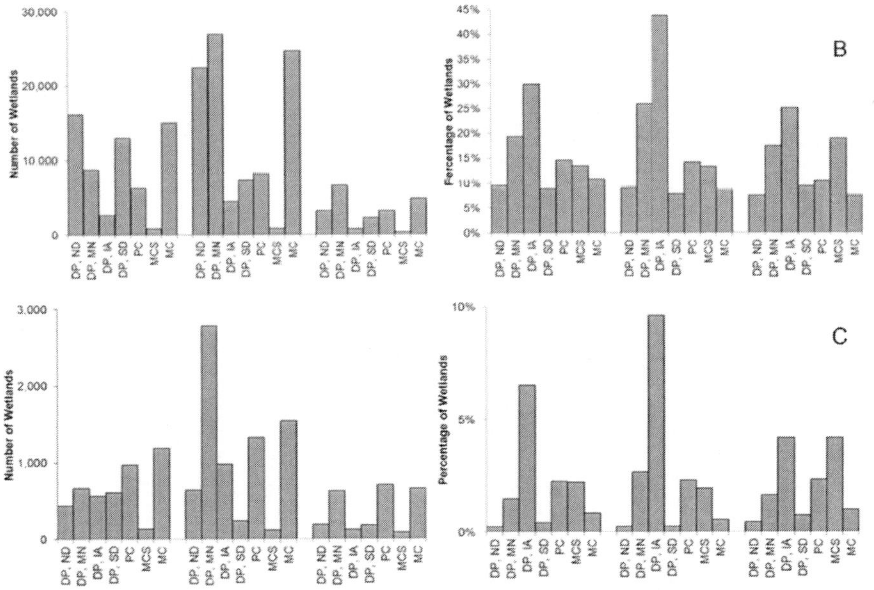

Figure. 4 Number *(left)* and percentage (*right*) of wetland basins within the U.S. Prairie Pothole Region (U.S. PPR) by subecoregion (defined in Fig. 1) and wetland type projected to a completely fill with sediments by 2100, b lose 50 % of their storage volume by 2100, and c accumulate at least 0.5 cm of sediment per year

3.2 Comparison with Empirical Studies

Our predicted annual sediment accumulation rates were similar to those reported by the two empirical field studies in the region (Table 2). The 95 % CIs for our comparative sample overlapped with the 95 % CIs reported by Gleason (2001) for seasonal and semipermanent across the Prairie Pothole Region (Montana, North Dakota, South Dakota, Minnesota, and Iowa) embedded in both grasslands and croplands; our point estimates were nonsignificantly lower. An empirical study conducted in the neighboring state of Nebraska reported similar annual sediment accumulation rates ranging from 0.18 cm yr^{-1} to 0.29 cm yr^{-1} (Tang et al. 2015). Wetlands embedded in agricultural fields had greater (non-overlapping 95 % CIs) annual sediment accumulation rates than wetlands surrounded by predominantly grassland in both Gleason's and our studies (Table 2).

Table 2 Comparison of vertical sediment accumulation rates for wetlands situated in cropland and grassland for this study and others

Method	Source	Grassland		Cropland	
		N (*n*) [a]	Accumulation rate (cm yr.$^{-1}$) (95 % CI)	N (*n*) [a]	Accumulation rate (cm yr.$^{-1}$) (95 % CI)
Semipermanent and seasonal wetlands across the PPR					
RUSLE$_{sdr}$	This study	1621 (7)	0.03 (0.01–0.07)	1730 (12)	0.16 (0.07–0.30)
^{210}Pb	Gleason 2001	(7)	0.08 (0.02–0.14)	(12)	0.27 (0.23–0.31)
Seasonal wetlands in the Montana Missouri Coteau[b]					
RUSLE$_{sdr}$	This study	596 (4)	0.03 (0.01–0.09)	354 (4)	0.23 (0.06–0.55)
^{210}Pb	Preston et al. 2013	(4)	0.06 (0.05–0.10)	(4)	0.21 (0.15–0.30)

[a]N reflects the population of wetlands and *n* the sample size drawn from the population. Confidence intervals were based on random draws using the sample size of the comparison study

[b]As our study did not extend into Montana, our comparison is with nearby North Dakota Missouri Coteau wetlands

The 95 % CIs of annual sediment accumulation rates for our comparative sample also overlapped with the intervals reported by Preston et al. (2013) for seasonal wetlands in the Montana Missouri Coteau embedded in both grasslands and croplands. Preston et al. (2013) also reported greater (non-overlapping 95 % CIs) annual sediment accumulation rates for wetlands embedded in agricultural fields than wetlands in grassland contexts.

3.3 Relative Importance of Covariates

Across the entire region, variation in rates of sediment accumulation were most strongly explained by three of six covariates (F = 129,344, df = 6, 142,252, p < 0.0001), the size of the catchment (t = 479.06; relative importance [t-value ratio] = 1), land use (C; relative importance =0.959); and slope (relative importance =0.930). Sediment accumulation was greater in wetlands in larger catchments and wetlands surrounded by steeper slopes; accumulation increased as coverage of tilled landscape within the catchment increased. Wetland basin size explained an intermediate amount of the variation in accumulation rates (relative importance =0.770), with greater rates for smaller wetlands. Annual precipitation (as expressed by R; relative importance 0.196) and soil erodibility (K, based on soil type; relative importance 0.132) were the least influential of the six factors in explaining variation in sediment accumulation rates. All covariates were statistically significant (p < 0.0001), and the amount of variance explained was high (R^2 = 0.845).

Depth of sediment accumulation in wetlands (n = 48,528) that did not fill prior to 2070 were estimated for periods of contemporary and projected climate to further evaluate the relative importance of land use and future changes in precipitation. This subset of wetlands experienced lower vulnerability than the population as a whole because one criterion for subset selection was lack of complete infilling by 2070. In this subset of less vulnerable wetlands, sediment accumulation in wetlands surrounded by croplands was 2–3 times more than in wetlands surrounded by in grasslands, and sediment accumulation was greater in 2041–2070 than 1981–2010 in cropland settings only (Table 3). Based on average precipitation, land use had a stronger influence on sediment accumulation than precipitation for all wetland regimes (Table 3). The differences attributable to land use were 10-fold greater than differences associated with changing precipitation.

Table 3 Average sediment accumulation depth (95 % CI, cm) by land use and wetland regime for a subset of wetlands predicted to fill after 2070 during periods of contemporary climate (1981–2010) and projected climate (2041–2070) using $RUSLE_{sdr}$

Regime	Time period	Surrounding upland: Grassland	Surrounding upland: Cropland	Difference between land uses
Temporary		n = 5950	n = 13,073	
	1981–2010	1.18 (1.12, 1.23)	2.26 (2.23, 2.30)	1.09 (1.00, 1.18) *
	2041–2070	1.22 (1.16, 1.27)	2.37 (2.34, 2.41)	1.16 (1.07, 1.25) *
Difference between time periods		0.40 (−0.07, 0.14)^{ns}	0.11 (0.04, 0.18)*	
Seasonal		n = 14,021	n = 10,739	
	1981–2010	1.34 (1.30, 1.39)	3.49 (3.44, 3.54)	2.14 (2.05, 2.23) *
	2041–2070	1.43 (1.38, 1.47)	3.71 (3.66, 3.76)	2.29 (2.20, 2.38) *
Difference between time periods		0.08 (0.00, 0.16)^{ns}	0.22 (0.13, 0.32)*	
Semipermanent		n = 3,285	n = 1,460	
	1981–2010	1.80 (1.67, 1.93)	6.26 (6.07, 6.45)	4.46 (4.16, 4.76) *
	2041–2070	1.89 (1.77, 2.02)	6.65 (6.45, 6.84)	4.75 (4.45, 5.05) *
Difference between time periods		0.09 (−0.14, 0.33)^{ns}	0.39 (0.03, 0.74)*	

*$p < 0.05$

4. DISCUSSION

Our study suggests that sedimentation warrants strong concern as a conservation issue because nearly one-third of wetlands in the U.S. Prairie Pothole Region may fill with sediments within the next century. The extent to which sedimentation may alter viability and function of individual wetlands across this region has not been explored previously, perhaps because sediment accumulation is a more subtle change agent than land and hydrologic alteration. Moreover, our estimates of sediment accumulation rates may be conservative if more annual precipitation manifests in the future than was predicted by the models used in this study. Our estimates of sediment accumulation rates were based on climate models that collectively predicted a range of future precipitation including both drier and wetter conditions. Climate predictions summarized by Ballard et al. (2014: Fig. 11), however, suggest increases in annual precipitation that are in agreement with our wettest model (CSIRO), with the greatest increases occurring in spring when newly tilled landscapes are susceptible to erosion.

The ability of prairie pothole wetlands to support vertebrate communities is influenced by land use modification, changing agricultural practices, pothole drainage systems, and climate change, which collectively result in altered patterns of wetland inundation (Johnson et al. 2005; Johnson et al. 2008; Oslund et al. 2010; Anteau 2012). Functionality will be lost by the sizeable percentage of wetlands likely to lose at least half of their volume. As basins of temporary and seasonal wetlands become shallower, rainfall will spread out over the landscape, yielding greater surface area for evaporation. The hydroperiod (the length of time a wetland holds water) of wetlands likely will be shortened due to increased evapotranspiration, thus potentially impacting the diversity of species dependent upon these shallow water habitats in the semi-arid landscapes of the Great Plains. Wetlands with short rather than intermediate hydroperiods typically have lower diversity in amphibian and avian species (Snodgrass et al. 2000; Balas et al. 2012; Tsai et al. 2012; Venne et al. 2012). These patterns may arise because of several mechanisms. Shorter hydroperiods may suppress the colonization and growth of aquatic invertebrates, thus decreasing the biomass of food available to amphibians, birds, and mammals. Shortened hydroperiods may be insufficient to allow larval amphibians with extended development periods (i.e., tiger salamanders *Ambystoma tigrinum*, northern leopard frogs *Lithobates pipiens*, and wood frogs *L. sylvatica*) to complete metamorphosis. Wetland-dependent birds require suitable habitat during critical periods in their annual cycles, including a lengthy time period for breeding birds that depend on wetlands for brood-rearing and a specific and sometimes narrow window of time for migrating birds (Murkin et al. 1997; Skagen et al. 2008).

Wetland functionality, habitat suitability, and food resources for vertebrate species can be significantly impacted not only by partial or complete infilling by sediments, but also by an annual sediment accumulation rate sufficient to suppress the emergence of aquatic invertebrates. A relatively small proportion but a sizeable number of wetlands experienced annual accumulation rates of >0.5 cm $yr.^{-1}$, the rate at which the emergence of plants and aquatic invertebrates from wetland seed and egg banks is 90–100 % suppressed (Jurik et al. 1994; Gleason et al. 2003). These wetlands may continue to provide the visual cues associated with suitable habitat and will likely attract birds, mammals, and amphibians that forage on emerging insects, but in actuality these wetlands may serve as ecological traps (sensu Schlaepfer et al. 2002).

The ability of landscape and wetland attributes to explain variability in sediment accumulation rates across the study area was greater than that of projected increasing precipitation with climate change. Across the study area, the wetlands most at risk of infilling were embedded in large catchments and a matrix primarily of cropland and often surrounded by steep slopes. Increased soil erosion in cropland relative to grassland was acknowledged by the developers of RUSLE based on empirical evidence and reflected in the C-factor (Renard et al. 1997). Our finding of greater sediment accumulation rates for wetlands within cropland matrices, possibly due to altered surface water flow, was consistent with earlier empirical studies (Luo et al. 1997, Gleason 2001; Preston et al. 2013) and not unexpected. Here we report that other landscape attributes (catchment size, steepness of the topography, and basin size) also play major roles in explaining variation in accumulation rates across the U.S. Prairie Pothole region.

This study provides a broadscale picture of projected sedimentation rates in the USPPR but does not account for all potential factors. We were unable to address the predicted increases in storm intensity (Karl et al. 1995) nor the potential changes in seasonality of storms because of the annual time-step used in calculating erosivity (from precipitation) in RUSLE and USPED. We were also unable to address the influence of hydrologic manipulations such as drainage ditches, tile drains, and resulting consolidation drainage on runoff and sedimentation patterns (Anteau 2012; McCauley et al. 2015) as well as wind erosion that may deliver sediments from unvegetated landscapes into wetlands. For our analyses, we assumed static land use patterns and did not attempt to project recent and future changes in land use or additional conversion of grassland to cropland (sensu Johnston 2013; Wright and Wimberly 2013).

Estimates of risk of conversion of existing grasslands to cultivated croplands range from 0.4 % to 1.33 % annualized loss (Stephens et al. 2008; Rashford et al. 2010). The sediment accumulation rates we present are likely conservative because the land cover data we used did not capture the rapid conversion of grassland to cropland that has occurred since 2000 (Rashford et al. 2010).

Sedimentation rates are determined not only by topographic features (catchment size, slope, basin size) and rainfall, but especially by land use practices. Our finding of greater sediment accumulation in wetlands embedded in tilled cropland rather than grassland matrices is in agreement with earlier studies here and in more southern regions of the Great Plains (Gleason 2001; Burris and Skagen 2012; Preston et al. 2013).

Several authors have recommended that conservation efforts be developed to strategically target grasslands and wetlands with high ecological value that are at high risk of conversion (Stephens et al. 2008; Rashford et al. 2010; Anteau 2012). Wetland-dependent biota may benefit if such efforts also incorporate the relative risk of wetland sediment infilling and thus emphasize conservation actions in areas of large watersheds and high topographic relief, especially in landscapes of predominantly cropland.

ACKNOWLEDGMENTS

Technical support was provided by Colorado State University, Department of Anthropology Geospatial Laboratory. This research was funded by the U.S. Geological Survey Science Support Program, the U.S. Geological Survey National Climate Change and Wildlife Science Center, the Plains and Prairie Potholes Landscape Conservation Cooperative, and the Department of Interior North Central Climate Science Center. Glenn Guntenspergen reviewed an earlier draft of this manuscript. Any use of trade, firm, or product names is for descriptive purposes only and does not imply endorsement by the U. S. Government. The findings and conclusions in this article are those of the author(s) and do not necessarily represent the views of the U.S. Fish and Wildlife Service.

REFERENCES

1. Anteau MJ (2012) Do interactions of land use and climate affect productivity of waterbirds and prairie-pothole wetlands? Wetlands 32:1–9
2. Balas CJ, Euliss NH Jr, Mushet DM (2012) Influence of conservation programs on amphibians using seasonal wetlands in the Prairie Pothole Region. Wetlands 32:333–345
3. Ballard T, Seager R, Smerdon JE, Cook BI, Ray AJ, Rajagopalan B, Kushnir Y, Nakamura J, Henderson N (2014) Hydroclimate variability and change in the Prairie Pothole Region, the "duck factory" of North America. Earth Interact 18:1–28. doi:10.1175/EI-D-14-0004.1
4. Batt BDJ, Anderson MG, Anderson CD, Anderson DC, Caswell DF (1989) The use of prairie potholes by North American ducks. In: van der Valk AG (ed) Northern Prairie wetlands. Iowa State University Press, Ames, Iowa, USA, pp. 204–229
5. Bring J (1994) How to standardize regression coefficients. Am Stat 48:209–213
6. Burris L, Skagen SK (2012) Modeling sediment accumulation in North American playa wetlands in response to climate change, 1940-2100. Clim Chang 117:69–83
7. Christensen JH, Hewitson B, Busuioc A, et al. (2007) Regional climate projections. In: Solomon S, Qin D, Manning M, et al. (eds) Climate change 2007: the physical science basis. Contribution of Working Group I to the fourth assessment report of the Intergovernmental Panel on Climate Change. Cambridge University Press, New York, pp. 849–940
8. Coulson DP, Joyce LA, Price DT, McKenney DW, Siltanen RM, Papadopo P, Lawrence K (2010) Climate scenarios for the conterminous United States at the 5 arc minute grid spatial scale using SRES scenarios A1B and A2 and PRISM climatology. US Department Of Agriculture Forest Service. Rocky Mountain Research Station, Fort Collins, CO, USA
9. Cowardin LM (1982) Some conceptual and semantic problems in wetland classification and inventory. Wildl Soc Bull 10:57–60
10. Cowardin LM, Carter V, Golet FC, LaRoe ET (1995) Classification of wetlands and deepwater habitats of United States. FWS/OBS-79/31. US Department of the Interior, Fish and Wildlife Service, Washington, DC
11. Dahl TE (1990) Wetland losses in the United States 1780s to 1980s. U.S. Department of the Interior, Fish and Wildlife Service, Washington DC
12. Dahl TE (2011) Status and trends of wetlands in the conterminous United States 2004 to 2009. US Department Of The Interior, Fish and Wildlife Service, Washington, DC

13. Daniel DW, Smith LM, Haukos DA, Johnson LA, McMurry ST (2014) Land use and Conservation Reserve program effects on the persistence of playa wetlands in the High Plains. Environ Sci Technol 48(8):4282–4288
14. Donald DB, Syrgiannis J, Hunter F, Weiss G (1999) Agricultural pesticides threaten the ecological integrity of northern prairie wetlands. Sci Total Environ 231:173–181
15. Environmental Systems Research Institute (ESRI) (1999–2008) ArcGIS version 9.3. Redlands, CA, USA
16. Foster GR, McCool DK, Renard KG, Moldenhauer WC (1981) Conversion of the universal soil loss equation to SI metric units. J Soil Water Conserv 36:355–359
17. Gesch DB (2007) The National Elevation Dataset. In: Maune D (ed) Digital elevation model technologies and applications: the DEM users' manual, 2nd edn. American Society for Photogrammetry and Remote Sensing, Bethesda, Maryland, USA, pp. 99–118
18. Gesch D, Oimoen M, Greenlee S, Nelson C, Steuck M, Tyler D (2002) The National Elevation Dataset. Photogramm Eng Remote Sens 68:5–11
19. Gleason RA (2001) Invertebrate egg and plant seed banks in natural, restored, and drained wetlands in the Prairie Pothole Region (USA) and potential effects of sedimentation on recolonization of hydrophytes and aquatic invertebrates. South Dakota State University, Brookings, South Dakota, USA, Dissertation
20. Gleason RA, Euliss NH Jr (1998) Sedimentation of prairie wetlands. Great Plains Res 8:97–112
21. Gleason RA, Euliss NH Jr, Hubbard DE, Duffy WG (2003) Effects of sediment load on emergence of aquatic invertebrates and plants from wetland soil egg and seed banks. Wetlands 23:26–34
22. Gleason RA, Tangen BA, Laubhan MK, Kermes KE, Euliss NH Jr (2007) Estimating water storage capacity of existing and potentially restorable wetland depressions in a subbasin of the Red River of the North. Reston. US Geological Survey Open File Report 2007–1159
23. Johnson RR, Higgins KF (1997) Wetland resources of Eastern South Dakota. South Dakota State University, Brookings, South Dakota, USA
24. Johnson WC, Millett BV, Gilmanov T, Voldseth RA, Guntenspergen GR, Naugle DE (2005) Vulnerability of northern prairie wetlands to climate change. Bioscience 55:863–872
25. Johnson RR, Oslund FT, Hertel DR (2008) The past, present, and future of prairie potholes in the United States. J Soil Water Conserv 63:84A–87A
26. Johnston CA (2013) Wetland losses due to row crop expansion in the Dakota Prairie Pothole Region. Wetlands 33:175–182
27. Jurik TW, Wang SC, van der Valk AG (1994) Effects of sediment load on seedling emergence from wetland seedbanks. Wetlands 14:159–165
28. Karl TR, Knight RW, Plummer N (1995) Trends in high frequency climate variability in the 20th century. Nature 377:217–220
29. Kehew AE, Teller JT (1994) History of late glacial runoff along the southwestern margin of the laurentide ice sheet. Quat Sci Rev 13:859–877
30. Luo H-R, Smith LM, Allen BL, Haukos DA (1997) Effects of sedimentation on playa wetland volume. Ecol Appl 7:247–252
31. McCauley LA, Anteau MJ, Post van der Burg M, Wiltermuth MT (2015) Land use and wetland drainage affect water levels and dynamics of remaining wetlands. Ecosphere 6(6):92. http://dx.doi.org/10.1890/ES14-00494.1
32. Millett B, Johnson WC, Guntenspergen G (2009) Climate trends of the North American prairie pothole region 1906-2000. Clim Chang 93:243–267
33. Mitas L, Mitasova H (1998) Distributed soil erosion simulation for effective erosion prevention. Water Resour Res 34(3):505–516
34. Mitasova HA, Mitas L (1999) Erosion/deposition modeling with USPED using GIS. University of Illinois, Urbana-Champaign, Illinois, USA. Available http://www4.ncsu.edu/~helena/gmslab/ denix/usped.html
35. Mitasova HA, Hofierka J, Zlocha M, Iverson LR (1996) Modeling topographic potential for erosion and deposition using GIS. Int J Geogr Inf Sci 10:629–641
36. Mitasova HA, Brown WM, Hohmann M, Warren S (2003) Using soil erosion modeling for improved conservation planning: a GIS-based tutorial. University of Illinois, Urbana-

Champaign, Illinois, USA. Available http://www4.ncsu.edu/~helena/gmslab/ reports/ CerlErosionTutorial/denix/default.html
37. Moore ID, Wilson JP (1992) Length-slope factors for the revised Universal soil loss equation: simplified method of estimation. J Soil Water Conserv 47:423–428
38. Murkin HR, Murkin EJ, Ball JP (1997) Avian habitat selection and prairie wetland dynamics: a 10-year experiment. Ecol Appl 7:1144–1159
39. Natural Resources Conservation Service (2000) National Resources Inventory, 1997. U.S. Department of Commerce, Washington, DC, USA
40. Niemuth ND, Estey ME, Reynolds RE, Loesch CR, Meeks WA (2006) Use of wetlands by spring migrant shorebirds in agricultural landscapes of the drift prairie of North Dakota. Wetlands 26:30–39
41. Oslund FT, Johnson RR, Hertel DR (2010) Assessing wetland changes in the Prairie Pothole Region of Minnesota from 1980 to 2007. J Fish Wildl Manag 1:131–135
42. Pachauri RK, Reisinger A (eds) (2007) Climate change: synthesis report. Contribution of working groups I, II, and III to the fourth assessment report of the Intergovernmental Panel on Climate Change, Intergovernmental Panel on Climate Change, Geneva, Switzerland
43. Preston TM, Sojda RS, Gleason RA (2013) Sediment accretion rates and sediment composition in Prairie Pothole wetlands under varying land use practices, Montana, United States. J Soil Water Conserv 68:199–211
44. Rashford BS, Walker JA, Bastian CT (2010) Economics of grassland conversion to cropland in the Prairie Pothole Region. Conserv Biol 25:276–284
45. Renard KG, Freimund JR (1994) Using monthly precipitation data to estimate the R-factor in the revised USLE. J Hydrol 157:287–306
46. Renard KG, Foster GA, Weesies GR, McCool DK, Yoder DC (1997) Predicting soil erosion by water: a guide to conservation with the revised soil loss equation (RUSLE). Agriculture Handbook No. 703. U.S. Department of Agriculture, Agricultural Research Service, Washington DC, 404 p
47. Reynolds RE, Shaffer TL, Loesch CR, Cox RR Jr (2006) The farm bill and duck production in the Prairie Pothole Region: increasing the benefits. Wildl Soc Bull 34:963–974
48. SAS Institute (2002–2010) SAS version 9.3. Cary, North Carolina, USA.
49. Schlaepfer MA, Runge MC, Sherman PW (2002) Ecological and evolutionary traps. Trends Ecol Evol 17:474–480
50. Skagen SK, Granfors DA, Melcher CP (2008) On determining the significance of ephemeral continental wetlands to North American migratory shorebirds. Auk 125:20–29
51. Smith LM, Haukos DA, McMurry ST, LaGrange T, Willis D (2011) Ecosystem services provided by playas in the High Plains: potential influences of USDA conservation programs. Ecol Appl 21:S82–S92
52. Snodgrass JW, Komoroski MJ, Bryan AL Jr, Burger J (2000) Relationships among isolated wetland size, hydroperiod, and amphibian species richness: implications for wetland regulations. Conserv Biol 14:414–419
53. Soil Survey Staff (2009) Soil survey geographic (SSURGO) database for Iowa, Minnesota, North Dakota, and South Dakota. USDA Natural Resources Conservation Service, available http://soildatamart.nrcs.usda.gov
54. Stephens SE, Walker JA, Blunck DR, Jayaraman A, Naugle DE, Ringelman JK, Smith AJ (2008) Predicting risk of habitat conversion in native temperate grasslands. Conserv Biol 22:1320–1330
55. Sylvester KM, Brown DG, Deane GD, Kornak RN (2013) Land transitions in the American plains: multilevel modeling of drivers of grassland conversion (1956-2006). Agric Ecosyst Environ 168:7–15
56. Tang Z, Gu Y, Drahota J, LaGrange T, Bishop A, Kuzila MS (2015) Using fly ash as a marker to quantify culturally-accelerated sediment accumulation in playa wetlands. J Am Water Resour Assoc. doi:10.1111/1752-1688.12347
57. Tangen BA, Gleason, RA (2008) Reduction of sedimentation and nutrient loading. In: Gleason RA, Laubhan MK, Euliss NH Jr (eds) Ecosystem services derived from wetland conservation practices in the United States Prairie Pothole Region with an emphasis on

the US Department of Agriculture Conservation Reserve and Wetlands Reserve Programs. US Geological Survey Reston, Virginia, USA, Professional Paper 1745:38–44

58. Thogmartin WE, Gallant AL, Knutson MG, Fox TJ, Suarez MJ (2004) A cautionary tale regarding use of the National Land Cover Dataset 1992. Wildl Soc Bull 32:970–978
59. Tiner RW Jr (1984) Wetlands of the United States: current status and recent trends. US Government Printing Office, Washington, DC
60. Tiner RW Jr (2003) Geographically isolated wetlands of the United States. Wetlands 23:494–516
61. Tsai J-S, Venne LS, Smith LM, McMurry ST, Haukos DA (2012) Influence of local and landscape characteristics on avian richness and density in wet playas of the Southern Great Plains, USA. Wetlands 32:605–618
62. US Department of the Interior (2009) Fish and Wildlife Service, National Wetlands Inventory Available http://www.fws.gov/wetlands/DATA/
63. USDA Soil Conservation Service (1983) National Engineering Handbook: section 3, sedimentation, chapter 6, sediment sources, yields, and delivery ratios. Washington, DC
64. van der Kamp G, Hayashi M (1998) The groundwater recharge function of small wetlands in the semi-arid northern prairies. Great Plains Res 8:39–56
65. Venne LS, Tsai J-S, Cox SB, Smith LM, McMurry ST (2012) Amphibian community richness in cropland and grassland playas in the Southern High Plains, USA. Wetlands 32:619–629
66. Warren SD, Mitasova H, Hohmann MG, Landsberger S, Iskander FY, Ruzycki TS, Senseman GM (2005) Validation of a 3-D enhancement of the Universal soil loss equation for prediction of soil erosion and sediment deposition. Catena 64(2–3):281–296
67. Winter TC, Rosenberry DO (1995) The interaction of ground water with prairie pothole wetlands in the cottonwood Lake area, east Central North Dakota, 1979-1990. Wetlands 15:193–211
68. Wright CK, Wimberly MC (2013) Recent land use change in the Western Corn Belt threatens grasslands and wetlands. Proc Natl Acad Sci 110:4134–4141

CHAPTER 6

Establishment and Application of Wetlands Ecosystem Services and Sustainable Ecological Evaluation Indicators

Han-Shen Chen [1,2]

[1]*School of Health Diet and Industry Management, Chung Shan Medical University, No. 110, Sec. 1, Jianguo N. Rd., Taichung City 40201, Taiwan*

[2]*Department of Medical Management, Chung Shan Medical University Hospital, No. 110, Sec. 1, Jianguo N. Rd., Taichung City 40201, Taiwan*

ABSTRACT

Gaomei wetlands are national Taiwanese coastal wetlands. Over the past few years, they have grown into an important water bird habitat and popular bird-watching location. However, the rapid growth in tourism has begun to affect the environmental quality in the Gaomei wetlands. This study combined ecosystem services (ES) and ecological footprint (EF) assessments to evaluate the sustainability status according to the features of each ecosystem service for the different Gaomei wetlands land uses. The results found that (a) the total Gaomei wetlands ecosystem service value increased from 59.24 million TWD in 2008 to 98.10 million TWD in 2015, and the ecosystem service function was continuously improving; (b) the EF increased by 56.12% over 8 years; and (c) there was a negative growth rate of 106.54% in the ecological deficit (ED) in the sustainable ecological evaluation indicators (SEEI). The ecological footprint index (EFI) in 2015 was at Level 4 at 1.02, and the environmental sustainability index (ESI) was at Level 3 at 0.49. Results show that Gaomei wetlands have a low sustainability; therefore, the local, regional, and national governments need to implement regulations to strictly control the Gaomei wetlands land use. This study demonstrated that ES and EF theory application can give an objective guidance to decision-makers to ensure that wetlands eco-security can be maintained at safe levels.

Keywords: wetlands; ecological footprint; ecosystem services; ecological security; sustainable ecological evaluation indicators (SEEI)

1. INTRODUCTION

As outlined in the Millennium Ecosystem Assessment of the United Nations, wetlands are one of the most threatened ecosystem in the world, with biodiversity loss being the major concern. With the continued loss and degradation of wetlands, ecological services are declining, negatively impacting human life. Factors such as climate change, rural poverty, and increased human population size have resulted in a wetlands loss of ~30%–50% in the last decade [1,2,3]. As well as providing ecosystem services such as flood control, coastline protection, nutrient recycling, carbon sequestration, and ecotourism, wetlands support many specialized plants and animal species [4,5,6,7,8].

This study was conducted in the Gaomei wetlands, located in Shimizu, Taichung, Taiwan, and designated national Taiwanese wetlands in 2007 (Construction and Planning Agency Ministry of the Interior, 2007). Recently, the Gaomei wetlands have become a major bird destination as a critical winter habitat. However, human activities such as highway construction have led to significant reductions in the sandbars and mangroves with a concurrent loss of biodiversity. Sustainable economic and societal development and reductions in the impact of tourism must be addressed when developing natural resources. This study sought to balance resource and biodiversity conservation with sustainable management and resource development. Because the Gaomei wetlands have similar topography and ecosystems to other Taiwanese wetlands systems, the evaluation model developed in this paper could be used for similar wetlands systems.

Chapin et al. (2000) suggested that ecosystem processes and biological diversity are crucial intermediaries in the overall economic and human systems' global environment [9]. Costanza et al. (1997) defined ecosystem services (ES) as ecosystems that "provide, directly or indirectly, the material and services to promote human welfare" [10]. The UN published the Millennium Ecosystem Assessment (MEA) in 2005, in which ecosystem services are divided into supply services, regulating services, cultural services, and support services [11]. Ecosystem service (ES) assessments have traditionally focused on identifying the individual monetary values for each ecosystem service [12,13]. However, the lack of theoretical frameworks has led to subjective judgments and criticisms [14,15,16]. The integration of deliberate and non-monetary valuation approaches to ES valuations has increasingly been advocated as a way of revealing the wider value concepts. Such methods, however, have had limited application in practice and have been mostly focused on localized case studies [17,18,19]. Barbier et al. (2011) evaluated the ecosystem service values in wetlands, mangroves, coral reefs, seagrass beds, and sandy beaches [20]. Bateman et al. (2011) explored the contribution of land use changes on ecosystem services and ecosystems [21]. Su et al. (2012) focused on four ecological zones in Hangzhou, China, to investigate the effect of landscape patterns and ecosystem service changes on urbanization [7]. To examine ecosystem services and biodiversity in Europe, Maes et al. (2012) used four supply function indicators, five adjustment function indicators, and a cultural function indicator to calculate ecosystem service values, and used average species richness and species diversity to measure biodiversity [22].

As previous research on land use and its impact on the environment has tended to focus more on exploring the single highest impact level, there has been less focus on the analysis and evaluation of the impact land-use development and the changes it has had on the natural environment. An econometric model [23,24], a statistical model [25,26,27,28,29,30] as well as a cellular automata model [31] have to date been the most commonly used evaluation models. Burkhard et al. (2013) believed that, for a more realistic ecosystem service status assessment, ecosystem services at different ecosystems, and the benefits that different ecosystems and land cover types provide, should also be considered [32]. Therefore, some studies have integrated the Millennium Ecosystem Assessment (MEA) and other assessment indicators into comprehensive evaluation indices [33,34,35]. Nelson et al. (2009) combined land use change ecosystem services with the integrated valuation of ecosystem services and tradeoffs (InVEST) mode to explore the relationship between biodiversity competition with other ecosystem services [36]. Polasky et al. (2011) also used the InVEST mode to quantify changes in ecosystem services, biodiversity, and land use in Minnesota from 1992 to 2001, and to assess the impact of different land use change scenarios on ecosystem services and biodiversity. Using the historical development (1964–2004) in

Leipzig, Germany [37], Lautenbach et al. (2011) developed regional scale indicators for different land use structure ecosystem services, such as water purification, pollination, food production, and outdoor recreation, and calculated the systemic functions and analyzed sensitivity tests under different land use types [34]. Geneletti (2012) simulated the impact generated by different land management policies on ecosystem services in the future based on historical land use [38]. In summary, using analysis and prediction modes for land use change along with mode simulations, decomposition, the analysis and synthesis of the complex socio-economic factors, and the interaction processes in natural ecosystems for given land uses to determine land-use change and spatial pattern trends [39,40,41,42] have become the focus of current research trends.

Sharp variabilities in the global climate have resulted in desertification, reduced ecosystem resilience, and loss of biodiversity. The 1972 United Nations Declaration on the Human Environment and Eco-Security raised concerns related to the preservation of food and ecosystems and outlined principles for sustainable human development projects, providing a new perspective on environmental resources, human survival and sustainable development reviews. With the development of ecological security theory and as ecological problems became increasingly prominent, researchers began using different indicators and measurement system models or methods to evaluate the ecological security of different regional scales, thus providing early warning models that could serve as vital references [43].

As using quantitative indicators to analyze complex information increases objectivity [44,45], a three-dimensional (economy, ecology, and society) indicator system was developed to study ecological security in Western Nepal [46]. Ecological security has also been measured using the Ecological Footprint Index (EFI) and environmental carrying capacity (ECC) [47].

The Ecological Footprint (EF) Model was proposed by Rees (1992) [48], with the primary feature being its ability to compare human demands on the environment with the biosphere's ability to regenerate resources and provide services. Wackernagel and Rees (2000) proposed that the EF magnitude was directly proportional to the environmental impact (the greater the EF, the greater the environmental impact), and was inversely proportional to the per-capita usable area of biologically productive land (the greater the EF, the smaller the per-capita usable area of biologically productive land) [49]. It is now a widely used measure in the field of ecological economics as it is a quantitative indicator that is easy to understand and calculate. Therefore, this paper uses sustainable ecological evaluation indicators (SEEIs) to measure regional ecological security on a per-unit ecological footprint basis.

In summary, this paper first combines the ecosystem services and ecological footprint models to evaluate the sustainability status based on each of the ecosystem service features for the different Gaomei wetlands land use covers, after which the ecosystem service values are calculated. Subsequently, the SEEI—the ecological remainder (ER), the ecological deficit (ED), the EFI and ESI—is used to analyze the resource utilization efficiency and ecological security in the Gaomei wetlands. The problems identified by the different indicator values are evaluated to develop a systematic measurement apparatus to encourage sustainable development and to review the evolution in sustainable development trends.

2. MATERIALS AND METHODS

2.1. Study Area

The Gaomei wetlands (24°18′35.07″ N, 120°33′08.21″ E) are located in Shimizu, Taichung, Taiwan. The Gaomei wetlands support diverse bird species and special habitats such as lagoons and sandbars, providing many possible tourist opportunities.

As the Gaomei wetlands have been evolving in the past few years from a primarily agriculturally based economy to a primarily tourism based economy, there have been several recent studies focused on tourist behavior and local tourism support initiatives [50,51,52].

2.2. Methods

This study combines ecosystem services and the EF model to reclassify the EF ecosystem according to the various land-use features for each ecosystem service system at the Gaomei wetlands.

The ecological footprint (EF) (demand) and ecological capacity (supply) at the Gaomei wetlands were first evaluated over various periods using an EF model developed from ecosystem services theory, after which an eco-security indicator system was established to estimate the Gaomei wetlands' eco-security, the steps for which were as follows:

2.2.1. Ecosystem Services Value Model

The methods used to estimate the ecosystem service value in the research area were based on the value assessment method models outlined in [10,53,54]. An equivalence factor for the ecosystem service value for the different land uses and land cover at the Gaomei wetlands was calculated using the assessment model, which took the ecosystem service value for the different land use and land covers as the basis for determining the ecosystem service value, as shown in Equation (1):

$$ESV_k = \sum_i^m \sum_j^n A_i \times f_{ij} \times E_a \times S_k \times T_k \quad (1)$$

where *ESV* is the total ecosystem service value; Ai is the distribution area for the *i*th type of land use and land cover (gha); *fij* is the equivalence factor for the *j*th ecosystem goods and services item provided by the *i*th ecosystem; E_a is the production per unit area or the ecosystem service value coefficient; S_k is a K coefficient for regional differences; T_k is a K regional service support coefficient; *i* is the land use and land covers in the different ecosystems; *j* is the ecosystem service category.

2.2.2. Accommodation Ecological Footprint

The EF concept [55,56] was used to evaluate the changes in the Gaomei wetlands from 2008 to 2015.

To examine the influence of the EF on the environment, the transportation ecological footprint (TREF), the activities ecological footprint (ACTEF), and the food and fiber consumption ecological footprint (FEF) were employed as the evaluation measures. The general formulas for calculating the EF and ECC are shown in Equations (2) and (3):

$$\mathrm{EF} = \mathrm{N} \times \mathrm{ef} = \mathrm{N} \times \sum_{i=1}^{n} (a \times a_i) = \mathrm{N} \times \sum_{j=1}^{6} \left(r_j \sum_{i=1}^{n} \frac{c_i}{p_i} \right) \; (i = 1, 2, 3, \ldots n;\ j = 1, 2, 3, \ldots, 6) \quad (2)$$

$$\mathrm{ECC} = \mathrm{N} \times \sum_{j=1}^{6} ec_j = \mathrm{N} \times \sum_{j=1}^{6} (A_j \times r_j \times y_j), \; (j = 1, 2, 3, \ldots, 6) \quad (3)$$

in which EF is the total EF (gha); N is the total population; ef is the EF per capita (gha); aa_i is the biologically productive area per capita (gha), converted to the *i*th traded commodity type; c_i is the consumption per capita (kg) of the *i*th commodity type; p_i is the average global productive capacity (kg/(t/gha)) of the *i*th consumer goods type; r_j and y_j are the equivalence factor and yield factor (YF) for the *j*th land type; *j* is the corresponding land use or land cover type; ECC denotes the total ecological carrying capacity; ecj is the ECC per capita; and A_j is the area per capita of the *j*th land type in the region.

2.2.3. Ecological Footprint Model

The traditional ecological footprint model converts the EF and ECC of the land ecosystems into biological resource consumption (agricultural land, forest land, grassland, and fisheries) and energy consumption (carbon footprint, completion land) and six other ecological system units. However, as this model includes systems and services ecological functions in the ECC calculation, it assumes that the ecosystem services of the various land ecosystem supply units can be used as fossil fuels. Therefore, the ECC evaluation of the ecosystem service value for the land ecosystems is reclassified in this paper to (1) agricultural ecosystem; (2) forest ecosystem; (3) grassland ecosystem; (4) ecosystem completions; (5) fisheries ecosystems (including lakes, rivers and wetlands); (6) unutilized land; and six other ecosystem units.

Traditional ecological footprint calculations consider only the production of food and raw materials to provide the two ecological functions. However, in the integrated EF model, an appropriate equivalence factor and a yield factor (YF) are used to determine the ecological functions for the biological land productivity and food production raw materials rather than the ecological functions and the ecosystem service value. YF is used mainly to reflect the differences in the different regions per unit area ecosystem services in the Gaomei wetlands, which is calculated as follows:

$$\mathrm{YF}_j = v_j / v_j^- \tag{4}$$

in which YF_j is the yield factor for the jth type of ecosystem unit, j = 1, 2, ..., 6, are the 6 ecosystem units; v_j is the per unit area value of the ecosystem service function for the jth type of ecosystem in a region; and v^-_j is the per unit area ecosystem service function value for the jth type of ecosystem in Taiwan.

Using Equation (4), the Gaomei wetlands yield factor was calculated from 2008 to 2015, as shown in Table 1.

2.2.4. Sustainable Ecological Evaluation Indicators (SEEI)

Multiple quantitative indicators (e.g., ED, ER, EFI and ESI) were employed to develop an indicator set to evaluate the Gaomei wetlands ecological sustainability and to establish the standards so as to be able to properly assess the ecological security.

The evaluation indicators utilized in this research are outlined in the following subsections.

Table 1. Yield factor (YF) for different land use and land cover ecosystems in the Gaomei wetlands from 2008 to 2015.

Year	Agricultural Ecosystem	Forest Ecosystem	Grassland Ecology System	Fisheries Ecosystems	Ecosystem Completions	Unutilized Land and Six Other Ecosystem Units
2008	1.51	0.92	1.15	1.88	0.62	1.92
2009	1.49	0.91	1.13	1.81	0.58	2.01
2010	1.48	0.88	1.12	1.80	0.54	2.15
2011	1.31	0.77	1.14	1.60	0.40	1.98
2012	1.06	0.60	0.87	1.28	0.15	1.62
2013	1.14	0.63	0.94	1.33	0.15	1.77
2014	1.07	0.57	0.88	1.21	0.11	1.64
2015	1.00	0.52	0.77	1.11	0.08	1.56

(a) ED or ER

When the EF is lower than the ECC of a region, there is an ecological remainder (ER), indicating that the corresponding development model is sustainable. When the EF is higher than the ECC of a region, there is an ecological deficit (ED), indicating that the corresponding development model is not sustainable. As the ecological deficit results from excessive human resource demands [57], demand must be reduced to achieve sustainable ecological development. The formulas for ED and ER are shown as Equations (5) and (6).

$$ER = ECC - EF \tag{5}$$

$$ED = EF - ECC \tag{6}$$

in which ER is the ecological remainder, ED is the ecological deficit, ECC is the ecological carrying capacity, and EF is the ecological footprint.

(b) EFI

The EFI compares the resources and energy expenditures to the region's ECC to assess resource utilization and determine development sustainability. The EFI formula is as in Equation (7), and the EFI levels are as shown in Table 2

$$EFI = EF/ECC \tag{7}$$

in which EFI is the ecological footprint index, EF is the ecological footprint, and ECC is the ecological carrying capacity.

Table 2. Ecological Footprint Index (EFI) levels and conditions.

Level	EFI	EFI Conditions
1	<0.5	Safe
2	0.5~0.8	Moderately safe
3	0.8~1.0	Threshold
4	>1.0	Unsafe

Resource: [49].

(c) ESI

The Environmental Sustainability Index (ESI), which was developed by the Center for International Earth Science Information Network (CIESIN), the Yale Center for Environmental Law and Policy (YCELP), and the World Economic Forum, assesses sustainability by measuring the degree to which the ecological systems in a region meet the human ecological demands. The ESI formula is shown in Equation (8) and the ESI levels are given in Table 3

$$ESI = ECC/(ECC + EF) \quad (8)$$

in which ESI is the environmental sustainability index, EF is the ecological footprint; and ECC is the ecological carrying capacity.

Table 3. Environmental Sustainability Index (ESI) levels.

Level	ESI	Regional Ecological Sustainability Extent
1	>0.7	High sustainability
2	0.50~0.70	Low sustainability
3	0.30~0.50	Low unsustainability
4	<0.30	High unsustainability

Resource: [58].

3. RESULTS AND DISCUSSION

3.1. Ecosystem Services Value Computation and Analysis Results

Equation (2) was used to calculate the ecological service value of each ecosystem type in the Gaomei wetlands from 2008 to 2015, as shown in Table 4. The total value of the ecosystem services in the Gaomei wetlands increased from 59.24 million TWD to 98.10 million TWD over the 8 years. From 2008 to 2013, the total Gaomei wetlands ecosystem service value increased by 42.27 million TWD due to the increased fisheries (including lakes, rivers, and wetlands) and agricultural land areas, both of which had large ecological service value coefficients. The total Gaomei wetlands ecosystem service value increased from 2013 to 2015; however, during the same period, the Gaomei wetlands total ecosystem service value decreased by 3.41 million TWD because the fisheries area, which had a large ecological service value coefficient decreased rapidly and the increased agricultural land area was not sufficient to compensate for the fisheries reduced ecological service value. Since 2013, therefore, the Gaomei wetlands total ecosystem service value has been decreasing.

Table 4. Ecosystem service values from 2008 to 2015 for each Gaomei wetlands ecosystem type.

	Year	Fisheries Ecosystems	Grassland Ecology System	Forest Ecosystem	Agricultural Ecosystem	The Completion of Ecosystems	Unutilized land and Six Other Ecosystem Units	Total
Ecosystem Services Values (10^6 TWD)	2008	40.93	6.17	0.32	10.68	0.01	1.13	59.24
	2009	48.85	6.28	0.34	10.74	0.01	1.27	67.49
	2010	53.12	6.35	0.34	10.82	0.01	1.35	71.99
	2011	60.89	6.39	0.36	10.10	0.01	1.44	79.19
	2012	65.11	6.62	0.55	12.08	0.01	1.39	85.76
	2013	83.79	3.34	0.04	12.76	0.01	1.57	101.51
	2014	78.18	4.37	0.05	14.56	0.01	1.44	98.61
	2015	74.12	4.11	0.05	18.49	0.02	1.31	98.10

From 2008 to 2015, the relative value of each Gaomei wetlands ecosystem type (Table 4) changed variably. The fisheries ecosystem services function value increased then decreased, the forest ecosystem services function value increased, decreased, and then increased, and the grassland ecosystem services function value had an M-shaped pattern: increase–decrease–increase–decrease; the agricultural ecosystem services function value decreased and then increased, with the overall ecosystem services function value increasing. Taken together, the total ecosystem service value in increased by a net of 65.60%, with the fisheries ecosystem service value increasing the most (81.09%), and that of the forest ecosystems decreasing the most (−84.38%) followed by the grassland ecosystems (−33.39%). The fisheries accounted for 39% of the total utilized land area in the study area; however, the fisheries ecological service value accounted for 75.56% of the total ecosystem service value. Because the fisheries in this area have a high ecosystem service value coefficient, this land use type has a higher total ecosystem service value, highlighting the importance of the fisheries in the Gaomei wetlands ecosystem.

3.2. EF Computation and Analysis Results

Table 5 details the three Gaomei wetlands activity types and the EF computations. The EF gradually increased from 244.03 global hectares (gha) in 2008 to 380.98 gha in 2015. Of the three EF activity types, the TREF had the largest proportion at an average of 70.18%, followed by the ACTEF at an average of 24.97% and the FEF at an average of 4.85%. Based on these empirical results, the TREF grew from 140.47 gha in 2008 to 267.36 gha in 2015 due to a significant growth in tourist numbers, which caused an increased demand for vehicles, further inflating liquefied fuel demands.

Table 5. Ecological Footprint (EF) for the three activities types and total Ecological Footprint (EF) (unit: gha).

Year	TREF	ACTEF	FEF	EF
2008	140.47	95.14	8.42	244.03
2009	183.70	95.14	13.52	292.36
2010	204.87	95.14	13.37	313.38
2011	184.57	95.14	13.77	293.48
2012	182.99	95.14	12.30	290.43
2013	220.37	95.14	15.54	331.05
2014	267.04	95.14	19.96	382.14
2015	267.36	95.14	18.48	380.98
Average proportion	70.18%	24.97%	4.85%	100.000%

Notes: TREF: transport ecological footprint; ACTEF: activities footprint; FEF: food & fiber consumption ecological footprint; EF: total ecological footprint.

As the Gaomei wetlands gross area did not change substantially between 2008 and 2015, the ACTEF was steady at 95.14 gha; however, the FEF increased from 8.42 gha in 2008 to 18.48 gha in 2015. The FEF indicates the tourist dietary demands inside the Gaomei wetlands and includes grains, coarse cereals, vegetables, fruit, meat, and fish; therefore, the substantial growth in the FEF was linked to the growing tourist numbers. This analysis found that the main tourism ecological resource consumption came from the fossil energy used by the vehicles traveling between residences and destinations and the increase in the land accessible to tourists for leisure activities.

3.3. SEEI Computation and Analysis Results

The *SEEI* indicated that the ED grew by about 106.54% from 2008 to 2015. In 2015, the EFI was rated Level 4 at 1.02, and the ESI was rated Level 3 at 0.49, indicating that the Gaomei wetlands were at an unsafe ecological security level within that time span.

(a) ER/ED

Table 6 gives the computation results for the ER/ED. The ER had a declining trend from 128.54 gha in 2008 to 41.52 gha in 2013. The Gaomei wetlands ecological deficit (ED) decreased from −9.57 gha in 2014 to −8.41 gha in 2015, primarily due to the increased tourist numbers.

Table 6. Ecological Footprint (EF), ecological carrying capacity (ECC), ecological remainder (ER)/ecological deficit (ED), ecological footprint index (EFI), and environmental sustainability index (ESI) (unit: gha).

Year	ECC	EF	ES/ED	EFI			ESI		
				Index	Level	Representation Condition	Index	Level	Representational State
2008	372.57	244.03	128.54	0.65	2	Moderately safe	0.60	2	Low sustainability
2009	372.57	292.36	80.18	0.78	2	Moderately safe	0.56	2	Low sustainability
2010	372.57	313.38	59.19	0.84	3	Threshold	0.54	2	Low sustainability
2011	372.57	293.48	79.09	0.79	2	Moderately safe	0.56	2	Low sustainability
2012	372.57	290.43	82.14	0.78	2	Moderately safe	0.56	2	Low sustainability
2013	372.57	331.05	41.52	0.89	3	Threshold	0.53	2	Low sustainability
2014	372.57	382.14	−9.57	1.03	4	Unsafe	0.49	3	Low unsustainability
2015	372.57	380.98	−8.41	1.02	4	Unsafe	0.49	3	Low unsustainability

(b) EFI

This study utilized the EFI to evaluate the Gaomei wetlands ecological security, the outcomes of which are shown in Table 6. The EFI increased from 0.65 in 2008 to 1.02 in 2015, indicating an unsafe ecological security level, which is predicted to further increase because of the rising tourist demand for resources and services (e.g., bus services, road construction, and waste production).

(c) ESI

Table 6 shows the computational results for the ESI. Between 2008 and 2013, the ESI remained at Level 2, signifying low sustainability. Yet from 2014 onward, the ESI has fallen to Level 3, signifying unsustainability. If this situation is not controlled and improved, the attainment of sustainable ecological development will be impossible.

4. CONCLUSIONS

This study employed the ecosystem service value, ecological capacity, EF, and sustainable ecological evaluation indicators (SEEIs) to assess the ecological security and the efficient use of resources in the Gaomei wetlands. We came to the following conclusions:

The total value of the ecosystem services in the Gaomei wetlands increased from 59.24 million TWD in 2008 to 98.10 million TWD in 2015. The EF gradually increased from 244.03 gha in 2008 to 380.98 gha in 2015. Of the three activity EFs, TREF had the biggest proportion (70.18%), with ACTEF (24.97%) and FEF (4.85%) following thereafter. The SEEI indicated that the ED grew by about 106.54% from 2008 to 2015. In 2015, the EFI was rated Level 4 at 1.02, and the ESI was rated Level 3 at 0.49. Therefore, as the Gaomei wetlands are predicted to become ecologically unsustainable over time, local, regional, and national governments need to implement regulations to strictly control the Gaomei wetlands land use.

According to the empirical analysis results, the primary factors influencing various types of activity EFs are presented below.

a) Tourists had a negative effect on the overall EF from all activities. Therefore, when tourist numbers increased, the EF increased and there was a greater environmental impact. Attempts should be made to increase the environmentally friendly behavior of tourists to decrease the impact of increasing tourist numbers.

b) The fossil fuels used for transportation had the greatest influence on the TREF. Therefore, strategies aimed at reducing energy use and the commensurate carbon footprints should be developed. Using public transportation and using environmentally friendly vehicles and services such as electric cars and motorcycles and bicycle rental services should be encouraged. Global positioning systems could be used in rental cars to monitor tourist activity, which can then be used to develop effective transportation systems aimed at decreasing overall fossil fuel use and minimizing the associated carbon footprints.

ACKNOWLEDGMENTS

I am thankful to the Ministry of Science and Technology (Republic of China, Taiwan) for financially supporting this research project (grant number MOST 102-2410-H-040-010).

REFERENCES

1. Rahman, M.M.; Islam, M.A. *Better Options for IFM: Uptake Promotion NRSP Project R8306 Final Technical Report: Annex A*; Center for Natural Resource Studies (CNRS): Dhaka, Bangladesh, 2005.
2. Valiela, I.; Fox, S.E. Managing coastal wetlands. *Science* **2008**, *319*, 321–323.
3. Miththapala, S. *Mangroves. Coastal Ecosystems Series (Volume 2)*; Ecosystems and Livelihoods Group Asia, International Union for Conservation of Nature: Colombo, Sri Lanka, 2008.
4. Fletcher, S.; Kawabe, M.; Rewhorn, S. Wetland conservation and sustainable coastal governance in Japan and England. *Mar. Pollut. Bull.* **2011**, *62*, 956–962.
5. Feng, M.; Liu, S.; Euliss, N.H., Jr.; Young, C.; Mushet, D.M. Prototyping an online wetland ecosystem services model using open model sharing standards. *Environ. Model. Softw.* **2011**, *26*, 458–468.
6. Hopkinson, C.; Cai, W.; Hu, X. Carbon sequestration in wetland dominated coastal systems: A global sink of rapidly diminishing magnitude. *Sustainability* **2012**, *4*, 186–194.
7. Su, S.; Xiao, R.; Jiang, Z.; Zhang, Y. Characterizing landscape pattern and ecosystem service value changes for urbanization impacts at an eco-regional scale. *Appl. Geogr.* **2012**, *34*, 295–305.
8. Ward, R.D.; Burnside, N.G.; Joyce, C.B.; Sepp, K.; Teasdale, P.A. Improved modelling of the impacts of sea level rise on coastal wetland plant communities. *Hydrobiologia* **2016**, *774*, 203–216.
9. Chapin, F.S.; Zavaleta, E.S.; Eviner, V.T.; Naylor, R.L.; Vitousek, P.M.; Reynolds, H.L.; Hooper, D.U.; Lavorel, S.O.; Sala, E.; Hobbie, S.E.; et al. Consequences of changing biodiversity. *Nature***2000**, *405*, 234–242.
10. Costanza, R.; d'Arge, R.; de Groot, R.; Farber, S.; Grasso, M.; Hannon, B.; Limburg, K.; Naeem, S.; O'Neill, R.V.; Paruelo, J.; et al. The value of the world's ecosystem services and natural capital. *Nature* **1997**, *387*, 253–260.
11. Millennium Ecosystem Assessment (MEA). *Ecosystems and Human Well-Being: Wetlands and Water Synthesis*; World Resources Institute: Washington, DC, USA, 2005.
12. The Ecological and Economic Foundations. *The Economics of Ecosystems and Biodiversity: The Ecological and Economic Foundations*; Earthscan: London, UK, 2010.
13. UK National Ecosystem Assessment. *UK National Ecosystem Assessment Follow-On Phase: Synthesis Report*; UNEP-WCMC: Cambridge, UK, 2014.
14. Kenter, J.O.; O'Brien, L.; Hockley, N.; Ravenscroft, N.; Fazey, I.; Irvine, K.N.; Reed, M.S.; Christie, M.; Brady, E.; Bryce, R.; et al. What are shared and social values of ecosystems? *Ecol. Econ.* **2015**, *111*, 86–99.

15. Scholte, S.S.K.; van Teeffelen, A.J.A.; Verburg, P.H. Integrating socio-cultural perspectives into ecosystem service valuation: A review of concepts and methods. *Ecol. Econ.* **2015**, *114*, 67–78.
16. Fish, R.; Church, A.; Winter, M. Conceptualising cultural ecosystem services: A novel framework for research and critical engagement. *Ecosyst. Serv.* **2016**, *21*, 208–217.
17. Bunse, L.; Rendon, O.; Luque, S. What can deliberative approaches bring to the monetary valuation of ecosystem services? A literature review. *Ecosyst. Serv.* **2015**, *14*, 88–97.
18. Hattam, C.; Böhnke-Henrichs, A.; Börger, T.; Burdon, D.; Hadjimichael, M.; Delaney, A.; Atkins, J.P.; Garrard, S.; Austen, M.C. Integrating methods for ecosystem service assessment and valuation: Mixed methods or mixed messages? *Ecol. Econ.* **2015**, *120*, 126–138.
19. Kenter, J.O. Integrating deliberative monetary valuation, systems modelling and participatory mapping to assess shared values of ecosystem services. *Ecosyst. Serv.* **2016**, *21*, 291–307.
20. Barbier, E.B.; Hacker, S.D.; Kennedy, C.; Koch, E.W.; Stier, A.C.; Silliman, B.R. The value of estuarine and coastal ecosystem services. *Ecol. Monogr.* **2011**, *81*, 169–193.
21. Bateman, I.J.; Mace, G.M.; Fezzi, C.; Atkinson, G.; Turner, R.K. Economic Analysis for Ecosystems Assessments. *Environ. Res. Econ.* **2011**, *48*, 177–218.
22. Maes, J.; Paracchini, M.L.; Zulian, G.; Dunbar, M.B.; Alkemade, R. Synergies and trade-offs between ecosystem service supply, biodiversity, and habitat conservation status in Europe. *Biol. Conserv.* **2012**, *155*, 1–12.
23. Liu, Y.; Huang, X.; Yang, H.; Zhong, T. Environmental effects of land-use/cover change caused by urbanization and policies in Southwest China Karst area—A case study of Guiyang. *Habitat Int.***2014**, *44*, 339–348.
24. Su, S.; Zhou, X.; Wan, C.; Li, Y.; Kong, W. Land use changes to cash crop plantations: Crop types, multilevel determinants and policy implications. *Land Use Policy* **2016**, *50*, 379–389.
25. Tomaz, C.; Alegria, C.; Monteiro, J.M.; Teixira, M.C. Land cover change and afforestation of marginal and abandoned agricultural land: A 10 year analysis in a Mediterranean region. *For. Ecol. Manag.* **2013**, *308*, 40–49.
26. Trincsi, K.; Pham, T.H.; Turner, S. Mapping mountain diversity: Ethnic minorities and land use land cover change in Vietnam's borderlands. *Land Use Policy* **2014**, *41*, 484–497.
27. Du, S.; Wang, Q.; Guo, L. Spatially varying relationships between land-cover change and driving factors at multiple sampling scales. *J. Environ. Manag.* **2014**, *137*, 101–110.
28. Xiong, X.; Grunwald, S.; Myers, D.B.; Ross, C.W.; Harris, W.G.; Comerford, N.B. Interaction effects of climate and land use/land cover change on soil organic carbon sequestration. *Sci. Total Environ.***2014**, *493*, 974–982.
29. Deng, Y.; Srinivasanb, S. Urban land use change and regional access: A case study in Beijing, China. *Habitat Int.* **2016**, *51*, 103–113.
30. Beyene, F. Land use change and determinants of land management: Experience of pastoral and agro-pastoral herders in eastern Ethiopia. *J. Arid Environ.* **2016**, *125*, 56–63.
31. Halmy, M.W.A.; Gessler, P.E.; Hicke, J.A.; Salem, B.B. Land use/land cover change detection and prediction in the north-western coastal desert of Egypt using Markov-CA. *Appl. Geogr.* **2015**, *63*, 101–112.
32. Burkhard, B.; Crossman, N.; Nedkov, S.; Petz, K.; Alkemade, R. Mapping and modelling ecosystem services for science, policy and practice. *Ecosyst. Serv.* **2013**, *4*, 1–3.
33. Posthumus, H.; Rouquette, J.R.; Morris, J.; Gowing, D.J.G.; Hess, T.M. A framework for the assessment of ecosystem goods and services; a case study on lowland floodplains in England. *Ecol. Econ.* **2010**, *69*, 1510–1523.
34. Lautenbach, S.; Kugel, C.; Lausch, A.; Seppelt, R. Analysis of historic changes in regional ecosystem service provisioning using land use data. *Ecol. Indic.* **2011**, *11*, 676–687.
35. Haines-Young, R.; Potschin, M.; Kienast, F. Indicators of ecosystem service potential at European scales: Mapping marginal changes and trade-offs. *Ecol. Indic.* **2012**, *21*, 39–53.
36. Nelson, E.; Mendoza, G.; Regetz, J.; Polasky, S.; Tallis, H.; Cameron, D.R.; Chan, K.M.; Daily, G.C.; Goldstein, J.; Kareiva, P.M.; et al. Modeling multiple ecosystem services, biodiversity

conservation, commodity production, and tradeoffs at landscape scales. *Front. Ecol. Environ.* **2009**, *7*, 4–11.
37. Polasky, S.; Calderone, G.; Duarte, K.E.; Goldstein, J.; Hannahs, N.; Ricketts, T.H.; Tallis, H. Putting ecosystem services to work: Conservation, management, and trade-offs. In *Natural Capital: Theory and Practice of Mapping Ecosystem Services*; Oxford University Press: Oxford, UK, 2011; pp. 249–263.
38. Geneletti, D. *Integrating Ecosystem Services in Land Use Planning: Concepts and Applications*; CID Research Fellow and Graduate Student Working Paper No. 54; Center for International Development, Harvard University: Cambridge, MA, USA, 2012.
39. Petus, C.; Lewis, M.; White, D. Monitoring temporal dynamics of Great Artesian Basin wetland vegetation, Australia, using MODIS NDVI. *Ecol. Indic.* **2013**, *34*, 41–52.
40. Julian, J.P.; Wilgruber, N.A.; Beurs, K.M.D.; Mayer, P.M.; Jawarnehd, R.N. Long-term impacts of land cover changes on stream channel loss. *Sci. Total Environ.* **2015**, *537*, 399–410.
41. Abrantes, P.; Fontes, I.; Gomes, E.; Rocha, J. Compliance of land cover changes with municipal land use planning: Evidence from the Lisbon metropolitan region (1990–2007). *Land Use Policy* **2016**, *51*, 120–134.
42. Chen, N.; Ma, T.; Zhang, X. Responses of soil erosion processes to land cover changes in the Loess Plateau of China: A case study on the Beiluo River basin. *Catena* **2016**, *136*, 118–127.
43. Chen, H.S.; Chen, C.Y.; Chang, C.T.; Hsieh, T. The Construction and Application of a Carrying Capacity Evaluation Model in a National Park. *Stoch. Environ. Res. Risk Assess.* **2014**, *28*, 1333–1341.
44. Warhurst, A. Mining, mineral processing, and extractive metallurgy: An overview of the technologies and their impact on the physical environment. In *Environmental Policy in Mining: Corporate Strategy and Planning for Closure*; Warhurst, A., Raton, L.N.B., Eds.; CRC Press LLC: Boca Raton, FL, USA, 2000.
45. Singh, R.K.; Murty, H.R.; Gupta, S.K.; Dikshit, A.K. An overview of sustainability assessment methodologies. *Ecol. Indic.* **2012**, *15*, 281–299.
46. Bhandari, B.S.; Grant, M. Analysis of livelihood security: A case study in the Kali-Khola watershed of Nepal. *J. Environ. Manag.* **2007**, *85*, 17–26.
47. Liu, R.Z.; Borthwick, A.G.L. Measurement and assessment of carrying capacity of the environment in Ningbo, China. *J. Environ. Manag.* **2011**, *92*, 2047–2053.
48. Rees, W.E. Ecological footprints and appropriated carrying capacity: What urban economics leaves out. *Environ. Urban.* **1992**, *4*, 121–130.
49. Wackernagel, M.; Rees, W. *Our Ecological Footprint-Reducing Human Impact on the Earth*; New Society Publishers: Gabriola Island, BC, Canada, 2000.
50. Chen, H.S. Establishment and Applied Research on a Wetland Ecosystem Evaluation Model in Taiwan. *Sustainability* **2015**, *7*, 15785–15793.
51. Chen, T.S.; Lin, H.J.; Huang, S.Y. A framework on habitat connectivity among Taiwan's wetlands for overwintering Black-faced Spoonbill. *Ocean Coast. Manag.* **2015**, *116*, 78–88.
52. Fang, G.; Huang, C.; Chang, C.; Han, J.; Kuan, H.; Jie, L.; Fu, Z.; Chen, X.; Yang, K.; Tsai, H.H. Atmospheric total arsenic (As), (As^{3+}) and (As^{5+}) pollutants study in central Taiwan. *Environ. Earth Sci.* **2016**, *75*, 124.
53. Mamattuesun, E.; Hamid, Y.; Anwar, M. Oasis land-use change and its effects on the oasis eco-environment in Keriya Oasia. *China* **2010**, *7*, 244–252.
54. Sawut, M.; Eziz, M.; Tiyip, T. The effects of land-use change on ecosystem service value of desert oasis: A case study in Ugan-Kuqa River Delta Oasis, China. *Can. J. Soil Sci.* **2013**, *93*, 99–108.
55. Gössling, S.; Hansson, C.B.; Hörstmeier, O.; Saggel, S. Ecological footprint analysis as a tool to assess tourism sustainability. *Ecol. Econ.* **2002**, *43*, 199–211.
56. Martin-Cejas, R.R.; Sanchez, P.P.R. Ecological footprint analysis of road transport related to tourism activity: The case for Lanzarote Island. *Tour. Manag.* **2010**, *31*, 98–103.
57. Rees, W.E. Getting serious about urban sustainability, eco–footprints and the vulnerability of twenty–first–century cities. In *Canadian Cities in Transition, New*

Directions in the Twenty-First Century; Oxford University Press: Toronto, ON, Canada, 2011; pp. 70–86.

58. World Economic Forum, Yale Center for Environmental Law and Policy, Center for International Earth Science Information Network (CIESIN) of Columbia University, and Joint Research Centre of the European Commission. 2005. Environmental Sustainability Index, 2005. Available online: http://www.ciesin.columbia.edu/indicators/ESI/ (accessed on 16 June 2016).

CHAPTER 7

Wetland Evapotranspiration: Eddy Covariance Measurement in the Biebrza Valley, Poland

Mariusz Siedlecki[1], Włodzimierz Pawlak[1], Krzysztof Fortuniak[1] & Mariusz Zieliński [2]

1 Department of Meteorology and Climatology, University of Łódź, Łódź, Poland

2 Department of Sustainability and Development of the Geographical Environment, Pedagogical University of Cracow, Cracow, Poland

ABSTRACT

This paper presents eddy covariance measurements of evapotranspiration and sensible heat flux in wetland area. The rate of evapotranspiration is characterized by a distinct seasonal pattern and assumes the highest values during summer. From April to September, the latent heat flux is characterized by a distinct diurnal pattern with the highest average values (at the noon hours) at a level of ca 200 $W \cdot m^{-2}$. Moreover, in the period from June to August the values of latent heat flux far outweigh those of sensible heat flux (the Bowen ratio in the noon hours ranging between 0.4 and 0.6) and account for 50–60 % of radiation balance. The average daily totals of evapotranspiration vary from about 0.5 $mm \cdot d^{-1}$ during the winter months to about 3–3.5 $mm \cdot d^{-1}$ in summer.

Keywords: Latent heat fluxes, Surface energy, budget, Water balance, Wetland

1. INTRODUCTION

Wetlands, occupying an area of ca 5.7 × 10^6 km^2, represent a considerable percentage of land area, especially in the zone of middle latitudes (Aselman and Crutzen 1989). Due to their characteristic hydrological conditions, these areas are also an important element of the water cycle. The high levels of groundwater and dense vegetation cause wetlands to have a surface heat balance different from e.g. agricultural or forest areas, and thus they can significantly influence the weather and climatic conditions. In addition, a large biodiversity of species, numerous breeding grounds of waterfowl and marsh birds as well as high biomass production make these areas an important component of the system of nature protection. Wetlands are also an element of the global carbon cycle. Moreover, the rate of exchange of carbon dioxide and of methane between the atmosphere and wetlands is strictly dependent on soil moisture conditions (Moore et al. 1998). They determine the course of aerobic and anaerobic processes of biomass decomposition (Fraser et al. 2001; Billett et al. 2004). Therefore, due to their specificity and diversity of environmental conditions, these areas are often subject to different forms of nature conservation. Therefore, the measures to preserve the original features of such areas require a deeper knowledge of the principles of wetland environment functioning, with particular emphasis put on their water balance.

The process of evapotranspiration mainly depends on there being sufficient energy resources which can be consumed to evaporate a certain amount of water. This quantity depends primarily on the net all-wave radiation Q* and other elements of surface heat balance, approximated by:

$$Q{*}\approx Qh + Qe + Qg \tag{1}$$

where Qh is sensible heat flux, Qe is latent heat flux and Qg is ground heat flux.

However, the high soil moisture, dense vegetation and thus a large surface area of foliage, as well as the roughness of surface result in the fact that wetlands are usually characterized by higher evaporation in relation to evaporation from an open water surface (e.g. Kim and Verma 1996; Herbst and Kappen 1999; Pauliukonis and Schneider 2001; Acreman et al. 2003). Evaporation in such an environment usually takes place by several parallel paths, including transpiration from vegetation, evaporation from surface water under the vegetation and evaporation from an open water surface (numerous lakes, rivers, canals characteristic of the landscape of wetlands). At the same time, the complexity of the process of evapotranspiration in wetlands can be evidenced by the fact that in many cases, however, higher evaporation from an open water surface was observed (Campbell and Williamson 1997; Burba et al. 1999) in relation to evapotranspiration from wetlands. In addition, a number of studies point to evapotranspiration as the main component of the water balance of marshland areas, and (the energy equivalent of evapotranspiration) the latent heat flux dominates in the surface heat balance (Priban and Ondok 1985; Drexler et al. 2004). Therefore, a detailed knowledge of evapotranspiration of wetlands can be crucial for the functioning of these ecosystems or for restoring the land transformed as a result of land reclamation or other human activities to its original condition. In the second half of the twentieth century, wetland areas in many countries were used for economic purposes and subjected to land amelioration. However, including such an area in the programme of nature conservation (e.g. the Birds Directive – 79/409/EEC, the Water Framework Directive – 2000/60/EC) involves restricting agricultural activities (such as grazing and mowing peat meadows), which results in the observed succession of forest and bushy vegetation onto the peat meadow areas (Grygoruk et al. 2014; Wassen et al. 2006). Therefore, the results of evapotranspiration evaluation are essential to assess the water balance of these areas in order to restore them to their original condition.

In the studies of wetland evapotranspiration, a number of methods to assess this phenomenon are applied. For example, in the direct measurement of evapotranspiration lysimeters (and more specifically, a set of several lysimeters) are used, enabling simultaneous evaluation of evaporation loss of water from individual bog formations and from an open water surface (Abtew and Obeysekera 1995; Abtew and Melesse 2013). In the direct measurements, microclimatic methods based on the Bowen ratio are also used (Burba et al. 1999; Campbell and Williamson 1997; Kellner 2001; Peacock and Hess 2004), as well as methods based on the measurements of fluctuations of groundwater levels (Gerla 1992; Lafleur et al. 2005; Grygoruk et al. 2011; Grygoruk et al. 2014; Mazur et al. 2014). Furthermore, the results of evaporation evaluation, obtained using the Penman equation, the Penman-Monteith equation or the water balance are also widely presented (Wessel and Rouse 1994; Campbell and Williamson 1997; Souch et al. 1998; Jacobs et al. 2002; Gasca-Tucker et al. 2007). In the last two decades, the eddy covariance method has been more and more commonly used. The method enables the direct measurement of the vertical flux of water vapour. An important advantage of the eddy covariance method is the fact that the obtained water vapour flux comes from a source area within a distance of 200–300 m from the sensor. At the same time, due to the high costs of measuring instruments, and also because of various technical factors affecting the quality of the data obtained, a number of studies devoted to the measurement of evapotranspiration present the results of measurement experiments lasting for several weeks or seasons. They mainly present the daily patterns of water vapour flux or the daily totals of evapotranspiration against the background of meteorological conditions (Souch et al. 1998; Kurbatova et al. 2002; Gasca-Tucker et al. 2007), as well as the results of impact assessment of parameters such as foliage surface area or surface resistance on the rate of evapotranspiration (Acreman et al. 2003; Gulden et al. 2007).

The primary aim of this study is to present the variability of daily and seasonal evaporation from the wetlands located in north-eastern Poland in the Biebrza National Park. Thanks to the simultaneous measurements of the radiation balance, the remaining elements of heat balance, as well as selected meteorological parameters, the results allow to present the following:

- a number of characteristics of the daily pattern of latent heat flux as compared to other components of the heat balance of bog surface,
- the variability of evapotranspiration in individual seasons of the year as well as the variability of the relationship of latent heat fluxes with the radiation balance values and sensible heat fluxes at various stages of development of vegetation.

Owing to the multi-seasonal measurements in the period 2013–15, it was possible to determine the daily and monthly evapotranspiration totals and their annual pattern. In addition, during the measurement period the average weather conditions in individual seasons were characterized by a fairly wide diversity. Therefore, the results of the study also present the long-term variability of evapotranspiration in wetland areas both under the conditions of a relatively humid growing season and of a drought.

The study results are also crucial from the point of view of works aimed at protecting the wetland areas of the Biebrza National Park. The knowledge of evapotranspiration is essential to assess the water balance of these areas in order to restore them to their original condition. In addition, despite the ever wider research into this issue using state-of-the-art measurement methods, there still are only a handful of studies on the complex characteristics of evapotranspiration in the wetlands of Poland (Kleniewska et al. 2009; Chojnicki et al. 2007; Urbaniak 2006; Fortuniak et al. 2013).

2. METHODS

2.1 Study Site

The Biebrza National Park (Polish: *Biebrzański Park Narodowy*) (BNP) is the largest national park in Poland (59 thousand ha) and occupies the biggest marshland areas in the valley of the Biebrza river in north-eastern Poland (ca 255 km^2). The Biebrza basin is characterized by a typical valley landscape with high levels of groundwater and strong paludification. Hydrogenic, mostly peat soils dominate, and in the areas subject to amelioration these are primarily peat-and-muck soils. A smaller percentage of the area of the national park consists of mineral soils, such as rusty soils, luvisols, brown soils and spodic soils (Banaszuk and Banaszuk 2004). In the wetlands, rush and sedge communities dominate, as well as wet and variably wet meadows. The transition zones between peat soils and mineral soils are overgrown with forests and thickets (Malinowska 2004).

The measurement station is located in the neighbourhood of the village of Kopytkowo (53°35′20″N, 22°53′31″E), within the so-called Central Basin, covering the area of the largest marshlands and peatlands (Fig. 1) which are named the Red Marsh. In the immediate vicinity of the measurement site (~ 500 m), there occur rush communities with the dominating reed sweet-grass (*Glycerietum maximae*) and the common reed (*Phragmitetum communis*), as well as moss and sedge communities represented by the *Carici canescentis-Agrostietum caninae*. These communities widely occur in the whole Biebrza valley (Matuszkiewicz 2004). During the measurements, in the immediate vicinity of the weather station, the average height of vegetation in the cold half of the year and in spring was approximately 0.4 m, whereas in summer and autumn the average height was approximately 1 m.

2.2 Eddy Covariance Measurement

The measurements of the turbulent flux of water vapour were performed in the period between June 2013 and September 30, 2015. In the eddy covariance method applied, the turbulent flux of water vapour E is calculated as follows:

$$E = \rho \cdot \overline{w' \cdot q'} \quad (2)$$

where: ρ – air density, w' – fluctuations in vertical wind velocity, q' – specific density fluctuations, the flux from the Earth's surface to the atmosphere being assumed as positive. The overbar and primes indicate the mean for the sampling interval (30-min frequency) and deviations from the mean, respectively.

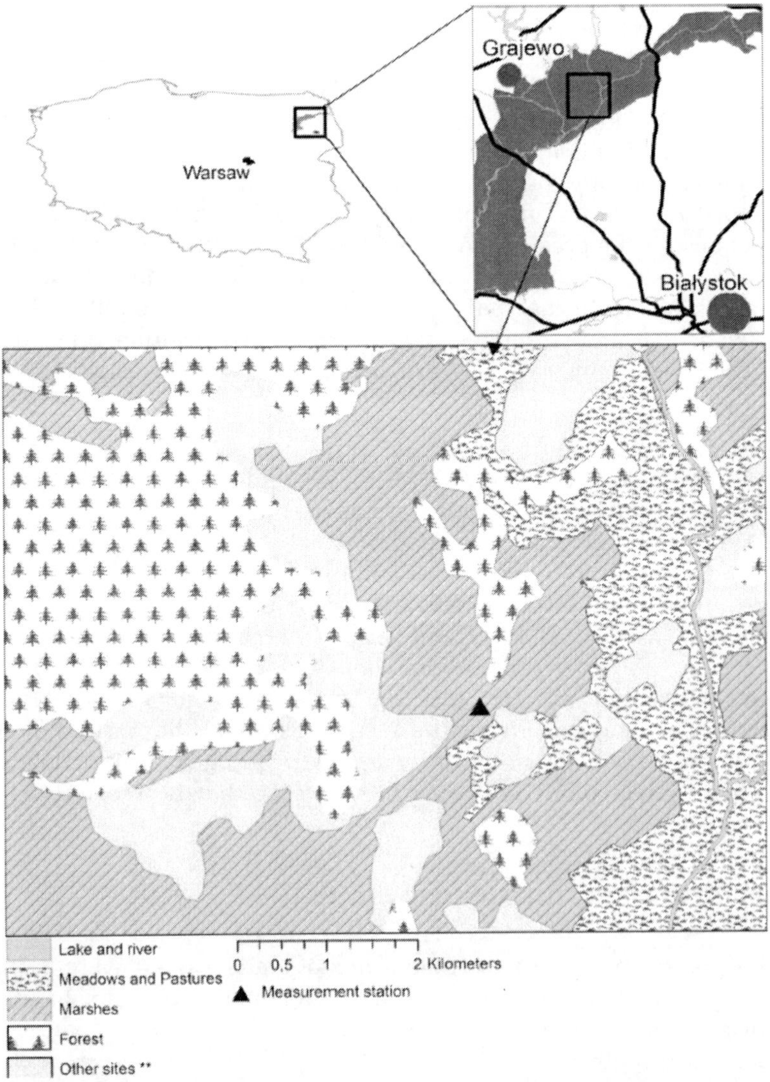

Figure. 1 Location map showing the Biebrza National Park with spatial distribution of different vegetation types near the measurement station (based on CORINE Land Cover data available from European Environment Agency (EEA 2006)). ** Other sites – mixed group included agricultural fields and areas covered by farm buildings

Then standard data-verification steps were used, including: elimination of artificial electrical impulses, rotations of coordinate systems, temperature and humidity corrections, the WPL correction (Webb et al. 1980), as well as corrections for spectral losses (Fortuniak 2010). The next step taken in the preparation of the data was to verify the requirement of series stationarity (Foken 2008; Fortuniak 2010). This issue was solved using the following three tests: the test proposed by Foken (Foken 2008), by Mahrt (Mahrt 1998), and by Dutaur as modified by Affre (Dutaur et al. 1999; Affre et al. 2000). In addition, the data were verified for precipitation and conditions of the presence of frost on instruments and the data from such situations were also omitted in further analysis. The criteria so adopted for the verification of data quality, as well as errors resulting from interference of measuring instruments cause gaps in the measurement series. With the aforementioned criteria adopted and measurements made in rainy or frosty weather eliminated, the percentage of good data was approximately 80 %.

To prevent underestimation of the actual evapotranspiration, data gap filled before determining the daily, monthly or annual totals. Gaps of several hours were filled using the average daily pattern of We values calculated over a period of 14 days (Falge et al. 2001; Foken 2008). On the other hand, in assessing the daily pattern of evapotranspiration, series with unfilled measurement gaps were used. Such approach allows to determine the basic statistics of variability only on the basis of data obtained from the eddy covariance measurements.

The eddy covariance method requires measurements made with a very short time step (10-20 Hz). The set of measuring instruments includes: a sonic anemometer (RM Young 81,000, USA) enabling the measurement of three components of wind velocity, and a gas analyzer (LI 7500, LI-COR, Inc., Lincoln, USA) which allows the identification of fluctuations of water vapour and carbon dioxide content. The instruments were installed on a mast at a height of 3.7 m above ground level. Data were recorded at a frequency of 10 Hz. The components of radiation balance were measured using a net radiometer CNR1 (Kipp&Zonen, the Netherlands) consisting of two pyrgeometers (CG3) for measuring the atmospheric infrared radiation, and two pyranometers (CM3), allowing the simultaneous measurements of sunlight received and reflected by the Earth's surface.

The measurements of temperature and humidity were performed at a height of 2.2 m and 0.5 m using HMP 60 sensors (Vaisala, Finland). In addition, the measurement station was equipped with instruments for the measurements of heat flux in the soil (HFP01SC, Campbell Scientific, Inc. USA), soil temperature (T107, Campbell Scientific, Inc. USA), soil moisture and the amount of precipitation.

2.3 Weather Conditions

The Biebrza basin, located in north-eastern Poland, is characterized by different meteorological conditions as compared to central and western Poland. In addition, due to a great diversity of its soil cover and plant communities, the area also presents a rich mosaic of microclimates (Kossakowska-Cezak et al. 1991), which requires thorough research, especially under conditions of progressing changes in soil moisture associated with the activities of nature conservation and restoration of the Biebrza marshes to their original features.

In the period under analysis, the thermal conditions were quite typical. In the winter period (both in 2014 and 2015), generally, the mean daily air temperatures oscillated between −5 °C and 5 °C. There were also periods of several days of very frosty air advection when the mean daily air temperature dropped below −20 °C (Fig. 2a). In spring, there was a gradual increase in air temperature and, for example, days with an average daily temperature above 20 °C were recorded at the measurement station in May 2014 (Fig. 2a). During the studied three summer seasons, the months of June, July and the first half of August were the period of the year with the highest air temperatures. There were periods of several days when the average daily air temperature was about 20 °C, while the maximum temperature exceeded 30 °C. From the second half of August, notably in 2013 and 2014, a gradual decrease in air temperatures began. In September, the average daily air temperature fell below 10 °C; moreover, there already occurred the first cases of ground frost.

The humidity conditions significantly affect the rate of evaporation. In the annual pattern of vapour pressure deficit (Fig. 2b), a marked seasonality is visible with the highest values occurring in summer. During winter, in turn, the vapour pressure deficit is low (less than 5 hPa), which indicates conditions similar to those of saturation of the air with water vapour. In the spring and summer periods, the average daily values of vapour pressure deficit grow. In particular, August 2015 was characterized by extremely low humidity, when several days with very high values of vapour pressure deficit occurred.

An important meteorological factor conditioning the rate of evapotranspiration is wind speed. During the measurement period, rather low wind speeds prevailed (Fig. 2c). In the warm season of the year, the daily average wind velocity oscillated around 2 m s^{-1}, and the maximum wind speed rarely exceeded 5 m s^{-1}. Higher wind speeds occurred during the winter months when several day periods with the daily mean wind speed exceeding 5 m s^{-1} were recorded.

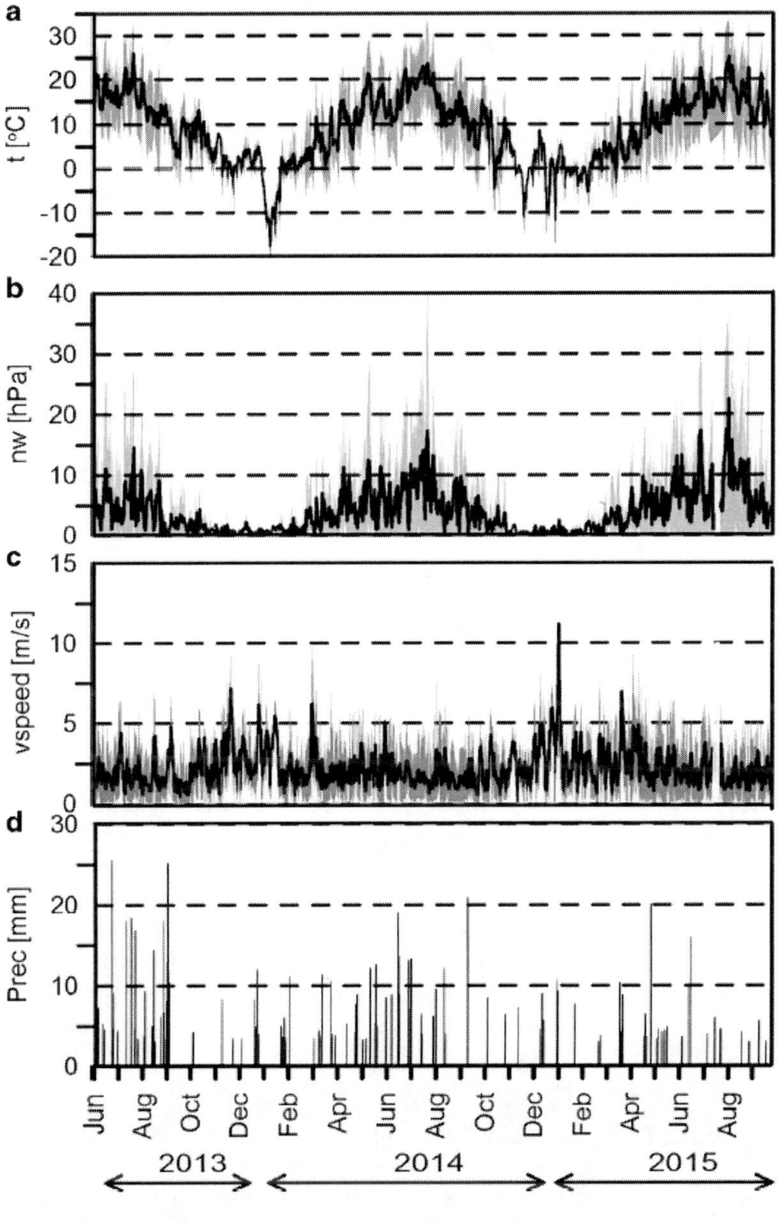

Figure. 2 Daily characteristics of selected weather conditions during study period: **A** air temperature, **B** vapour pressure deficit (VPD), **C** wind speed, **D** daily precipitation total. Solid line represents daily average, shaded area indicates daily minimum and maximum values

Precipitation is a major component of water balance conditioning the level of surface waters, as well as soil moisture. Seasonal precipitation during the summer of 2013 and the spring and summer of 2014 at the study site was similar to the long-term average values (Ustrnul et al. 2013; Ustrnul et al. 2014). High precipitation (110–140 % of the norm from the years 1971–2000) in north-eastern Poland occurred in September 2013 and during the winter and spring seasons of 2015 (Ustrnul et al. 2015a). In contrast, the summer of 2015 was characterized by very low precipitation at a level of 30–70 % of the average seasonal total of the period 1971–2000 (Ustrnul et al. 2015b). In the pattern of daily precipitation totals at the measurement station, longer periods without precipitation occurred mainly in the autumn season. The spring and summer periods were characterized by a greater number of days with precipitation; in addition, there also was precipitation with the daily total exceeding 10 mm. Against this background, the summer of 2015 stands out because in August and September there were only a few days with rainfall, and its daily totals did not exceed 5 mm (Fig. 2d).

3. RESULTS AND DISCUSSION

3.1 Mean Daily Pattern of Surface Energy Balance Components

Between April and September, the net all-wave radiation at the measurement station is characterized by a clear daily pattern with the highest values at around midday. The average value of Q^* in the afternoon from May to August reach a value of 400 $W \cdot m^{-2}$, the Q^* at the same time being characterized by a high variability due to more dynamic weather conditions in summer (Fig. 3a). In the night hours, in turn, the surface of the marsh loses heat through radiation. After sunset, the average radiation balance values in the summer months assume negative values at a level of -10 to -40 $W \cdot m^{-2}$.

In the warm half of the year, an increase in solar radiation and higher air temperatures lead to the development of turbulent sensible heat fluxes Qh and latent heat fluxes Qe. From April to September, with an increase in Q^*, the pattern of the Qh and Qe values show a distinct diurnal cycle (Fig. 3 b, c); at the same time, in the spring season, i.e. in the months of March, April and May an increase in the values of sensible heat flux is clearly visible. Around noon, the average values of Qh vary from about 100 $W \cdot m^{-2}$ in March to about 150 $W \cdot m^{-2}$ in May (Fig. 3b) and account for 30–40 % of the net all-wave radiation (Fig. 4a). The average ratio of sensible heat flux and radiation balance (Qh/Q^*) in the noon hours assume values of 0.29 to 0.36

(Table 1). The latent heat flux Qe is lower than Qh and in the noon hours the average values of Qe vary from ca 50 $W·m^{-2}$ in March to 100 $W·m^{-2}$ in April (Fig. 3c) which is approximately 30 % of the Q* (Fig. 5a). A sensible heat flux exceeding the values of Qe in spring was also detected in the multi-season measurements in the wetlands in north-eastern China (Zhou and Zhou 2009). The recorded mean values of sensible heat flux in the noon hours reach a level of 300 $W·m^{-2}$, while the values of Qe range between 80 and 100 $W·m^{-2}$. A similar trend is also characteristic of evapotranspiration from the surface of marshland areas in the Central United States (Lenters et al. 2011) where, for example, in April the average Bowen ratio is 3.3.

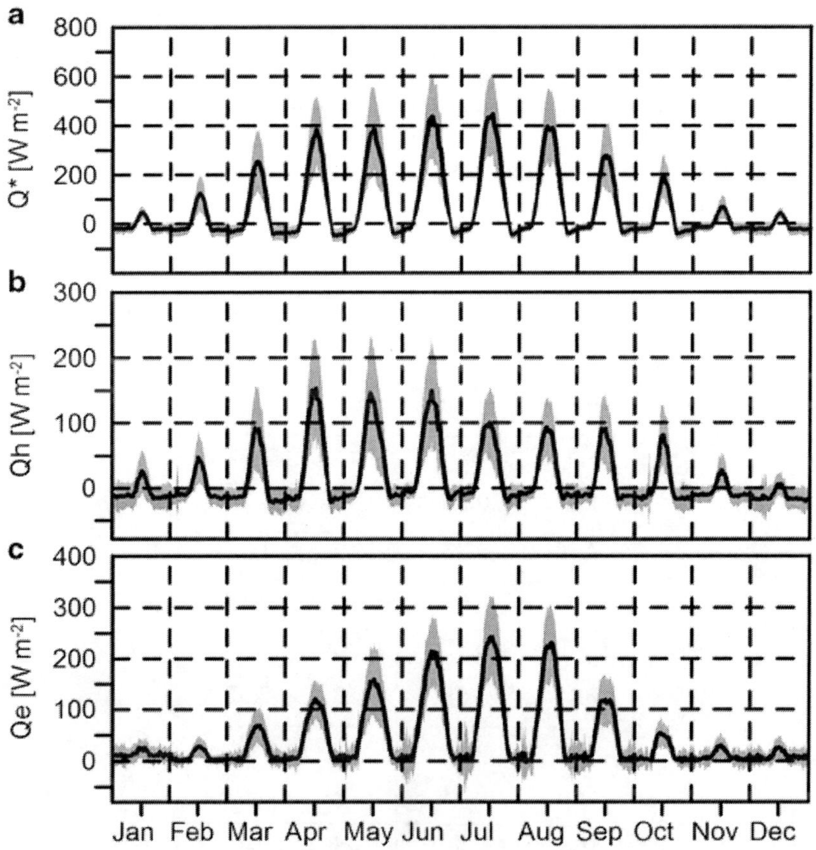

Figure. 3 Mean daily pattern of surface energy balance components (*solid line*) and its mean variance described as ± standard deviation (*shaded area*)

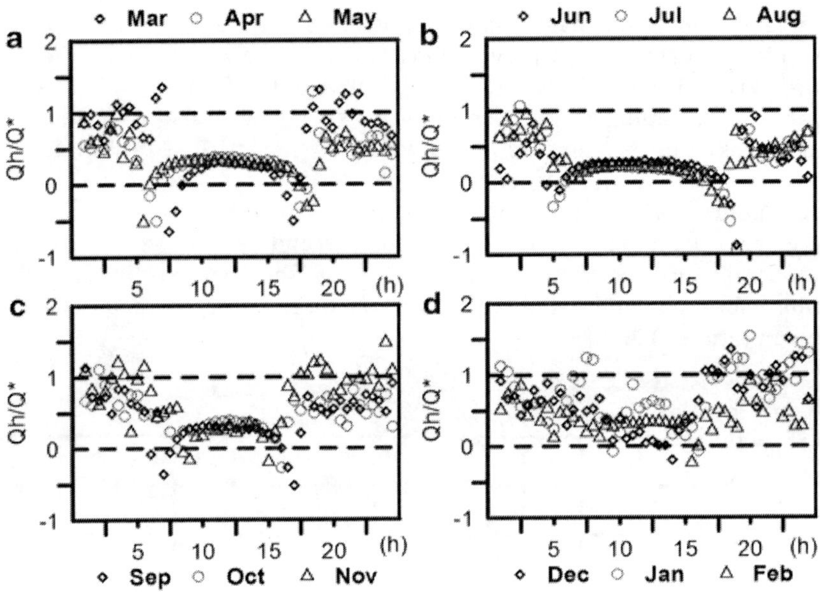

Figure. 4 Mean daily patterns of Qh and Q* ratios

Table 1 Average daytime (from 10.00 a.m. to 2.00 p.m.) flux ratio

Month	Qe/Q*	Qh/Q*	Qh/Qe
Jan	0.69	0.48	1.14
Feb	0.28	0.33	1.59
Mar	0.30	0.29	1.28
Apr	0.35	0.36	1.18
May	0.48	0.29	0.78
Jun	0.53	0.28	0.59
Jul	0.59	0.18	0.35
Aug	0.59	0.19	0.35
Sep	0.51	0.27	0.62
Oct	0.38	0.34	1.47
Nov	0.54	0.21	0.78
Dec	0.58	0.09	0.16

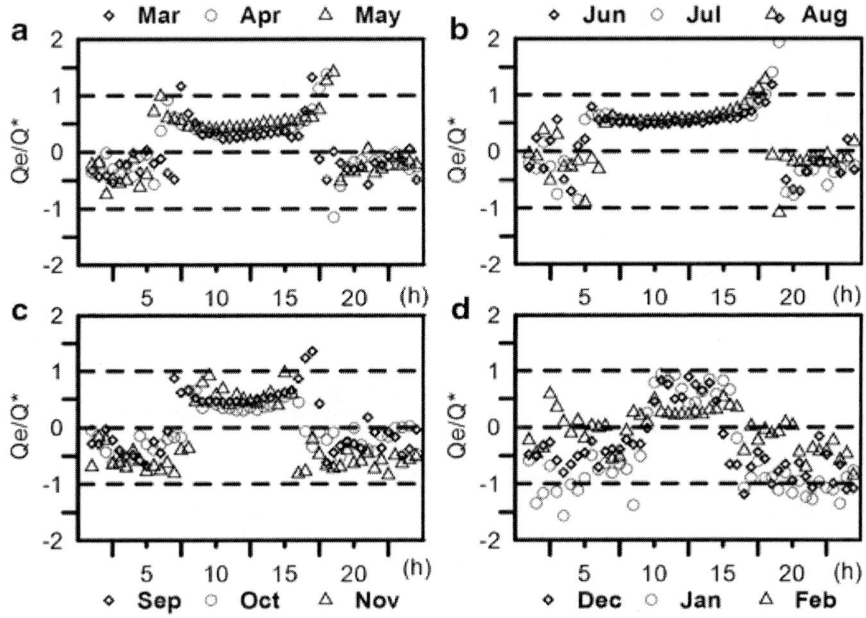

Figure. 5 Mean daily patterns of Qe and Q* ratios

In May, the sensible and latent heat fluxes achieve similar values. In the noon hours, the average values of Qh and Qe reach a level of 150 W·m^{-2}. In July and August, the high value of solar radiation in conjunction with fully developed vegetation and high precipitation lead to high values of the latent heat flux. The highest average values of Qe in the noon hours reach a level of 250 W·m^{-2} (Fig. 3c), while the values of the sensible heat flux at the same time amount to 100 W·m^{-2} (Fig. 3b). Despite the different climatic conditions (warm temperate climate conditions with the monsoon variant), quite similar values of turbulent latent heat flux are characteristic of evapotranspiration from the wetland area overgrown mainly with reeds in north-eastern China. For example, in the period of strong growth of vegetation (May and June), as well as in summer, the average values of Qe in the noon hours vary between 180 and 200 W·m^{-2} (Zhou and Zhou 2009). The average values of Qe at a level of about 200 W·m^{-2} are also characteristic of the marshland areas of central Sweden (Kellner 2001); the average value of Qe for the noon hours at a level of 300 W·m^{-2} was found, in turn, in the case of evapotranspiration from the surface of wetlands in the subtropical zone (Jacobs et al. 2002; Wu and Shukla 2014) or coastal marshes (Souch et al. 1998).

In summer, the values of Qe obtained in the daytime account for an average of 50–60 % of the radiation balance (Fig. 5b), while the ratio of

Qh/Q* has a value of 0.2 to 0.3 (Fig. 4b). In the diurnal pattern of the Qe/Q*, a clear u-shaped pattern is characteristic for the daylight hours (Fig. 5b). After sunrise in the summer months, the ratio Qe/Q* assumes values close to 1 (solar radiation energy received is consumed to evaporate dew), and next these values fall to about 0.5 in the noon hours to reach values equal to or greater than 1 in the evening, which means that the energy accumulated in the soil and the downward sensible heat flux can be used for evaporation (Fig. 5b). At night, due to the fact that the measured values of the turbulent latent heat flux are low, the result of the ratio Qe/Q* assumes values rather randomly (without a distinct diurnal pattern), ranging between −1 and 1.

September is another month in which the average pattern of Qh and Qe achieve similar values (around 100 $W \cdot m^{-2}$). In October, like in early spring, Qh exceeds Qe. In this season, the values of Qe obtained in the daytime account for an average of 35–50 % of the radiation balance (Fig. 5c), while the ratio of Qh/Q* has a value of 0.2 to 0.4 (Fig.4c). In winter, when the Q* is low, the turbulent sensible and latent heat fluxes thereby assume low values, and moreover, without a clear diurnal cycle. Furthermore, in the daily patterns of the relation of Qh and Qe to Q* it is difficult to identify a clear diurnal cycle (Fig. 4d and Fig. 5d). The low measured values, close to the accuracy of measuring instruments, in combination with the conditions of weak turbulence mean that the results obtained during this period may be subject to a considerable error.

The presented comparison of the average daily patterns of the sensible and latent heat fluxes in the warm half of the year shows that in spring the increase of solar radiation leads in the first place to the development of the sensible heat flux. It is only in summer, thanks to well-developed flora, evaporation heat losses are greater than those consumed for the sensible heat flux. This is also confirmed by the average pattern of the value of the Bowen ratio (Fig. 6). In the three summer seasons of study, the quotient Qh/Qe assumed values smaller than 1. The autumn period, namely the end of September and early October, showed a clear increase in the value of the Bowen ratio (> 1). Changes in the vegetative cycle lead to reduced transpiration of water vapour through the stomata of plants. At the same time, due to the accumulation of thermal energy in the soil after the summer period, and at relatively high values of solar radiation, the sensible heat flux assumes high values which exceed evaporation heat losses. The high values of the ratio Qh/Qe (> 1) are characteristic of the winter period, the visible high variability of the observed values being likely caused by the fact that the obtained values of Qh and Qe are small, within the accuracy of the measurement method (Fig. 6).

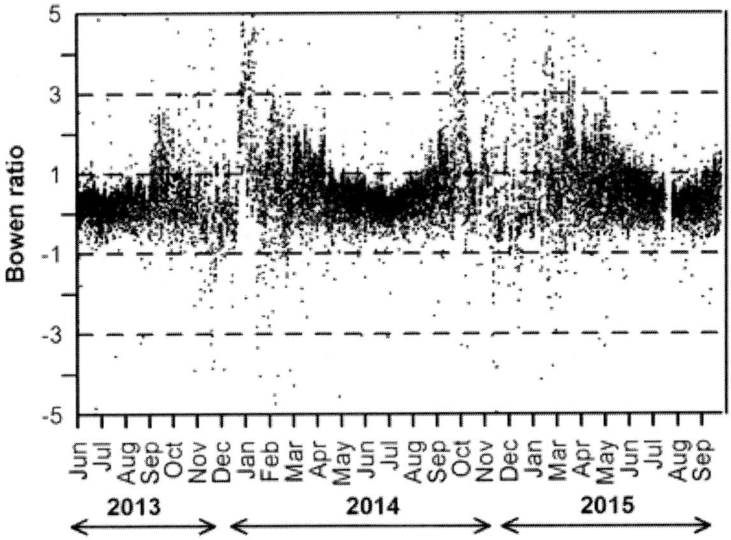

Figure. 6 Pattern of Bowen ratio values in daytime hours (Kd > 5 $W \cdot m^{-2}$)

The high variability of the Bowen ratio (Qh/Qe) values in winter is also confirmed by the average daily patterns for the winter months with highest values in the middle of the day and smaller or negative values around sunrise and sunset (Fig. 7a). However, in the spring, summer as well as autumn seasons a distinct average daily cycle is visible, especially for the daylight hours. For the months of March and April, in the morning hours the mean values of the Bowen ratio range between 0 and 1 to assume values greater than 1 in the afternoon (Fig. 7b). In May and in the summer months (Fig. 7c), the mean values of the Bowen ratio in the afternoon are in the range 0.35–0.6 (Table 1). In addition, from May to August a marked asymmetry is observed in the daily profile, where the values of afternoon and evening Qh/Qe quotient are lower than those of the morning. It is probably heat stored in the ground which may be an additional source of energy intensifying evapotranspiration at this time of day. However, due to difficulties in measuring Qg in an environment of lush vegetation and high moisture content, these results are not presented in this paper.

3.2 Evapotranspiration

The conversion of the latent heat flux values from Wm^{-2} to $mm \cdot h^{-1}$ allows to determine water losses associated with evaporation and transpiration. In the winter months, the mean hourly evapotranspiration totals are not big and the highest mean hourly evapotranspiration totals slightly exceed 0.05 mm/h, while the maximum values oscillate around 0.2 mm/h. From March to September, evapotranspiration it is characterized by a marked diurnal cycle (Fig. 8). In the spring months, the mean hourly totals of evapotranspiration at noon change from 0.1 mm·h^{-1}to 0.3 mm·h^{-1} in June. Equally high mean hourly totals of evapotranspiration are typical of the summer months.

The highest hourly totals of evapotranspiration observed so far (using the adopted criteria for data verification) hovered around 0.5 mm·h^{-1} (Fig. 8). Similar values of hourly evapotranspiration totals in the noon hours are characteristic of the wetlands constituting the west coast of Lake Huron, Michigan, USA (Mazur et al. 2014). Based on the water table fluctuation method, the presented daily patterns of evapotranspiration recorded during the measurement sessions of several days in the summer season show a clear pattern with maximum values of 0.5 to 0.8 mm·h^{-1}.

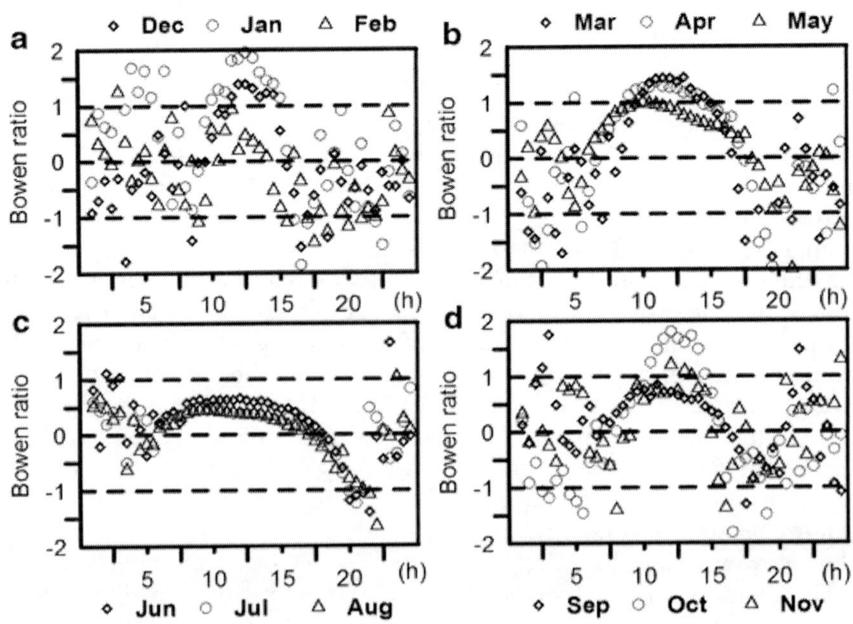

Figure. 7 Mean diurnal patterns of Bowen ratio

The results of measurements of hourly evapotranspiration totals with gaps in the data filled were used to determine the daily and monthly evapotranspiration totals. In the period of study, in the winter months the daily evapotranspiration totals (except in a few cases) ranged between 0.1

and 1 mm (Fig. 9). The median of daily totals varied from 0.20 mm in December to 0.48 mm in January. At the end of March, a significant increase in the daily evaporation totals is seen. The mean daily total both in 2014 and 2015 amounted to 0.8 and 0.9 mm, respectively, evaporation also being characterized by a greater variability due to more dynamic weather conditions. For example, as compared to January or February, in March the inter-quartile range of daily totals of evapotranspiration is determined by the values 0.1 and 0.6 mm.

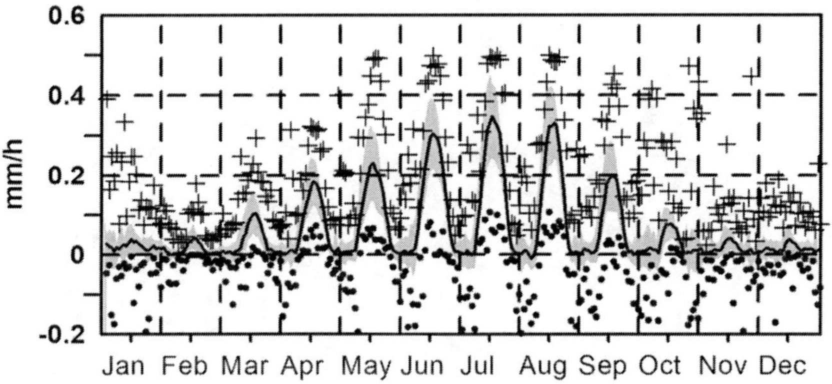

Figure. 8 Mean hourly evapotranspiration totals, shaded area – mean +/– standard deviation of hourly evapotranspiration totals, dots and cross – observed minimum and maximum values of evapotranspiration

In the subsequent months, the obtained daily evapotranspiration totals increase, to achieve the highest values in June and July when their average value ranges between 3.0 and 3.5 mm. In addition, the summer season, especially in July, is characterized by the highest daily totals with values of 4–5 mm (Fig. 9). This was particularly evident in the summers of 2013 and 2014. However, in 2015, when summer was relatively dry, the highest daily evapotranspiration totals (only in a few cases) reached a level of 4–4.4 mm. Another characteristic feature of evaporation in summer is the fact that the daily totals may assume values within a very wide range. In June and July, the lowest daily totals were 1.2 and 1.7 mm, while the absolutely highest daily totals reached values of 5.0 and 4.6 mm (Fig. 10). It is worth mentioning that the maximum evapotranspiration was observed during sunny and hot weather (incoming solar radiation reached 800–900 W m^{-2} and the maximum daily temperature exceeded 25 °C), whereas very low daily totals were associated with cloudy weather and advection cold air mass (maximum daily temperature below 15 °C). At the same time, with a large difference between the minimum and maximum values, the inter-quartile range of daily totals changed by about 1 mm. The value of the 1st quartile varied from about 2.5 mm in June and August to 3 mm in July, while the values for the 3rd quartile were, respectively, 3.5 and 4.0 mm. In the autumn season, in the period under analysis the daily totals steadily decreased. The average daily totals for September, October and November amounted to 1.6 mm, 0.8 mm and 0.4 mm, respectively.

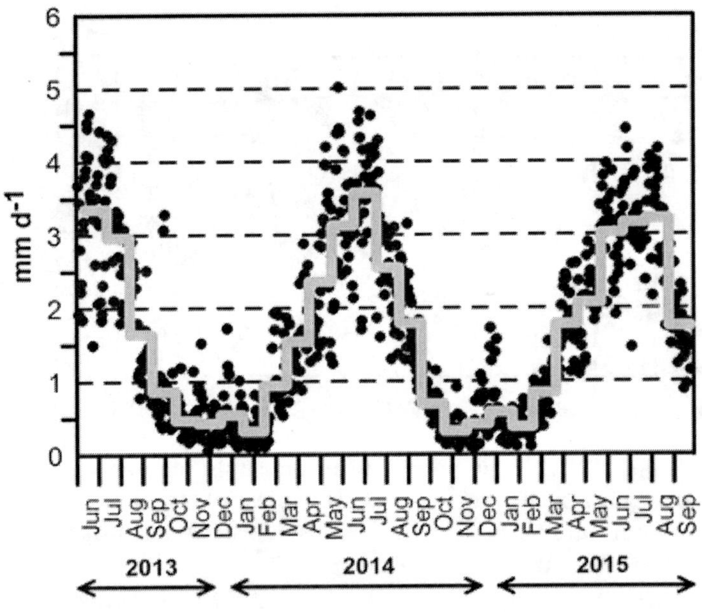

Figure. 9 Patterns of daily totals (*dots*) and mean daily total (*gray step plot*) of evapotranspiration during the study period

The results of the daily evapotranspiration totals in summer, obtained using the eddy covariance methods, are quite similar to the results obtained for marshland areas in Europe, Siberia and Canada. Daily totals at a level of 2–3.5 mm were obtained during a few weeks of summer measurements in the wetlands overgrown by reeds (Reed Belt) (Peacock and Hess 2004), or also in wet meadows (Gasca-Tucker et al. 2007) in southern England. Also in northern Germany (Herbst and Kappen 1999) and Canada (Lafleur et al. 2005), evapotranspiration in summer from the surfaces of marshlands where the dominant vegetation is reed is characterized by daily totals at a level of 2–4 mm·d^{-1}. Similar values of daily evapotranspiration totals are characteristic of wetlands in northwestern Russia (Kurbatova et al. 2002) or western Siberia (Shimiyama et al. 2004), but in both the cases the dominant vegetation is Sphagnum moss.

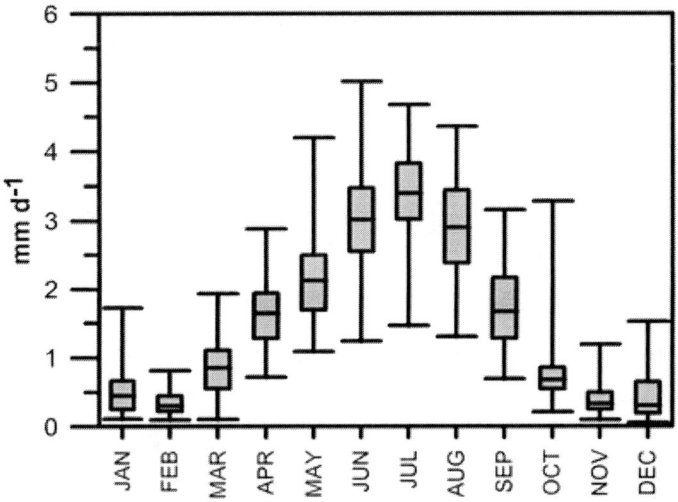

Figure. 10 Median, inter-quartile ranges, maximum and minimum daily values of evapotranspiration

In hydrological practice, in assessing the water balance of a specific area from the point of view of evaporation and evapotranspiration water losses, monthly or seasonal totals are normally used. In the annual pattern, the period from October to February is characterized by the lowest values of the monthly totals of evapotranspiration (Fig. 11c). Both in the winter season of 2013/14 and that of 2014/2015, the radiation balance and humidity conditions were similar. The marshland area radiated out more energy than it gained from solar radiation and diffuse sky radiation, i.e. the radiation balance from December to February was negative (Fig. 11a). In addition, low air temperatures (monthly averages in January and February below 0 °C) also resulted in high relative humidity. The average vapour pressure deficit in the winter seasons of 2013/14 and 2014/2015 was low and amounted to 1–2 hPa. Under such weather conditions, evaporation water losses were low, at a level of 9–15 mm per month. At the same time, it is also the period when precipitation exceeds evapotranspiration, resulting in an increase in soil moisture content. This is particularly evident in the pattern of the mean values of the volumetric water content in the soil (VWC). For instance, the winter season 2014/15 in the months from November to March marked an increase of VWC from 0.7 to about 0.9. In contrast, the low values of VWC in February and March 2014 were due to problems with the measurements due to very low temperatures in January 2014 and frozen soil (Fig. 11a).

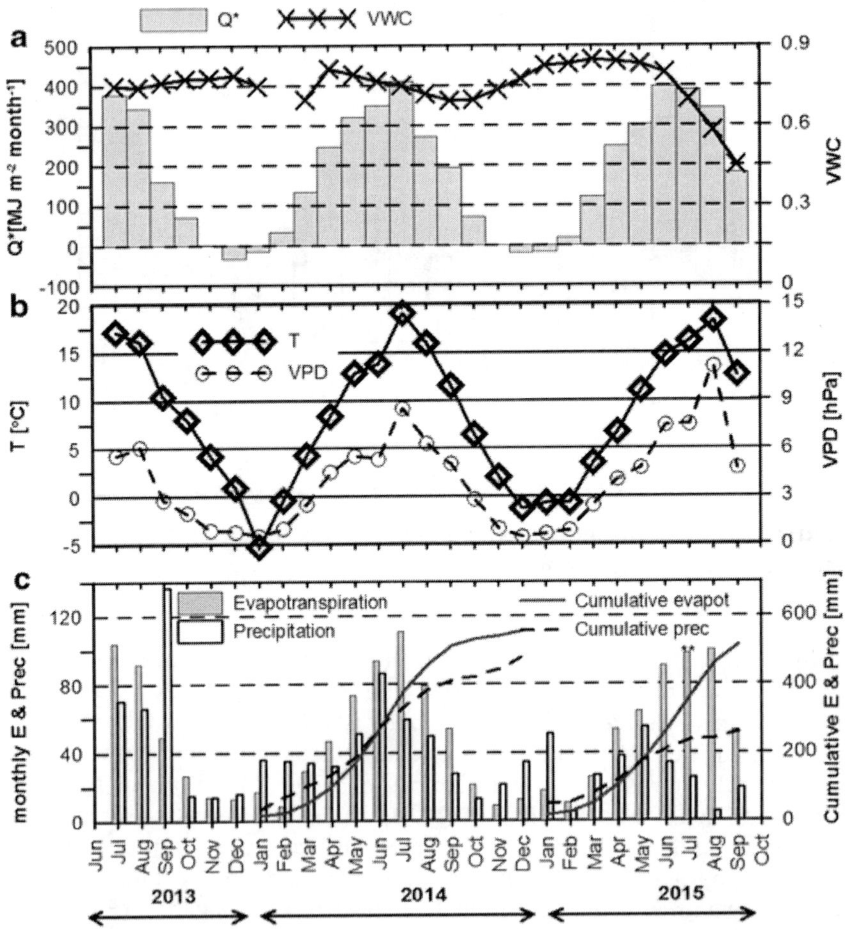

Figure. 11 Patterns of monthly evapotranspiration (E) and precipitation (P) totals during the study period with cumulative evapotranspiration and precipitation totals, mean monthly air temperature (T), vapour pressure deficit (VPD) and volumetric water content (VWC), monthly total of net radiation (Q*). ** Monthly sum of evapotranspiration taken from 20 days

In spring, as a result of increased turbulent water vapour flux, the monthly totals are also higher. In both the analyzed spring seasons, the weather conditions were very similar. The obtained monthly totals varied from about 25 mm in March to about 70 mm in May. The highest monthly evapotranspiration totals are characteristic of summer. In the three summer seasons studied, the monthly evapotranspiration totals from June to August exceeded 80 mm. The highest evaporation occurred in July 2014, with a monthly total of 110 mm. That month was exceptionally warm (a monthly average of 19.1 °C)(Fig. 11b), and the surface of the wetland area received, via radiation exchange, a total of 410 MJ m^{-2} of energy. Another factor contributing to the high evaporation that month was high rainfall which led to high soil moisture values (VWC monthly averages in June and July amounted to ca 0.75). In the summer months of 2015, evapotranspiration was not so high, although the summer was exceptionally dry. The monthly rainfall did not exceed 40 mm and the average monthly value of vapour pressure deficit was the highest(11.2 hPa) throughout the measurement period (Fig. 11b). Under such conditions, the monthly evapotranspiration totals increased from 90 mm in June to 100 mm in August due to water resources in the soil (average monthly values of the VWC index dropped from 0.8 in May and June to 0.4 in September). Such high evaporation with very low rainfall confirms the pivotal role played by wetlands in the water balance of a specific area as well as in its surface heat balance. In the autumn months, the obtained monthly totals decreased from about 50–60 mm in September to about 10–15 mm in November.

A characteristic feature of the seasonal pattern of evapotranspiration in Poland is that in the warm half of the year it exceeds precipitation (Daniela and Lenart 1989; Musiał and Rojek 2007). A similar pattern was observed for the marshland area studied. From April to November, evapotranspiration was higher than the recorded precipitation (except September 2013, when there was exceptionally high rainfall). For example, from April to October 2014 evaporation losses amounted to more than 450 mm, while total precipitation at the measurement station during this period was 320 mm. The annual precipitation and evapotranspiration totals, in turn, amounted to 475 mm and 555 mm, respectively (Fig.11c).

4. CONCLUSION

Evapotranspiration in the marshland areas of north-eastern Poland is characterized by a distinct diurnal cycle, particularly in the growing season. The average values of latent heat in the noon hours vary from about 100 Wm^{-2} in early spring to about 250 Wm^{-2} in summer. During this season, the losses of radiation energy available for evaporation vary from about 30 % in March and April to 60 % in July and August. At the same time, with the growth of wetland vegetation during the growing season, the redistribution of radiation energy received between the sensible and latent heat fluxes changes quite significantly. The average Bowen ratio in the spring season assumes values greater than 1, while in the summer season they are in the range 0.35 to 0.7.

The obtained average daily totals in the summer months at a level of 3–3.5 mmd^{-1} are similar to the results obtained when characterizing evapotranspiration in other wetland areas in Europe, Canada and Siberia. At the same time, multi-seasonal measurements under different weather conditions in individual years confirmed the role which the marshlands play for the catchment water balance and the surface heat balance. For example, the soil moisture reserves are able to maintain evapotranspiration rates under drought conditions that are only slightly lower than under standard seasonal conditions.

ACKNOWLEDGMENTS

We would like to thank the Biebrza National Park for the cooperation and the permission to conduct the study in the Park. The study was supported by the Polish National Science Centre under projects UMO-2011/01/B/ST10/07550 and N N306717740.

REFERENCES

1. Abtew W, Melesse A (2013) Evaporation and evapotranspiration. Springer, Measurement and estimation
2. Abtew W, Obeysekera J (1995) Lysimeter study of evapotranspiration of cattails and comparison of three estimation methods. Transactions of the ASAE 38(1):121–129
3. Acreman MC, Harding RJ, Lloyd CR, McNeil DD (2003) Evapotranspiration characteristics of wetlands: experience from a wet grassland and a reedbed using eddy correlation measurement. Hydrology and Earth System Sciences 7:11–21
4. Affre C, Lopez A, Carrara A, Druilhet A, Fontan J (2000) The analysis of energy and ozone flux data from the LANDES experiment. Atmospheric Environment 34:803–821
5. Aselman I, Crutzen PJ (1989) Global distribution of natural fresk-water wetlands and rice paddies, their net primary productivity, seasonality and possible methane emissions. Journal of Atmospheric Chemistry 8:307–358
6. Banaszuk H, Banaszuk P (2004) Charakterystyka gleb Biebrzańskiego Parku Narodowego. [in:] Banaszuk H, (ed.) Kotlina Biebrzańska i Biebrzański Park Narodowy. Aktualny stan walory zagrożenia i potrzeby czynnej ochrony środowiska, Wydawnictwo Ekonomia i Środowisko, Białystok, pp 265–290
7. Billett MF, Palmer SM, Hope D, Deacon C, Storeton-West R, Hargreaves KJ, Flechard C, Fowler D (2004) Linking land–atmosphere–stream carbon fluxes in a lowland peatland system. Global Biogeochemical Cycles 18:GB1024. doi:10.1029/2003GB002058
8. Burba GG, Verma SB, Kim J (1999) Energy fluxes of an open water area in a mid-latitude prairie wetland. Boundary-Layer Meteorology 91:495–504
9. Campbell DI, Williamson JL (1997) Evapotranspiration from a raised peat bog. Journal of Hydrology 193:142–160

10. Chojnicki BH, Urbaniak M, Józefczyk D, Augustin J (2007) Measurement of gas and heat fluxes at Rzecin wetland. In: Okruszko et al. (eds) Wetlands: Monitoring, Modelling and Management. Taylor & Francis Group, London, pp. 125–131
11. Daniela D, Lenart W (1989), Wyznaczanie parowania terenowego w Polsce metodą kompleksową, Roczniki Akademii Rolniczej w Poznaniu, Melioracje i Inżynieria Środowiska, 8:41–51
12. Drexler JZ, Snyder RL, Spano D, Kyaw Tha Paw U (2004) A review of models and micrometeorological methods used to estimate wetland evapotranspiration. Hydrological Processes 18:2071–2101
13. Dutaur L, Cieslik S, Carrara A, Lopez A (1999) The detection of nonstationarity in the determination of deposition fluxes. Proceedings of EUROTRAC Symposium 98, t 2, WIT Press, Southampton, pp. 171–176
14. EEA [European Environmental Agency] (2006) CORINE Land Cover database. Online: http://www.eea.europa.eu/data-and-maps/data/clc-2006-vector-data-version-3
15. Falge E, Baldocchi D, Olson RJ, Anthoni P, Aubinet M, Bernhofer C, Burba G, Ceulemans R, Clement R, Dolman H, Granier A, Gross P, Grunewald T, Hollinger D, Jensen N-O, Katul G, Keronen P, Kowalski A, Ta Lai C, Law BE, Meyers T, Moncrieff J, Moors E, Munger JW, Pilegaard K, Rannik U, Rebmann C, Suyker A, Tenhunen J, Tu K, Verma S, Vesala T, Wilson K, Wofsy S (2001) Gap filling strategies for defensible annual sums of net ecosystem exchange. Agricultural and Forest Meteorology 107:43–69
16. Foken T (2008) Micrometeorology. Springer, Berlin
17. Fortuniak K (2010) Radiacyjne i turbulencyjne składniki bilansu cieplnego terenów zurbanizowanych na przykładzie Łodzi. Wyd. Uniwersytetu Łódzkiego, Łódź, 232 s
18. Fortuniak K, Pawlak W, Siedlecki M, Zieliński M (2013) Surface energy balance and exchange of greenhouse gases in Eastern Poland wetland – A new EC site in Biebrza National Park. [in:] Reckermann M., Köppen S. (ed.) – 7th Study Conference on BALTEX, 10–14 June 2013, Borgholm, Conference Proceedings, pp. 18–19
19. Fraser CJD, Roulet NT, Moore TR (2001) Hydrology and dissolved organic carbon biogeochemistry in an ombrotrophic bog. Hydrological Processes 15:3151–3166
20. Gasca-Tucker DL, Acreaman MC, Agnew CT, Thompson JR (2007) Estimating vaporation from a wet grassland. Hydrology and Earth System Sciences 11(1):270–282
21. Gerla PJ (1992) The relationship of water-table changes to the capillary fringe, evapotranspiration and precipitation in intermittent wetlands. Wetlands 12(2):91–98
22. Grygoruk M, Mirosław-Świątek D, Okruszko T, Batelaan O, Szatyłowicz J (2011) Estimation of actual evapotranspiration of birch thickets on a fen based on diurnal fluctuations of shallow groundwater level. Water-Environmnet-Rural Areas 11(4):121–136
23. Grygoruk M, Batelaan O, Mirosław-Świątek D, Szatyłowicz J, Okruszko T (2014) Evapotranspiration of bush encroachments on a temperate mire meadow – a nonlinear function of landscape composition and groundwater flow. Ecological Engineering 73:598–609
24. Gulden ML, Litvak M, Miller SD (2007) Factors that control Typha marsh evapotranspiration. Aquatic Botany 86:97–106
25. Herbst M, Kappen L (1999) The ratio of transpiration versus evaporation in a reed belt as influenced by weather conditions. Aquatic Botany 63:113–125
26. Jacobs JM, Mergelsberg SL, Lopera AF, Myers DA (2002) Evapotranspiration from a wet prairie wetland under drought conditions: Paynes Prairie Preserve, Florida, vol 22 (2), pp. 374–385
27. Kellner E (2001) Surface energy fluxes and control of evapotranspiration from Swedish sphagnum mire. Agricultural and Forest Meteorology 110:101–123
28. Kim J, Verma S (1996) Surface exchange of water vapor between an open sphagnum fen and the atmosphere. Boundary-Layer Meteorology 79:243–264
29. Kleniewska M, Gołaszewski D, Majewski G, Siedlecki P (2009) System pomiaru głównych składników bilansu cieplnego łąki bagiennej w dolinie dolnej Biebrzy. Przegląd Naukowy. Inżynieria i Kształtowanie Środowiska 18(1):32–41

30. Kossakowska-Cezak U, Olszewski K, Przybylska G (1991) Klimat Kotliny Biebrzańskiej (climate of the Biebrza Valley). Zeszyty Problemowe Postępów Nauk Rolniczych 372: 119–160
31. Kurbatova J, Arneth A, Vygodskaya NN, Kolle O, Valralgin AV, Milyukova IM, Tchebakova NM, Schulze E-D (2002) Comparative ecosystem – atmosphere exchange of energy and mass in a European Russian and central Siberian bog I., Interseasonal and interannual variability of energy and latent heat fluxes during the snowfree period. Tellus 54B:497–513
32. Lafleur PM, Hember RA, Admiral SW, Roulet NT (2005) Annual and seasonal variability in evapotranspiration and water table at a shrubcovered bog in southern Ontario, Canada. Hydrological Processes 19:3533–3550. doi:10.1002/hyp.5842
33. Lenters JD, Cutrell GJ, Istanbulluoglu E, Scott DT, Herrman KS, Irmak A, Eisenhauer DE (2011) Seasonal energy and water balance of a *Phragmites australis*-dominated wetland in the Republican River basin of south-Central Nebraska (USA). Journal of Hydrology 408:19–34
34. Mahrt L (1998) Flux sampling errors for aircraft and towers. Journal of Atmospheric and Oceanic Technology 15:416–429
35. Malinowska E (2004) Krajobraz Biebrzańskiego Parku Narodowego [in:] Banaszuk H (ed.) Kotlina Biebrzańska i Biebrzański Park Narodowy. Aktualny stan walory zagrożenia i potrzeby czynnej ochrony środowiska, Wydawnictwo Ekonomia i Środowisko, Białystok, pp 327–364
36. Matuszkiewicz AJ (2004) Roślinność nieleśna. [in:] Banaszuk H (ed.) Kotlina Biebrzańska i Biebrzański Park Narodowy. Aktualny stan walory zagrożenia i potrzeby czynnej ochrony środowiska, Wydawnictwo Ekonomia i Środowisko, Białystok, pp 423–424
37. Mazur MLC, Wiley MJ, Wilcox DA (2014) Estimating evapotranspiration and groundwater flow from water-table fluctuation for a general wetland scenario. Ecohydrology 7:378–390
38. Moore TR, Roulet NT, Waddington MJ (1998) Uncertainty in predicting the effect of climatic change on the carbon cycling of Canadian peatlands. Climatic Change 40:229–245
39. Musiał E, Rojek MS (2007) Potential evapotranspiration and precipitation in Wrocław-Swojec during the period 1965-2000. Acta Agrophisica 9(3):685–698
40. Pauliukonis N, Schneider R (2001) Temporal patterns in evapotranspiration from lysimeters with three common wetland plant species in the eastern United States. Aquatic Botany 71:35–46
41. Peacock CE, Hess TM (2004) Estimating evapotranspiration from a reed bed using the Bowen ratio energy balance method. Hydrological Processes 18:247–260
42. Priban K, Ondok JP (1985) Heat balance components and evapotranspiration from a sedge-grass marsh. Folia Geobotanica Phytotaxon 20:41–56
43. Shimiyama K, Hiyama T, Fukushima Y, Inoue G (2004) Controls on evapotranspiration in a eest Siberian bog. Journal of Geophysical Research 109:D08111. doi:10.1029/2003JD004114
44. Souch C, Grimmond CSB, Wolfe CP (1998) Evapotranspiration rates from wetlands with different disturbance histories: Indiana dunes National Lakeshore. Wetlands 18(2):216–239
45. Urbaniak M (2006) Ocena sezonowej zmienności strumieni dwutlenku węgla i pary wodnej na terenach podmokłych metodą kowariancji wirów. Doctoral thesis. Poznań. Institute of Agrometeorology, Faculty of Land Reclamation and Environmental Engineering, AR
46. Ustrnul Z, Marosz M, Limanówka D, Biernacik D, Czekierda D, Kilar P, Czarnecki B, Mizera M (2013) Monthly Climate Monitoring Bulletin, Summer 2013, Warszawa (2013) Available via http://www.imgw.pl/images/stories/biuletyn_monitoringu/ 2013/lato_2013.pdf
47. Ustrnul Z, Limanówka D, Biernacik D, Czekierda D, Kilar P, Mizera M (2014) Monthly Climate Monitoring Bulletin, Summer 2014, Warszawa 2014. Available via http://www.imgw.pl/images/stories/biuletyn_monitoringu/2014/lato2014.pdf

48. Ustrnul Z, Limanówka D, Biernacik D, Czekierda D, Kilar P, Mizera M (2015a) Monthly Climate Monitoring Bulletin, Winter 2014–2015, Warszawa 2015. Available via http://www.imgw.pl/images/stories/biuletyn_monitoringu/2015/wiosna2015.pdf.
49. Ustrnul Z, Limanówka D, Biernacik D, Czekierda D, Kilar P, Mizera M (2015b) Monthly Climate Monitoring Bulletin, Summer 2015, Warszawa 2015. Available via http://www.imgw.pl/images/stories/biuletyn_monitoringu/2015/lato2015.pdf
50. Wassen MJ, Okruszko T, Kardel I, Chromanski J, Swiatek D, Mioduszewski W, Bleuten W, Qerner EP, Kahloun ME, Betelaan O, Meire P (2006) Eco-Hydrological Functioning of the Biebrza Wetlands: Lessons for the Conservation and Restoration of Deteriorated Wetlands. [In:] Bobbink R, Beltman B, Verhoeven JTA, Whigham DF (eds.) Wetlands: Functioning, Biodiversity Conservation, and Restoration. Ecological Studies, 191:285–310
51. Webb KF, Pearman GI, Leuning R (1980) Correction of flux measurements for density effects due to heat and water vapor transfer. Quarterly Journal of the Royal Meteorological Society 106:85–100
52. Wessel DA, Rouse WR (1994) Modeling evaporation from a wetland tundra. Boundary-Layer Meteorology 68:109–130
53. Wu C-L, Shukla S (2014) Eddy covariance-based evapotranspiration for a subtropical wetland. Hydrological Processes 28:5879–5896
54. Zhou L, Zhou G (2009) Measurement and modeling of evapotranspiration over a reed (*Phragmites australis*) marsh in Northeast China. Journal of Hydrology 372:41–47

CHAPTER 8

A Novel Protection and Utilization Pattern for Urban Wetland—As an Example of Hemu Wetland in Hangzhou, China

Nongnong Shi[1*#], Minger Ye[2#], Tongfei Lai[1], Zhiming Yu[1], Yunli Pan[2], Mengying Yu[2], Xiaoqi Zhou[2]

[1]*School of Life and Environmental Sciences, Hangzhou Normal University, Hangzhou, China*

[2]*School of Agriculture and Biological Technology, Zhejiang University, Hangzhou, China*

ABSTRACT

This paper proposes an innovative plan and its implement strategy of Hemu urban wetland. The plan takes into consideration its historical background and current status, its protection significance, strategies, solutions and support conditions. It highlights the landscape characteristics of rivers and lakes, and restores its historical flood prevention functions and climate regulation for the city. Especially, this paper proposes to achieve the restoration of "Mulberry-sided Fish Pond" farming style—an important agricultural heritage created by the ancestors inside the wetland, which is a unique and creative way to use low land for ecological agriculture and cycling economy. The proposed wetland sewage treatment schemes would be of practical importance to improve and enhance the ecological environment and promote the construction of ecological civilization in the region. The "Two zones—One block and Three complexes" protection and utilization plan is reasonable and feasible. It fully presents the functions of the wetland for its production, living, life forms and ecology, and is characterized by joint protection of the wetland by original residents and social public. This will result in active, instead of passive, wetland protection, like turning a "blood transfusion protection" to "hematopoietic protection", which reflects the sustainable development. The paper offers valuable information and strategies for planning and development of urban wetland.

Keywords: Urban Wetland, Protection and Utilization, Mulberry-Sided Fish Pond Ecosystem, Two Zones—One Block and Three Complexes

1. INTRODUCTION

Hemu wetland is located in the Wanjing Subdivision, Yuhang District in the western Hangzhou city, Zhejiang Province, China. The total area is about 8.51 square kilometers. It is adjacent to the famous Hangzhou Xixi National Wetland Park. At present, it is an undeveloped and original urban wetland [1]. Located in ancient Taihu basin, Hemu wetland is a typical plain depression in the network of rivers and ponds. In the ancient times, irrigation systems consisting of dikes and fields were constructed, where the dikes were built around the low areas to prevent flooding and protect polders, and the paddy fields surrounded by the dikes were for crop cultivation.

1.1. Background Review of Protection and Utilization

Along with the migrated silk weaving technology, the silk industry in the region resulted in prosperity. Ancestors creatively planted mulberry trees on the dikes, grew rice in elevated flat and fertile polders, and cultured fishes in low and swamp-like ponds. The thick and fertile sludge in the bottom of pond was dug and moved to the dikes around the ponds to fertilize mulberry trees, and the excess nutrients in the dike soil were washed down with rain and flowed into the fish ponds. Excess pupae and silkworm excrement from the sericulture was used as fish feed. These activities resulted in "Mulberry-sided Polder" and "Mulberry-sided Fish Pond" systems which fitted well with the principles of cycling economy and ecological agriculture, as well as with agricultural development of the region [2] [3] . These systems greatly promoted the development of fishery, sericulture and silk industries in the Taihu River basin and reached a peak period in the late Ming and early Qing Dynasty. The mulberry-sided fish pond system provides an example of solution to agricultural source pollution due to overuse of chemical fertilizers and pesticides. This system is a unique and creative farming style in low land use and ecological and cycling economy [4] . Therefore, for protection and utilization of the Hemu wetland, what needs to be done first is to protect and utilize the "Mulberry-sided Fish Pond" system, an important and valuable agricultural heritage created by ancestors for the wetland.

Hemu wetland is located along the northern margin of subtropical monsoon climate zone with great biological diversity. It has plants of 85 families, 182 genera and 221 species, including 207 species of angiosperm, 9 species of pteridophytes, and 5 species of gymnosperm. Plants on river banks are wingnut, willow, early bamboo, persimmon, *Sapium sebiferum, Paulownia, camphor, elm, hibiscus, peach, gardenia,* etc. *Reed, Arundo donax* and *silvergrass* are the common plants on the riversides. In the water *Eichhornia crassipes, Alternanthera philoxeroides,* lotus, water chestnut, duckweed and *Acorus calamus* are growing [5] .

Approximately 12 orders, 26 families and 89 species of birds including herons, egrets, herons, green winged teals, spot billed duck, white-fronted Goose, kingfishers, woodpeckers, swallows, thrushes and tits, 3 orders and 6 families of mammals such as rabbits and weasels, 3 orders and 6 families of reptiles like toad, limnocharis, *Rana nigromaculata*, money frogs and *Rana rugulosa* were once observed to live in the Hemu wetland zone [6] .

1.2. Current Problems with the Wetland Protection

Since the beginning of the 1990s, the city of Hangzhou has been expanding continuously and a lot of farmlands and wetlands have been replaced with real estate and factories. Right now, the Hemu wetland area is equivalent to only half of that original. Industrial wastewater is discharged directly to the wetland, resulting in serious pollution of rivers and lakes, and killing large amounts of aquatic organisms and accumulation of harmful substances such as heavy metals in aquatic organisms.

In addition, a large number of residential housings developed in the wetland release a large volume of sewage into the wetland, leading to the eutrophication of the wetland. Sometimes, some algae proliferate explosively, heavily degrading the ecological environment of the wetland and severely damaging the biodiversity [7] [8] . According to a survey made by *Ni et al.* in July and November 2011 and January and April 2012, the data on average contents of ammonia nitrogen, nitrite nitrogen, nitrate nitrogen, total nitrogen, total phosphorus, chemical oxygen demand (COD), biochemical oxygen demand (BOD), chlorophyll, temperature, turbidity, dissolved oxygen, pH, electronic conductivity (EC), oxidation and restore potential (ORP) and other indexes taken from 15 sites show that water environment quality is below the standards for agriculture and landscape in the towns inside the wetland, such as Hemu bridge, Chenjia bridge, Nanxing bridge, Shanghe road, Yujiadoumen, Jiuqu bridge, Xingfu bridge, Jiuqu bay, Dongdun pond, Xidun lock, Yanyun bridge, Wutong bridge, Xidun pond, Ruanjia farm, and Aquatic product village. The water was classified as low-grade.

The monitoring results showed that N and P pollution were serious in four time periods at all sites. In addition, the water in the Ruanjia farm and Aquatic product village sites was most heavily polluted. This might be because they are geographically close to the Xianlin industrial zone [9] .

Field investigation by the project team found that the traditional and ecological agriculture systems "Mulberry- sided Polders" and "Mulberry-sided fish Ponds" in the wetland no longer existed. What left over is an extensive and unmanaged natural ecological landscape. Urban wetland is different from wetlands in remote and sparsely populated areas, because it is located in the urban areas with low position. This makes it prone to nutrient gatherings as a result of flowing of sewage and other polluted water. If there is no adjustment by artificial cycling ecological agriculture systems with high output and high nutrient consumption, and such large amount of nutrients are handled only by natural vegetation in ecological system in the urban wetland, eutrophication of water will occur easily and frequently.

This will result in the degradation of the structure and functions of wetland ecosystems, and in turn further reduce the self-purification ability of water and consequently exacerbate deterioration of water quality, forming a vicious circle and eventually leading to the degradation of wetland landscape. These indicate that for the urban wetland, as the city expands gradually, the sources of pollution not only continue to exist but also to expand [10] [11]. If we continue to use the protection modes for remote or sparsely populated natural wetlands for the urban wetland, it will be difficult to play its ecological functions.

A survey showed that the Wanjing village located within the jurisdiction area of the Hemu wetland was completely flooded due to Typhoon "Feite" outbreak on 7 October 2013. All the houses had the water of 1 m deep, and crops and fields were largely flooded.

Therefore, without solid control of upstream floods and perfect drainage system of downstream floods, Hemu wetland will be flooded and its ecosystem will be severely damaged, if the wetland is managed naturally, when extreme disasters such as flooding occur [12] - [14] .

1.3. Protection and Utilization: Necessity and Significance

Hemu wetland was historically responsible for flood discharge for the rivers between Tianmu Mountain in western Hangzhou and Dongtiao brook. In recent years, with the rapid development of the region, the functions of wetland gradually are weakening for its flood discharge. We should not only keep its beautiful landscape, but also the flood discharge function. Therefore, it is very important to figure out how to protect and enhance the flood discharge function of the Hemu wetland. This will prevent disasters resulting from extreme weather and benefit sustainable development of regional economy [15] . Historically, the "Mulberry-sided Polder" and "Mulberry- sided Fish Pond" systems within the Hemu wetland were cycling ecological agriculture systems. It played an important role in maintaining good regional ecological environment within the water systems. Restoration and development of these systems will improve and enhance the regional ecological environment and promote the management of five different types of water (sewage treatment, flood control, drainage of flooded water, water supply and water saving) carried out by local government.

1.4. The Goals of Protection, Utilization and Development

The goals are to let the wetland play important roles in reducing river eutrophication due to city life and preventing city flooding due to extreme weather by restoration and utilization of the "Mulberry-sided Fish Pond" wetland ecosystem created by ancestors. Meanwhile, through exploration of sericulture, fish culture and already formed "sheep-pot" culture in the local region, Hemu wetland has been developed into an integrated eco-agricultural demonstration area with high productivity, technology development, scientific education, tourism and leisure, vacation and health, and cultural heritage.

2. PROTECTION AND UTILIZATION PLAN

2.1. Space Layout

We propose to have a "Two districts—One block and Three complexes" overall planning for Hemu wetland protection and utilization.

Two districts: to divide the Hemu wetland into the original ecological protection zone and the restoration protection zone (Figure 1).

One block: to construct a block possessing wetland culture.

Three complexes: to construct sericulture complex, "sheep pot" culture complex and fish culture complex.

Figure 1. The original ecological protection zone and the restoration protection zone in protection planning of Hemu wetland.

2.2. Protection Planning

The original ecological protection zone will maintain the current wetland natural ecological landscape. At the east and west ends of the zone, there will establish two ecological isolation bands for sewage. The center original ecological protection area of the zone is the ecological wetland area for "Mulberry-sided Fish Pond" system (Figure 2).

2.2.1. Restoration Plan for the "Mulberry-Sided Fish Pond" Ecosystem

In the system, mud from the rivers and ponds were usually dug up and used to fertilize the mulberry trees every year, thus preventing eutrophication of rivers and ponds and elevating the dikes every year, and made the river clean and smooth. This also continuously increased the overall water storage capacity in the whole system and played important role in preventing or mitigating flood disaster.

High standard "Mulberry-sided Fish Pond" system is planned by restructuring the paddy fields and some fish ponds in the center original ecological protection area. The dike to pond ratios in the planned system will be 6:4 or 5:5 to ensure a balanced ecological system to maximize the economic and social benefits. The crops for the dike can be selected from mulberry tree, fruits, sesame, rapeseed, flowers and Bahia grass to form an interspersed, changing and beautiful landscape of the fish ponds with mulberry-, fruits-, rapeseeds-, and flowers-sided. Such landscape will be very scenic and attractive to tourists.

2.2.2. Water Purification Plan

The following three measures are planned to purify the water in the protected areas.

1) Establishment of ecological isolation zone for sewage

At the east and west ends in southern restoration protection area, ecological isolation bands for sewage will be built to prevent the upstream polluted water from industrial parks on the Xianlin road and large residential areas from entering the center wetland area directly through the Xianlin and Wutong harbors without any purification. For better purification of the upstream polluted water, three layers of biological water purification systems including aquatic plant purification, aquatic animal purification and reed purification are planned to ensure water quality being at standard II and III before entering the protected zone.

Aquatic plant purification system: Proper deployment of floating, submerged and emergent plants in the water of the wetland ecological isolation zone and irrigation rivers will have both ornamental value and the biological purifying function. Plants of various heights can be planted beside the water for better landscape purpose, such as tall plants like reed, and middle plants like cattail and windmill grass, and short plants like arrowhead. Over the water surface, floating plants (such as eye lotus) or floating carriers (such as aquatic biological floating island, in which terrestrial flowers such as beauty banana, or water spinach, celery and other vegetables are cultured in water). These plants will reduce light flux into the lakes, inhibit the growth of phytoplankton and increase the transparency of water. Submerged plants (such as Vallisneria spiralis, Potamogeton, etc.) or the leaf-floating plants (such as lotus) can be planted at the bottom of the water to absorb and transform organic matters and nutrients in sediment and water. They will reduce nutrient concentration in water, inhibit the growth of phytoplankton, prevent sediment from re-suspension to increase the water transparency and generate a clean river.

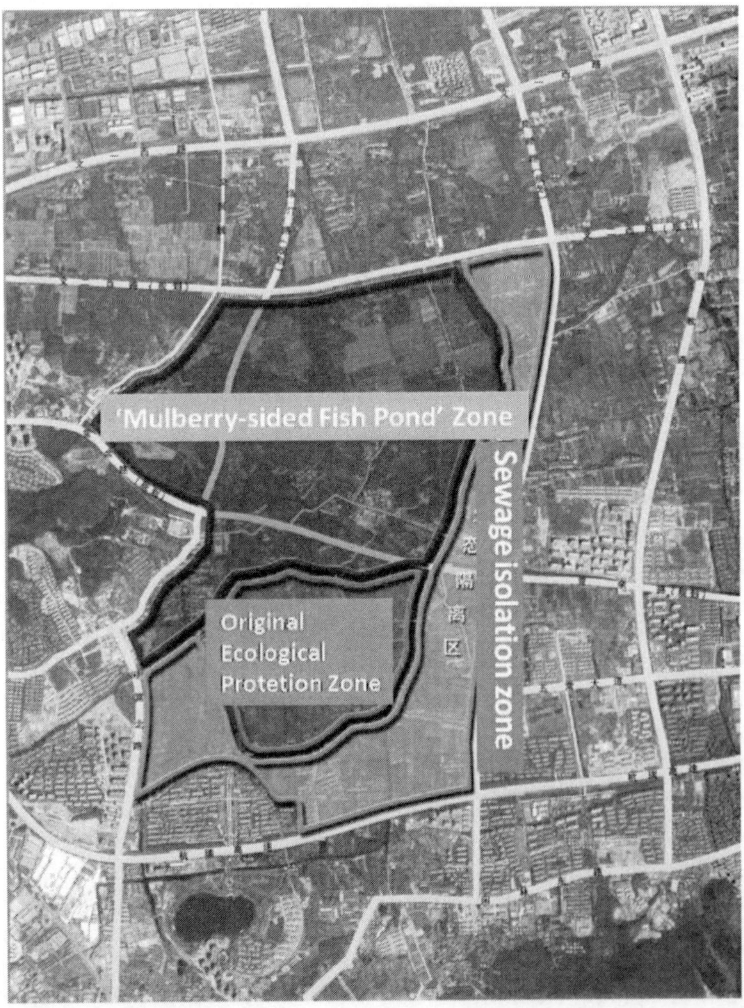

Figure 2. The "Mulberry-sided Fish Pond" area and the ecological isolation bands for sewage in protection planning.

Aquatic animal purification system: appropriate culture of aquatic animals can effectively remove the surplus nutrients in water. For example, mandarin fish can be raised to inhibit wild fish, resulting in increased planktonic animals and reduced phytoplankton, and improved water transparency and self-purification ability of wetland and rivers. At the same time, mussels, snails and other aquatic animals can be raised to further purify the water. Mussels can filter and consume the suspended algae and organic debris, and improve water transparency. Snails eat mainly periphyton, and secrete flocculent materials to aggregate the suspended matters in lake water, and clean up the water.

Reed purification system: In the protected area, reed water purification system should be greatly developed. With reed, the removal rate of BOD5 can be above 90%, suspended solids (SS) more than 91%, ammonia nitrogen (NHO-N) over 76%, total nitrogen (T-N) over 84%, total phosphorus (T-P) over 87%, chlorobenzenes over 81%, chlorophenols over 82%, pesticides over 89%, other benzenes over 95%, enterovirus over 94%, total bacteria above 99%, worm eggs over 99%, E. coli over 99%, Salmonella over 99% and bacteriophages over 99%. Reed purification system can also maximize the absorption and adsorption of the inorganic pollutants (heavy metals) and organic matters in the sewage, thus eliminating the water pollution risks, improve and restore the natural ecological environment.

2) Ecological purification system with fish pond

Herring (150 - 200), grass carps (10 - 20), chub (bighead carps 30 - 50, silver carps 200), flat fish (50) and crucian (50) are planned to be raised in fishpond per mu, a well-controlled water system can be established. The abundant snail and silkworm excrements can be used as fish feeds, and excrements from herrings are good fertilizer for planktonic animals and plants. The plants grown on water surface can be controlled by grass carps to ensure adequate lighting for better proliferation of planktons, which in turn provide large amount of feed for chubs. The left-over from herrings will be eaten by carps, flatfishes and crucian carps to form excellent ecological food chain. This will result in balanced pond ecology and prevent the water quality from deterioration [16] . If the water quality is "cared" by the integrated biological measures, ecologically balanced water will be achievable and water quality will be well purified [17] .

3) River and pond dredging systems

Dredging is planned for the rivers and ponds in the protected area every year in the winter time, particular drainage channels such as Xianlin and Wutong harbors will be built to prevent eutrophication and ensure smooth water flow.

4) Establishment of a monitoring platform for regional water environment

Automatic remote monitoring system of water quality in the protected area will be set up to provide the real- time, all dimensional monitoring of water quality to ensure that the water quality being at grade II and III standards all year around. This will be necessary for good ecological environment and biological diversity. Water sampling sites and video monitoring systems are planned for Hemu bridge, Chenjia bridge, Nanxing bridge, Shanghe road, Yujiadoumen, Jiuqu bridge, Xingfu bridge, Jiuqu bay, Dongdun pond, Xidun lock, Yanyun bridge, Wutong bridge, Xidun pond, Ruanjia farm, and Aquatic product village in the protected area. The signals will be wirelessly transmitted to water environment monitoring platform in the management center. The monitored information will include algal concentration, water turbidity, pH, dissolved oxygen, turbidity, water temperature, electronic conductivity, concentration of potassium permanganate, total phosphorus, total nitrogen and other water quality parameters. The fully automatic and real-time monitoring system will provide early warning for water quality to ensure the excellent condition in water network.

3. Plan for Development and Utilization

According to the dynamic protection idea that development while protected and protection while under development [18] [19] , we propose a "One block-Three complexes" overall development and utilization plan for the Hemu wetland (Figure 3).

3.1. One Block

A block reflecting wetland culture is planned to display the history of farming and agriculture, low-carbon life and environmental protection as the themes. This requires to fully explore the farming culture, Liangzhu culture, heritages left behind by famous ancient scholars. Exhibition halls can be set up to show farming methods, life styles and customs in history. At the same time, corresponding products of low energy consumption, low pollution, and low emission that are related to low-carbon daily life can be displayed, sold, experienced (including DIY room) along the block, such as energy-saving houses, energy-saving appliances, energy-saving building materials, environment-friendly daily necessities, and resources recycling technology and products, the solar energy technology and products, energy-saving and intelligently controlling technology and products. The consumers can learn what carbon reduction means. In addition, intangible cultural heritage products (Mulberry silk skills, Yuhang rolling lights, Wuchang dragon-boat, 18 Martial arts, Jinshan tea party et al.) and stores with a hundred year of history from local district and city (Fuyi Warehouse—selling rice; A Hundred Huichang—selling preserves;

Tongfuyong—selling liquor et al.) can be actively introduced to the block in the style of traditional specific crafts shops and local snacks. This will make the block special full of farming culture with local favors, and integrated with the surrounding wetland ecological system, showing a beautiful scene of suburban in south of Yangtze River (Figure 4).

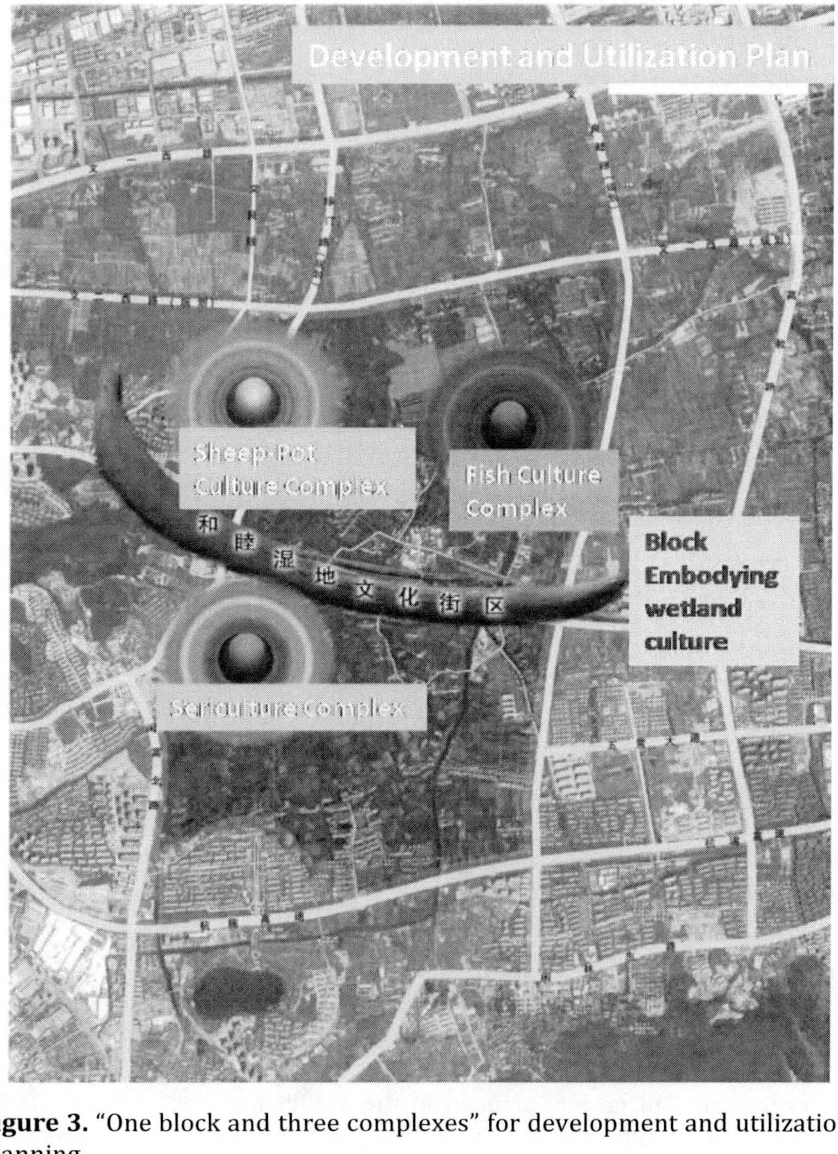

Figure 3. "One block and three complexes" for development and utilization planning.

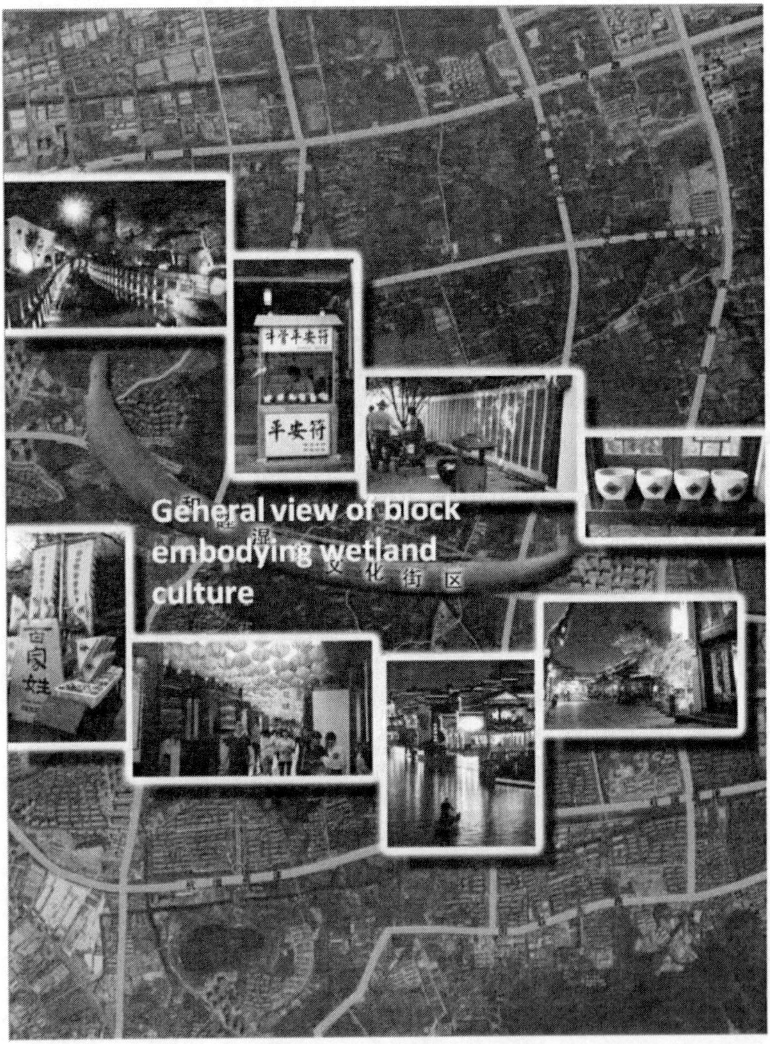

Figure 4. General view of block embodying wetland culture.

3.2. Three Complexes

3.2.1. Sericulture Complex

A sericulture museum is planned in the Wanjing village of the protected area. The museum will have exhibition halls to display the history and development of sericulture in China, using the materials and pictures to show the distribution and breeding of mulberry trees, wild silkworm and domesticated silkworm, silk reeling, and silk products.

Figure 5. General view of sericulture complex.

At the same time, old resident houses in the village will be rebuilt to show traditional sericulture customs including worshiping silkworm, pleasing silkworm, burning silkworm fireworks, covering silk flowers, scaring silk ghosts, burning silkworm in field, predicting silk production, and thanks for silkworm. Also planned is a "mulberry leaf tea" museum that will show products made from mulberry leaf, such as products with functions of weight reduction, beauty and blood sugar reduction. Mulberry leaf and fruit will be used to make special food products such as biscuit, noodle, cake, ice cream, steamed bun, wine, drink and jam. Beauty shops selling cosmetic products made mainly with silk fibroin as raw materials and the beauty finger gloves made of cocoons will be built to enhance the cultural taste of the village to attract tourists (Figure 5).

3.2.2. Sheep Pot Culture Complex

The neighbor Gexiang village is planned to be the center of the traditional sheep pot food culture originated from unique folk food festival "sheep pot". The sheep culture will be investigated and explored from the aspects of history, language, diet, product, moral, etiquette, religion and aesthetic to promote the Gexiang village from "sheep pot village" to "sheep culture village". The final goal is to develop a "sheep pot culture complex" that has remarkable and regional characteristics of the Hemu wetland with its seasonal and dynamically changing wetland landscape—"Mulberry-sided Fish Pond" (Figure 6).

3.2.3. Fish Culture Complex

A fish culture complex is planned in the "Gonglaohan" recreational fishery Park. The complex will offer unique organic fishes from the "Mulberry-sided Fish Pond" system. The complex will be used to investigate, display and explore the fishery history and development in the wetland through wetland fish species, fishing boats, fishing gear, fishing methods, aquaculture and processing technology in different periods. It will also be used to show the fishermen living habits, local customs, fish and fishermen's legends, fish-related arts, and fish cooking, derivatives from fusions of fish with religion (Figure 7).

4. CONCLUSION

Hemu wetland is low-lying in the western suburb of Hangzhou, China, which has the function of water-log- ging prevention in history. Mulberry-sided Fish Pond is its important agricultural cultivation pattern and landscape in ancient time. Nowadays, with the expansion of the city and the surrounding real estate development, the water quantity in Hemu wetland is decreased and it is polluted severely. Based on its background, historical function and current status, this study suggests protecting and restoring the original function and sustainable utilization of Hemu wetland by re-constructing and developing "Mulberry-sided Fish Pond" model, treating polluted river, clearing silt in order to be the restoration protection zone and original ecological protection zone, and by establishing a block area indicating local wetland culture and products, three complexes of sericulture, sheep-pot culture and fish culture. This is the novel protection and development model for urban wetland—"Two zones—One block and Three complexes" presented in this study. The results provide valuable reference in planning and utilizing urban wetland.

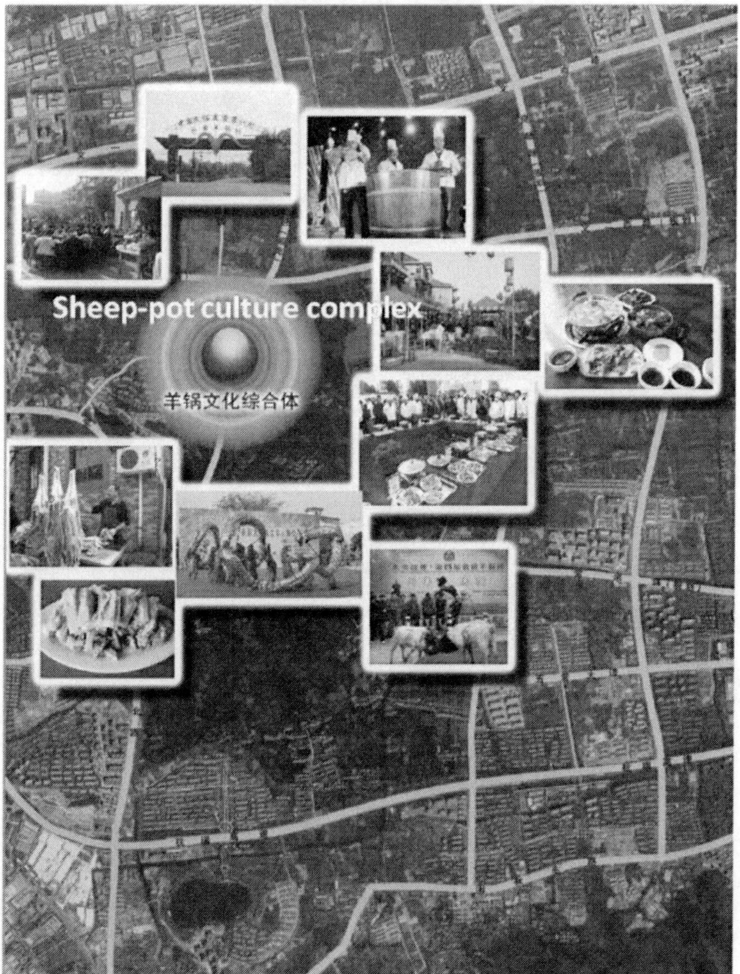

Figure 6. General view of sheep-pot culture complex.

Figure 7. General view of fish culture complex.

ACKNOWLEDGEMENTS

Nongnong Shi#, Minger Ye#, Tongfei Lai, Zhiming Yu, Yunli Pan, Mengying Yu, Xiaoqi Zhou (2012R10039-19) of "The Program for Zhejiang Leading Team of Science & Technology Innovation". We thank Yuhang Management Committee for offering us the Hemu Wetland CAD Blueprint; we also thank Dr. Weimin Ni in School of Life and Environmental Sciences of Hangzhou Normal University for offering us the data and information of water quality of Hemu Wetland.

REFERENCES

1. Wu, W.Z. and Fang, L. (2013) Measures of the Situation and Protection of Wetland Resources in Zhejiang Province. Zhejiang Science and Technology of Forestry, 33, 81-84.
2. Zhou, Q. (2013) The Mulberry-Sided Fish Pond and Soil-Water Environment in the Dongtiao Downstream of Qing and Ming Dynasty. China History of Agriculture, 32, 80-90.
3. Zhang, J., Zhong, G.F. and Wu, H.S. (1993) Analysis on Coordination Relationship of Human-Earth in Dike-Pond economic Ecosystem. Ecological Science, 12, 55-59
4. Ding, J.H., Wen, Y.M. and Shu, Q. (2001) Research on the Situation, Problem and Strategy for the Sustainable Development of Dike-Pond System. Environmental Sciences Chongqing, 23, 12-14.
5. Li, Z.F., Zhang, F. and Hua, G.F. (2007) Research on Variations of Vascular Plants in Xixi Wetland of Hangzhou. Zhejiang Agricultural Sciences, 6, 655-658.
6. Kong, Y.Y. and Xia, Y.P. (2006) Protection of Biodiversity and the Formation of Ecological Landscape in Xixi Wetland Park. Modern Gardern, 2, 11-13.
7. Yang, L.X. (2012) Problems and Protection Measures on Urban Wetland. Forum on Science and Technology, 6, 448- 449.
8. Xu, J.L. and Xu, J.L. (2011) Problems and Strategies on Wetland Protection and Management in China. Economy and Law, 4, 193-194.
9. Ni, W.M., Zhao, L. and Hua, W.G. (2014) Analysis and Evaluation of Surface Water Quality in Hemu Wetland, Hangzhou China. Wetland Protection and Co-managing for Five Types Water, 8, 12-15.
10. Liu, Y.J. (2013) Principles and Way to Protect Landscape of Urban Wetland. Science and Technology Wind, 7, 228- 232.
11. Wang, J.N., Hu, A. and Wei, J.J. (2013) Research on Protection and Utilization Model of Urban Wetland and Compound Strategy. City Plan, 212, 137-143.
12. Cheng, J. and Han, C. (2012) Wetland Ecological Function and Protection. Anhui Agricultural Science, 40, 9851-9854.
13. Hong, D.Z. (2012) Research on Management Model for the Stable Development of the Wetland Ecosystem. Innovation of Science & Technology and Application, 7, 94.
14. Li, J. (2012) Investigation and Analysis of the Protection of Xixi Wetland Park Hangzhou. Integrated Forum, 4, 336.
15. Li, H.H. and Zhang, H.P. (2002) Management Mode and Evaluation for Wetland Protection Zone in China. Journal of Zhongnan Forestry College, 22, 53-56.
16. Pu, Y.F., Zhang, F. and Yuan, W. (2006) Development Situation and Direction of Tourism Agriculture in Zhejiang Province. Zhejiang Agricultural Sciences, 6, 714-716.
17. Wang, H. (2003) Planning and Running on Agricultural Tourism Park. China Forestry Press Inc., Beijing, 53-56.
18. Zhao, L.X., Li, Y., Kong, F.L., Qu, L.L. and Sun, P. (2013) Research on the Development Strategy of Community- Joint Eco-Trip in Luyang Lake Wetland. Wetland Science and Management, 9, 31-34.
19. Wang, Y. (2010) Research on the Mechanisms of Community-Joint Eco-Trip in Wetland Development. Travel Resources, 26, 1138-1147.

CHAPTER 9

Applying Process-Based Models for Subsurface Flow Treatment Wetlands: Recent Developments and Challenges

Guenter Langergraber

Institute of Sanitary Engineering andWater Pollution Control, University of Natural Resources and Life Sciences Vienna (BOKU), Muthgasse 18, A-1190 Vienna, Austria

ABSTRACT

To date, only few process-based models for subsurface flow treatment wetlands have been developed. For modelling a treatment wetland, these models have to comprise a number of sub-models to describe water flow, pollutant transport, pollutant transformation and degradation, effects of wetland plants, and transport and deposition of suspended particulate matter. The two most advanced models are the HYDRUS Wetland Module and BIO-PORE. These two models are briefly described. This paper shows typical simulation results for vertical flow wetlands and discusses experiences and challenges using process-based wetland models in relation to the sub-models describing the most important wetland processes. It can be demonstrated that existing simulation tools can be applied for simulating processes in treatment wetlands. Most important for achieving a good match between measured and simulated pollutant concentrations is a good calibration of the water flow and transport models. Only after these calibrations have been made and the effect of the influent fractionation on simulation results has been considered, should changing the parameters of the biokinetic models be taken into account. Modelling the effects of wetland plants is possible and has to be considered when important. Up to now, models describing clogging are the least established models among the sub-models required for a complete wetland model and thus further development and research is required.

Keywords: horizontal flow; HYDRUS Wetland Module; numerical simulation; treatment wetland; vertical flow

1. INTRODUCTION

Treatment Wetlands (TWs) are engineered systems that optimize treatment processes found in natural environments. Processes in TWs are very complex, there is a large number of active physical, chemical, and biological processes occurring simultaneously that mutually influence each other. TWs are used worldwide and are able to efficiently treat different types of polluted water [1,2].

During the last few decades, the interest in modelling removal of pollutants in TWs steadily increased. However, most of the models published to date are "simple" or "black-box" models, i.e., models where data from measurements are used to derive the model equations. Examples of "black-box" models are

- correlation models that correlate influent and effluent concentration,
- first-order rate equations, and
- more sophisticated correlation models, such as artificial neural networks and fuzzy logic models.

The first two model types are typically used for designing horizontal flow (HF) wetlands and free water surface wetlands [1]. For these models, model parameters are derived from experimental data. Only for TWs operating under similar conditions—i.e., wastewater composition, porous filter material, climate, plant species, etc.—can good designs be obtained [3].

In contrast, the governing equations of process-based models are derived from the processes occurring in wetlands. These types of models have various degrees of complexity and are predominantly based on balance equations (e.g., for energy, mass, charge). Experimental data are thus not required for deriving the model equations but are used for calibration and validation of the model. During the calibration and validation of process-based models, experimental data are compared with simulation results. The better predictions that should be possible using these models should make them more applicable for wetland design [3].

For a process-based wetland model, a number of sub-models for different wetland processes have to be considered [4]:

The water flow model:

Different types of TWs require different models to describe water flow. In subsurface flow (SSF) wetlands, water flows either horizontally or vertically through the porous filter media whereby no free water level is visible. HF wetlands can be simulated when only saturated water flow conditions are considered. For modelling vertical flow (VF) wetlands with intermittent loading, transient variably saturated flow models are required. The intermittent loading makes VF wetlands highly dynamic and makes modelling these systems even more complex. Models applicable to VF wetlands use either the Richards equation or various simplified approaches to describe variably saturated flow [5].

The transport model:

Wastewaters contain particulate and solute compounds that are transported with the flowing water. The transport model generally considers convective-dispersive transport in the liquid phase, diffusion in the gaseous phase, as well as adsorption and desorption processes between the solid and liquid phases.

The biokinetic model:

Biokinetic models describe the transformation and degradation processes of the pollutants. The complexity of biokinetic models ranges from considering a single process affecting the concentration of one compound to multiple processes affecting multiple compounds. The reaction rates are often assumed to be constant, i.e., independent of environmental conditions. However, reaction rates are dependent on a number of environmental conditions (e.g., concentrations of oxygen, substrate and nutrients). These dependencies are often modelled with Monod-type equations.

The plant model:

Plants are an essential part of wetland treatment systems and thus plant related processes have to be included in wetland models. Root growth influences the water flow pattern in the subsurface, and via evapotranspiration the water balance. During growth and decay, plants take up nutrients and release organic matter and nutrients, respectively. Additionally, some plants release substances via roots (e.g., oxygen and/or specific organic substances). Plant models describe uptake and release of substances either associated with water uptake or as a process dependent on concentration gradients.

The clogging model:

Clogging models for SSF wetlands need to be able to describe transport and deposition of suspended particulate matter and processes that reduce the hydraulic capacity/conductivity of the filter medium (i.e., reduction of free pore volume due to deposition of particulate matter, bacterial and plant growth). This is important for simulations of the long-term TWs performance and predictions of the potential failure of TWs due to clogging [6].

After briefly describing the two most advanced process-based wetland models, the HYDRUS Wetland Module [7] and BIO-PORE [8], the main aim to the paper is to present typical applications of and important details that have to be considered when applying process-based wetland models. Firstly, typical simulation results for VF wetlands using the HYDRUS Wetland Module are presented as an example. Secondly, experiences and challenges using process-based wetland models are discussed in general and in relation to the five sub-models described above.

2. MATERIALS AND METHODS

2.1. Process-Based Models for Treatment Wetlands

During the last two decades, a couple of process-based models have been developed for simulating mainly HF and VF wetlands [5,9,10,11,12]. The two most advanced models are:

- The HYDRUS Wetland Module [7] implemented in the HYDRUS simulation software [13], and
- BIO_PORE [8] implemented in the COMSOL Multiphysics™ platform.

Both, the HYDRUS Wetland Module and BIO_PORE use multi-component biokinetic models, i.e., CW2D (Constructed Wetland 2D; [14]) and/or CWM1 (Constructed Wetland Model #1; [4]), respectively. Only aerobic and anoxic transformation and degradation processes are taken into account in the CW2D biokinetic model; organic matter, nitrogen and phosphorus are components modelled. To provide a widely accepted model formulation for biochemical transformation and degradation processes in both HF and VF wetlands was the main goal when developing the CWM1 biokinetic model, thus aerobic, anoxic and anaerobic processes are considered. Components modelled in CWM1 are organic matter, nitrogen and sulphur. Table 1 lists processes and components described by the biokinetic models CW2D and CWM1 as well as the typical application of these models. Table 2 compares sub-models implemented in the HYDRUS Wetland Module and BIO_PORE. The HYDRUS Wetland Module is more applicable for VF wetlands and both biokinetic models are available. BIO_PORE was mainly developed and applied for HF wetlands and also includes a clogging model.

Table 1. Processes and components (a number of components/processes in brackets) and applications of the biokinetic models CW2D (Constructed Wetland 2D) and CWM1 (Constructed Wetland Model #1) [7].

Biokinetic Model	CW2D	CWM1
Processes	Aerobic and anoxic (9 processes)	Aerobic, anoxic, and anaerobic (17 processes)
Components	Oxygen, organic matter, nitrogen, and phosphorus (12 components)	Oxygen, organic matter, nitrogen, and sulphur (16 components)
Application	Vertical flow (VF) wetlands Lightly loaded horizontal flow (HF) wetlands	VF and HF wetlands

Table 2. Comparison of implemented sub-models in the HYDRUS Wetland Module and BIO_PORE (adapted from [6]).

Simulation Tool	HYDRUS Wetland Module [7]	BIO_PORE [8]
Flow model	Richards equation (variably saturated flow)	Variable water table (saturated flow)
Transport model	Advection, dispersion, adsorption	Advection, dispersion, adsorption
Biokinetic model	CW2D + CWM1	CWM1
Influence of plants	Evapotranspiration, uptake and release of substances	Evapotranspiration, uptake and release of substances
Clogging model	Not considered	Included

2.2. The HYDRUS Wetland Module

In the HYDRUS software, the Richards equation is solved numerically for saturated/unsaturated water flow and the convection–dispersion equation for heat and solute transport. The flow equation incorporates a sink term to account for water uptake by plant roots. Convective-dispersive transport in the liquid phase, diffusion in the gaseous phase, as well as non-linear non-equilibrium reactions between the solid and liquid phases are considered in the solute transport equations [13,15]. Version 2 of the HYDRUS Wetland Module [7] includes two biokinetic model formulations: (1) CW2D and (2) CWM1.

Both CW2D and CWM1 have been developed to describe processes in TWs treating domestic wastewater. A number of applications of the

HYDRUS Wetland Module show that a good match between measured data and simulation results can be achieved [16,17,18,19,20]. Besides the application to municipal wastewater, the HYDRUS Wetland Model has also been used to model

- TWs treating combined sewer overflow (CSO) [21,22,23,24,25],
- TWs treating effluents of the wastewater treatment plant for irrigation purposes [26,27],
- TWs treating artificial wastewater [28,29,30],
- TWs treating greywater [31], and
- TWs treating run-off from agricultural sites [32,33].

Besides the Wetland Module, the HYDRUS software also includes other modules. These include the colloid-facilitated solute transport C-Ride module, the HP1/2/3 multicomponent solute transport modules, the colloid-facilitated solute transport C-Ride module, the major ion UnsatChem module, and a module for fumigants. Water flow and solute/heat transport processes in two-dimensional transport domains are simulated for these modules.

All these modules are supported by the HYDRUS graphical user interface. Many processes of these specialized modules are also available as part of the public domain HYDRUS-1D software [34].

3. RESULTS AND DISCUSSION

3.1. Simulation Results for Vertical Flow Wetlands Treating Domestic Wastewater

This section shows simulation results for VF wetlands using the HYDRUS Wetland Module. Simulations have been carried out for a VF pilot-scale wetland (Figure 1) operated in the technical laboratory hall of the Institute of Sanitary Engineering at BOKU University in Vienna, Austria [5]. The systems had a 60 cm main layer consisted of sand with a grain size of 0.06–4 mm (d_{10} = 0.2 mm; d_{60} = 0.8 mm) and a surface area of 1 m^2. The intermediate layer (gravel 4–8 mm) prevents fine particles from being washed out into the drainage layer (gravel 16–32 mm). The systems were loaded four times per day with mechanically pre-treated municipal wastewater (each loading 10 L) and were planted with giant cane (*Arundo donax*).

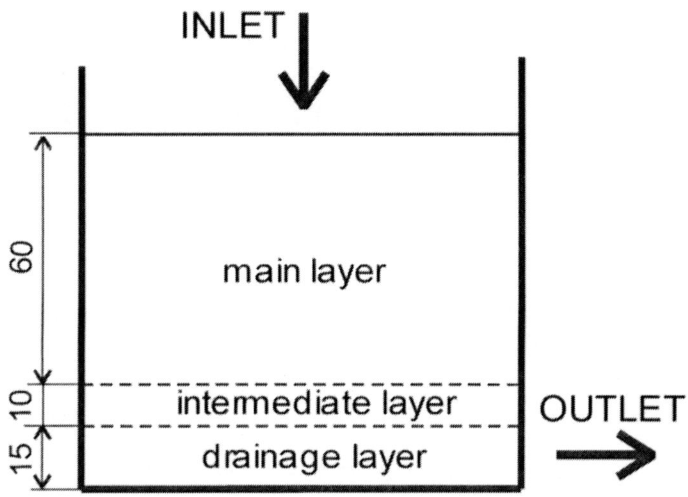

Figure 1. Schematic representation of the indoor VF pilot-scale wetland (values are in cm).

For the numerical simulations, only the main layer was considered. The transport domain, i.e., a vertical cross-section of the main layer, was discretized into 11 columns and 40 rows. Using this discretisation, the two-dimensional finite element mesh consists of 440 nodes and 780 finite elements. An atmospheric boundary condition is assigned to the top of the system, a constant pressure head boundary condition with a constant head of −2 cm [5].

Table 3 shows three parameter sets of the van Genuchten–Mualem model [35] of the soil hydraulic properties used in simulations of water flow. The *"HYDRUS parameter set for 'Sand'"* uses default soil hydraulic properties for sand provided by HYDRUS. For the *"Measured K_s and porosity"* parameter set, measured values of porosity and saturated hydraulic conductivity K_s were used whereas the default residual water content and shape parameters were used for sand. The third parameter set was obtained by inverse simulation. Measured effluent flow rates were used to estimate the soil hydraulic parameters (i.e., residual water content and shape parameters) using the inverse simulation routine provided by HYDRUS [13]. Measured values of porosity and K_s were kept constant. Figure 2 compares the measured and simulated effluent flow rates using these three parameter sets. The simulated effluent flow rate is rather constant when using the default parameter set for sand due to the low K_s value. When using the *"Measured K_s and porosity"* parameter set, measured and simulated effluent flow rates match better, however, the peak of the effluent flow rate is over-predicted and occurs too early.

Using the *"Fitted parameter set"*, a good match between simulated and measured data can be achieved. The results show that by measuring the porosity and the saturated hydraulic conductivity of the filter material, the results can be greatly improved and thus these two parameters should be measured when modelling VF wetlands.

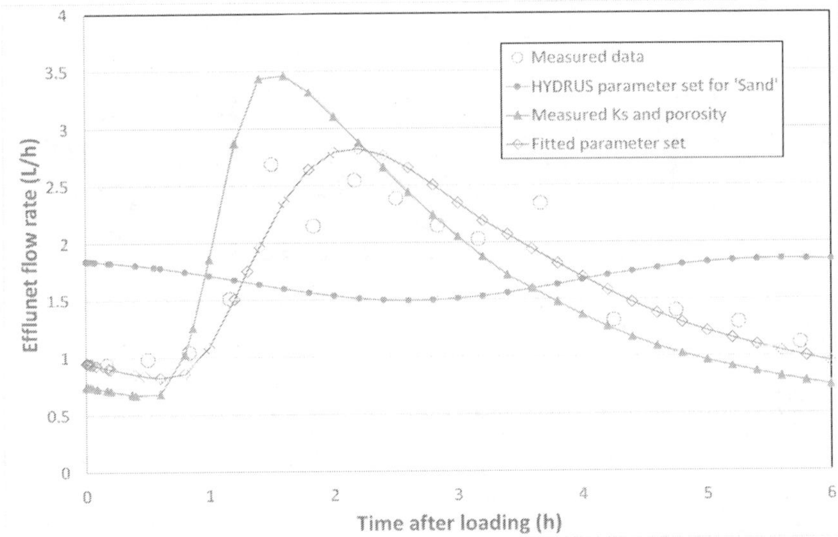

Figure 2. Measured and simulated effluent flow rates of the VF pilot-scale wetland for a single loading of 10 L (*"HYDRUS parameter set or 'Sand'"*, *"Measured Ks and porosity"* and *"Fitted parameter set"* represent the parameter sets from Table 3).

Table 3. Soil hydraulic parameters of the van Genuchten–Mualem model [35] for the 0.06–4 mm main layer.

Parameter	Residual Water Content θ_r	Saturated Water Content θ_s	Shape Parameters			Saturated Hydraulic Conductivity K_s
			α	N	L	
Unit	(-)	(-)	(1/cm)	(-)	(-)	(cm/h)
HYDRUS parameter set for 'Sand'	0.045	0.43	0.145	2.68	0.5	29.7
Measured K_s and porosity	0.045	0.30	0.145	2.68	0.5	117
Fitted parameter set	0.013	0.30	0.147	2.42	0.636	117

Table 4 shows measured and simulated concentrations of ammonia and nitrate nitrogen (NH_4-N and NO_3-N, respectively) for the VF pilot-scale wetlands with two different sandy filter materials used for the main layer. In addition to the sand with a grain size of 0.06–4 mm, in a parallel-operated VF wetland, sand with a grain size of 1–4 mm was also used. The different levels of the NO_3-N effluent concentrations for the two filter materials could be simulated. For the VF wetland with 1–4 mm sand, the effluent concentrations of NO_3-N have been higher. This can be explained by the fact that due to the higher flow rates, less readily degradable organic matter is produced by hydrolysis and thus denitrification is limited by lack of available organic matter [16].

Table 4. Median values of measured and simulated influent and effluent concentrations for NH_4-N and NO_3-N in mg/L (adapted from [16]).

Parameter	Influent	Effluent (0.06–4 mm Sand)		Effluent (1–4 mm Sand)	
	Measured	Measured	Simulated	Measured	Simulated
NH_4-N	60.0	0.15	0.01	1.20	0.16
NO_3-N	3.0	38.5	41.1	50.0	63.0

For the VF pilot-scale wetlands with a main layer of 0.06–4 mm sand, the distribution of microbial and bacterial biomass in different depths of the filter was also measured [36]. Converting the measured data into biomass COD results in mean values of 3400 to 5100 μg·COD/g·DW (dry weight of the substrate, i.e., sand) for the first cm of the main layer, 1100 to 2600 μg·COD/g·DW from 1 to 5 cm, and 640 to 1400 μg·COD/g·DW from 5 to 10 cm, respectively. Most of the biomass could be found in the upper 10 cm of the VF bed [18].

When using heterotrophic lysis rates of between 0.25 and 0.35 day^{-1}, the simulated microbial biomass COD is between 5600 and 3400 μg·COD/g·DW (the range of the measured values) in the first cm of the main layer. Calculated and simulated microbial biomass COD in different depths of the main layer showed a good fit for a heterotrophic lysis rate of 0.30 day^{-1}(Figure 3). When comparing measured and simulated biomass COD in different depths of the main layer, simulations seem to over-predict biomass COD in 1–5 cm depth and under-predict biomass COD in 5–10 cm depth. This could be an indication that the influence of biomass growth on the hydraulic properties has to be considered in the model [18].

Additionally, outdoor experiments have been carried out at experimental wetlands located at the wastewater treatment plant Ernsthofen, Lower Austria [38]. Three VF wetlands with a surface area of about 20 m^2 each have been operated in parallel. The organic loading rates for bed 1 through 3 have been 20, 27 and 40 $g \cdot COD \cdot m^{-2} \cdot day^{-1}$, which correspond to hydraulic loading rates of 32.2, 43.0, and 64.7 mm/day, respectively. The main layer of the filter consisted of 50 cm sandy substrate (gravel size 0.06–4 mm, d_{10} = 0.2 mm; d_{60} = 0.8 mm). All three VF wetlands have been planted with common reed (*Phragmites australis*). The data gained from this experiment have been used to verify the temperature model incorporated into CW2D [17].

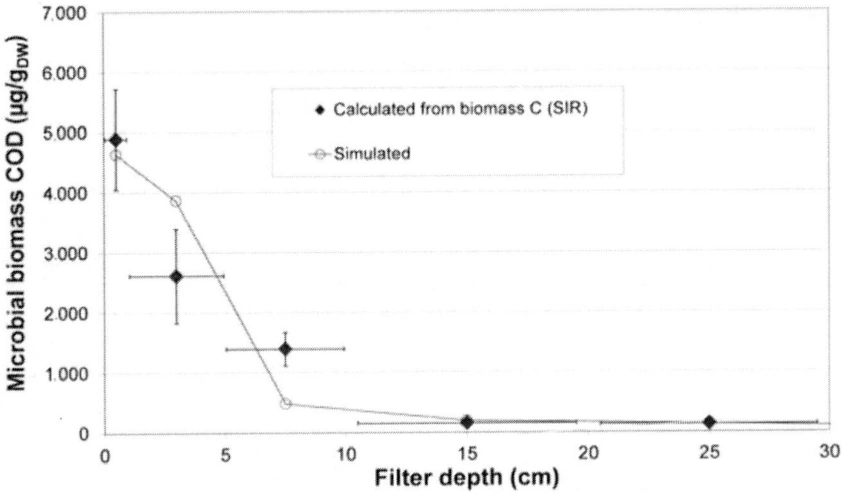

Figure 3. Calculated and simulated microbial biomass COD (Chemical Oxygen Demand) in different depths of the main layer of the VF pilot-scale wetland (adapted from [18,37]).

The width of the transport domain in the numerical simulations was 4 m, and its depth 0.8 m, while the transport domain itself was discretized into 31 columns and 33 rows. The two-dimensional finite element mesh consisted of 1023 nodes and 1920 triangular finite elements. An atmospheric boundary condition was assigned to the top of the system representing the influent distribution system, and a constant pressure head boundary condition (constant head of −4 cm) was assigned to one side of the drainage layer [17].

Measured and simulated effluent flow rate and cumulative effluent of the experimental wetland showed good match [17]. Using the calibrated water flow model and using the original CW2D parameter set [14], measured effluent concentrations during summer could be simulated. However, at lower temperatures, measured COD and NH_4-N effluent concentrations could not be simulated because hydrolysis and nitrification at low temperatures were over predicted. The standard CW2D parameter set considers temperature dependencies only for maximum growth, decay and hydrolysis rates. Using a modified parameter set (that includes temperature dependencies for the half-saturation constants of hydrolysis and nitrification) measured and simulated COD and NH_4-N effluent concentrations showed a good fit (Figure 4). With this modified parameter set, simulating the COD and NH_4-N effluent concentrations at low temperatures was possible [17]. Temperature dependencies for the half-saturation constants of hydrolysis and nitrification are now included in the standard parameter sets of CW2D and CWM1 [7].

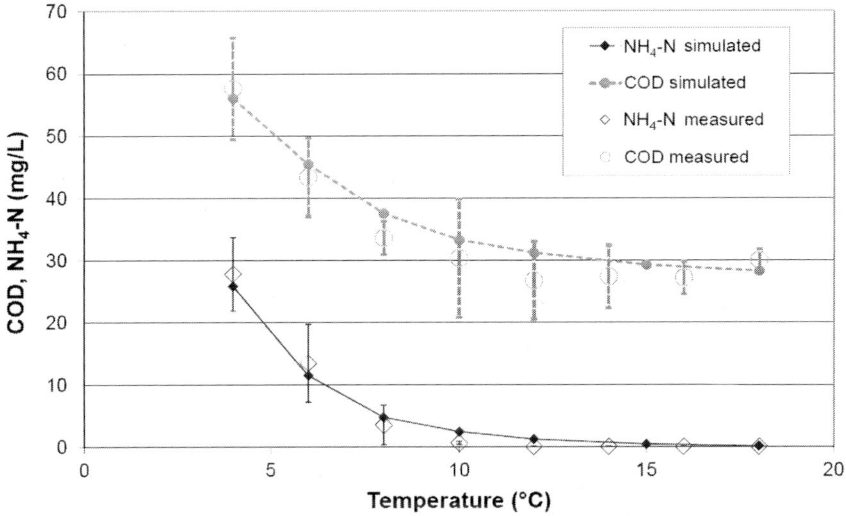

Figure 4. Measured and simulated COD and NH_4-N effluent concentrations of the experimental VF wetland in Ernsthofen using the modified parameter set (adapted from [17]).

3.2. Experiences and Challenges Using Process-Based Wetland Models

3.2.1. Water Flow Model

It is generally agreed that the hydraulics have a high impact on the treatment performance of TWs. When simulating SSF wetlands, the first step thus should be a good calibration of the water flow model. Experience shows that simulated effluent concentrations can match the measured data only when the hydraulic behaviour of the system is well described.

For VF wetlands, the influence of the parameters describing the hydraulic properties of the filter material (such as for the van Genuchten–Mualem model [35] in HYDRUS [13]) is much higher than the influence of the parameters of the biokinetic model [31]. To achieve a good calibration of the water flow model, it is advised to measure at least the porosity and the saturated hydraulic conductivity of the filter material and, if possible, the volumetric effluent flow rate between two intermittent loadings [3]. With these measurements, it is possible to estimate the remaining parameters describing the hydraulic properties of the filter material using the inverse simulation implemented in HYDRUS [13] (see also Figure 2).

Preferential flow paths in VF wetlands can be caused by macropores/cracks in some layers. Through this effect, water can bypass most of the soil porous matrix in an unpredictable way [39]. This is especially true for the sludge layer in French VF wetlands [40,41]. Water flow in such systems cannot be modelled with uniform flow models (such as using the Richards equation combined with the van Genuchten–Mualem function in HYDRUS [13]). A non-equilibrium approach uses a model that separately describes flow and transport in preferred flow paths and slow or stagnant pore regions. This approach seems to be more appropriate for simulating preferential flow in VF wetlands [39]. Such a dual-porosity model therefore also needs to be incorporated into the software tools to accurately describe water flow and solute transport in French VF wetlands.

For HF wetlands in which gravel is usually used as filter material, the saturated hydraulic conductivity of gravel is much larger than the actual water flow in the filter. Tracer experiments are usually applied to calibrate the water flow and single solute transport models.

3.2.2. Transport Model

Tracer experiments are used to calibrate the transport model, i.e., the transport of solutes through the filter bed of VF and HF wetlands. When calibrating tracer experiments, i.e., one tries to match measured and simulated breakthrough curves, the longitudinal dispersion coefficient is the most sensitive parameter and has the highest influence on the simulation results.

When simulating tracer breakthrough curves, the selection of the proper water flow model is important. If sand is used as filter media, single-porosity models such as the van Genuchten–Mualem model [35] are sufficient and measured breakthrough curves from tracer experiments can be fitted well [16,26]. However, if filter media with high porosity such as woodchips and/or crushed seashells are used [33], the use of dual-porosity models might be required (similar as for simulating preferential flow in the sludge layer of French VF wetlands). Dual-porosity models consider mobile and immobile water zones and exchange of solutes between these two phases. A number of dual-porosity models are implemented in the HYDRUS software package [13]. Currently, the use of the HYDRUS Wetland Module is limited to single-porosity models.

For a number of substances (e.g., ammonia, phosphate), adsorption/desorption is a main retention mechanism. It is advised to use measured adsorption isotherms for the filter material used. For some applications, adsorption also needs to be considered for some COD fractions, e.g., when simulating wetlands treating CSO [22]. Also, slow release of carbon from soils or specific filter media can be modelled by considering adsorption/desorption for different fractions of COD [32].

3.2.3. Biokinetic Model

Bacteria in TWs are similarly to those in activated sludge systems. Based on this assumption, it was proposed that parameters of the biokinetic models developed for activated sludge systems should be applicable to describe processes in TWs as well [42]. This assumption was confirmed by a number of applications of the HYDRUS Wetland Module that showed that a good match between measured and simulated concentrations could be achieved when the hydraulic behaviour of the system is well described (see above). Additionally, Table 5 shows good agreement between measured and calculated volumetric nitrification rates [43].

Table 5. Comparison of measured and calculated volumetric nitrification rates (adapted from [43], SD = standard deviation).

Method	Measured with Solid Respirometry	Calculated from Simulation Results *
Results ($mg \cdot O_2/L_{sample}/h$)	32–50 (mean = 41, SD = 9; 2 values)	30.5

Notes: * From simulations using parameters for the biokinetic model from activated sludge systems.

When using the biokinetic models, such as CW2D or CWM1, it is therefore advised not to change the default parameters of the biokinetic model, except for good reasons. However, parameters describing treated wastewater and that are related to the chosen biokinetic model have to be amended and shall be used for calibration process. Influent fractionation has high impact on the simulation results and needs to be adapted for each simulation study as treated wastewaters are different. Parameters describing the influent fractionation include (1) fractionation of influent COD (i.e., estimation of different COD model fractions from measured total COD) and (2) organic N content of different COD fractions.

Rizzo et al. [29] described the set-up of a model to simulate experimental data from a HF wetland fed with artificial wastewater. During the experiments, the only nitrogen parameter measured in influent was TKN (Total Kjehldal Nitrogen). As the model requires influent concentrations of ammonia nitrogen, the organic N content of different COD fractions had to be adapted from standard values for the type of used artificial wastewater. This results in different ratios of ammonia nitrogen to TKN. Figure 5 shows the impact of different influent concentrations of ammonia nitrogen on simulated effluent concentrations. When all influent TKN is assumed to be ammonia (i.e., no organic nitrogen is present), simulated effluent concentrations of ammonia nitrogen are higher than measured effluent concentrations. On the contrary, when all influent TKN is assumed to be organic nitrogen (i.e., no ammonia is present), simulated effluent concentrations are too low.

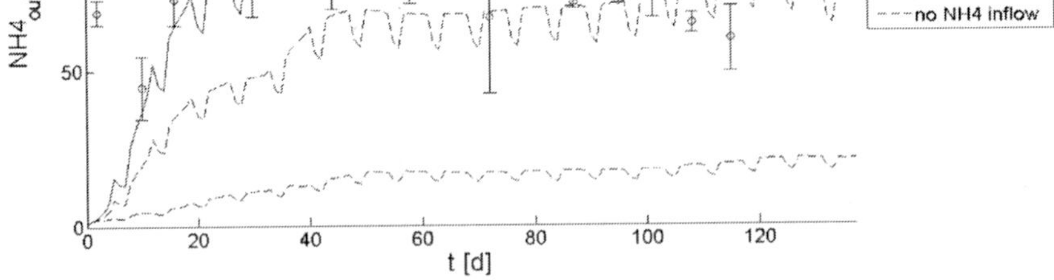

Figure 5. Influence of inflow NH4-N concentrations on effluent NH4-N concentrations of a HF wetland (adapted from [6]).

Pucher and Langergraber [44] simulated VF filters treating domestic wastewater. The main layer of two parallel operated filters consisted of sand with a grain size of 1–4 mm or zeolite with a grain size of 2–5 mm. Both filters had an impounded drainage layer. After calibrating the flow model for the sand and zeolite filters (Table 6), the procedure for calibrating the biokinetic model was as follows: the standard parameter set of the CW2D biokinetic model [7] was used in the first simulation of the sand filter ("Effluent (Sand)—Simulated 1" in Table 7). This resulted in over-prediction of nitrification, i.e., simulated ammonia nitrogen concentrations were much lower than measured data. After reducing the maximum nitrification rate from 0.9/day to 0.175/day, the simulated effluent concentrations of ammonia nitrogen matched measured data well ("Effluent (Sand)—Simulated 2" in Table 7). After adding the measured adsorption isotherms [45] and a first-order rate coefficient for non-equilibrium adsorption of 0.01/day in the zeolite filter simulation compared to the sand filter simulation, a good match between simulated and measured date could be obtained ("Effluent (Zeolite)—Simulated" in Table 7).

Table 6. Estimated soil hydraulic parameters of the van Genuchten–Mualem model [35].

Parameter	Residual Water Content θ_r	Saturated Water Content θ_s	Shape Parameters			Saturated Hydraulic Conductivity K_s
			α	N	L	
Unit	(-)	(-)	(1/cm)	(-)	(-)	(cm/min)
Sand (1–4 mm)	0.063	0.37	0.124	3.2	0.49	55
Zeolite (2–5 mm)	0.045	0.42	0.096	4.6	1.27	154

Table 7. Measured and simulated influent and effluent concentrations of VF filters (in mg/L, measured data: average values and standard deviations in brackets, simulated data: average values over 1 day; CR = readily biodegradable COD, CS = slowly biodegradable COD, and CI = inert COD as defined in the CW2D biokinetic model [7]; bold data are used for comparison).

Parameter		CR	CS	CI	COD	NH_4-N	NO_2-N	NO_3-N
Influent	Measured	-	-	-	294	93	0.008	0.59
	Simulated	274	100	20	-	93	0.008	0.59
Effluent (Sand)	Measured	-	-	-	**35 (5)**	**14.5 (4.6)**	0.026 (0.040)	**70.8 (8.4)**
	Simulated 1	0.17	0.08	31.2	31.5	0.42	<0.003	85.7
	Simulated 2	0.14	0.01	29.4	**30.0**	**15.3**	0.016	**76.2**
Effluent (Zeolite)	Measured	-	-	-	**29 (9)**	**0.06 (0.05)**	0.52 (1.10)	**51.8 (4.6)**
	Simulated	0.16	0.01	29.1	**29.3**	**0.27**	<0.003	**43**

Pálfy and Langergraber [28] simulated experimental results from batch-fed column experiments using the CWM1 biokinetic model. They described the need to adjust some parameters of the biokinetic model, i.e., the inhibition coefficients of acetotrophic methanogenic and sulphate reducing bacteria for dissolved oxygen and nitrate, respectively, to be able to simulate anaerobic, anoxic, and aerobic processes to occur in parallel. The local effect of root zone re-aeration can explain these phenomena. Figure 6 shows measured and simulated sulphate concentrations before (**a**) and after (**b**) the adjustment of parameters of the CWM1 biokinetic model. Batch experiments can be used to calibrate the biokinetic model parameters, as there is no impact of water flow on the treatment performance.

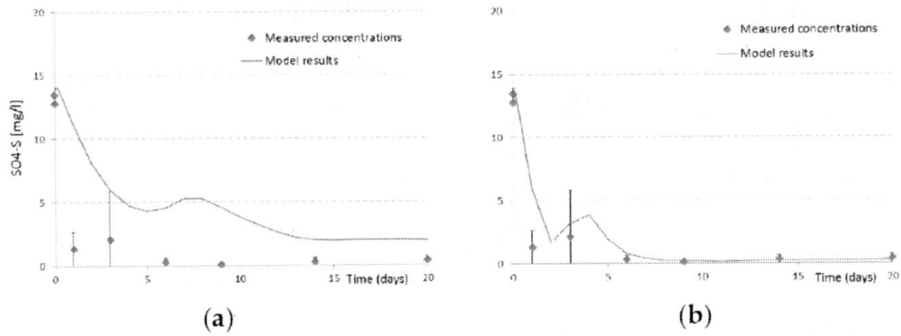

Figure 6. Measured and simulated sulphate concentrations for a batch-fed column at 24 °C planted with *Carex* spp. ((**a**): with a standard parameter set of the CWM1 biokinetic model; (**b**): after adjusting CWM1 biokinetic model parameters; adapted from [46]).

3.2.4. Plant Model

It is generally agreed that plants play an indispensable factor in treatment wetlands. The plant model that can be used with the HYDRUS Wetland Module describes nutrient uptake coupled to water uptake (which is modelled by the evapotranspiration rate). This approach has been shown to be sufficient for high loaded systems; however, for water qualities with lower nutrient contents, the simulated potential nutrient uptake is far over-estimated [47].

Toscano et al. [26] simulated pilot-scale two-stage subsurface flow treatment wetlands in Eastern Sicily. HF and VF wetlands were used for secondary and tertiary treatment of municipal wastewater. Each of the wetland cells had a surface area of 4.5 m^2 (1.5 m × 3.0 m). The HF beds were filled with 0.6 m volcanic gravel having a size of 10–15 mm. The VF beds were filled with two different filter media: volcanic sand as the main layer for a depth of 0.5 m (about 0.06–4 mm; d_{10} = 0.13 mm; d_{60} = 2.0 mm) at the top of the filter bed and coarse volcanic gravel as the drainage layer (about 16–32 mm, 25 cm depth). The average influent flow rate for each line was about 40 ± 5 L/h. Planted and unplanted VF and HF beds have been operated in parallel. Planted wetland cells were planted with reed (*Phragmites* sp.). For the simulations, an oxygen release of 5 g O_2/m^2/day and an evapotranspiration rate of 7.4 mm/day were used. Additionally, N and P uptake by plants was considered. Using these input data for the plant model, simulated effluent concentrations matched the measure concentrations well (Table 8). The effect of the plants can be best seen for the HF wetlands: the different NH_4-N effluent concentrations in planted and unplanted HF wetlands could be simulated. In the model, the difference was caused by uptake of ammonia by wetland plants through their roots as well as ammonia removal through nitrification that was possible due to the oxygen input via the roots. The simulated NO_3-N effluent concentrations are a clear indication of nitrification in the HF filter bed. For VF wetlands, no significant difference between planted and unplanted wetland cells was observed or simulated.

Table 8. Measured and simulated influent and effluent concentrations of planted and unplanted VF and HF wetlands for secondary treatment of municipal wastewater (in mg/L, standard deviations in brackets, adapted from [26]).

Parameter		COD	NH_4-N	NO_2-N	NO_3-N
Influent	Measured	63.9 (3.3)	22.1 (1.9)	0.18 (0.38)	0.34 (0.10)
Effluent HF (planted)	Measured Simulated	47.5 (2.8) 44.4	16.6 (1.2) 17.2	0.06 (0.05) 0.02	0.18 (0.17) 2.82
Effluent HF (unplanted)	Measured Simulated	48.2 (4.1) 50.6	20.4 (0.6) 21.5	0.02 (0.01) 6×10^{-5}	0.12 (0.11) 1×10^{-4}
Effluent VF (planted)	Measured Simulated	23.0 (2.5) 24.6	2.0 (0.1) 0.54	0.10 (0.07) 0.01	16.2 (1.1) 26.9
Effluent VF (unplanted)	Measured Simulated	23.2 (2.3) 24.0	2.0 (0.1) 0.96	0.19 (0.08) 0.01	16.4 (1.3) 25.0

As described above, Pálfy and Langergraber [28] simulated batch-fed column experiments. Figure 7 shows the measured and simulated COD, NH_4-N and SO_4-S concentrations at 12 °C. The different behaviour of unplanted columns as well as those planted with *Carex* and *Schoenoplectus* could be simulated. Parameters of the biokinetic model were adjusted using the unplanted columns. For modelling the effect of the wetland plants, only the parameters' oxygen release and evapotranspiration rate were adjusted. These experiments clearly showed that the different effects of wetland plants on the treatment performance of wetland systems can be modelled with existing plant models.

3.2.5. Clogging Model

Up to now, modelling of clogging processes is not possible using the HYDRUS Wetland Module. In a first trail, the colloids attachment/detachment model implemented in the HYDRUS C-Ride module [34,48] was coupled with the CW2D biokinetic model [49]. It was assumed that CS, the slowly biodegradable organic matter (as defined by [14]), consists of a particulate matter XS (slowly biodegradable organic matter as defined in the CWM1 biokinetic model [14]) that can be divided into five different classes according to different particle sizes (<1, 1–3, 3–10, 10–30, and 30–100 μm, respectively).

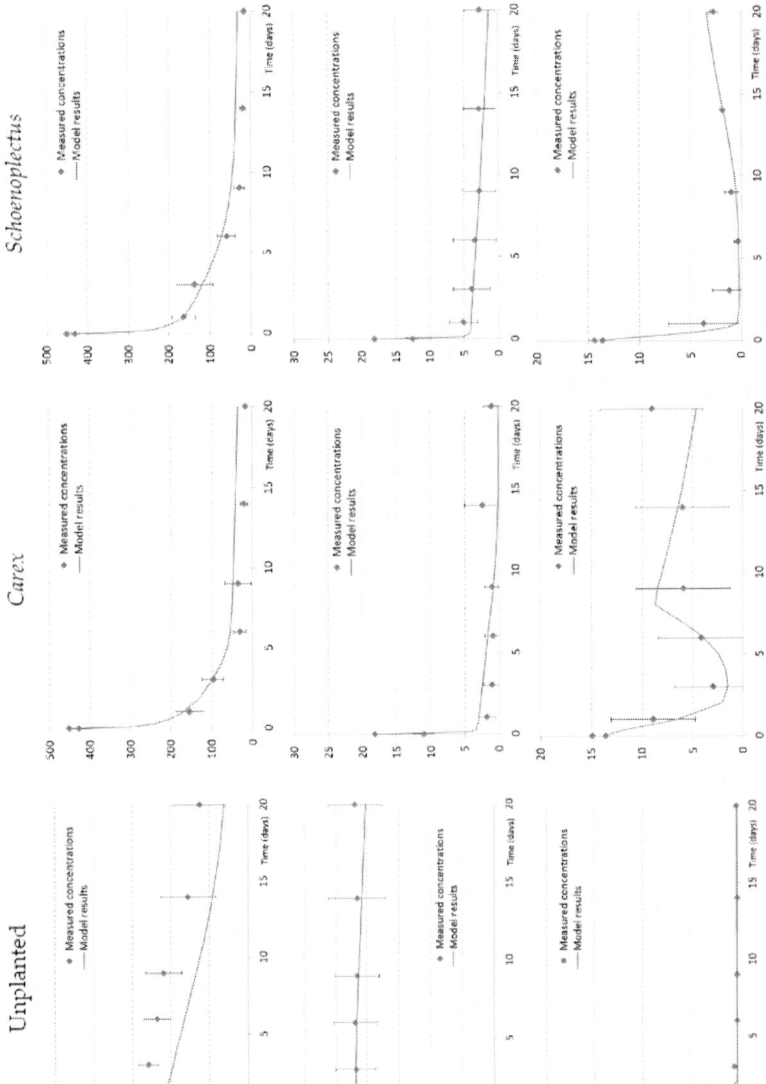

Figure 7. Simulation results for COD (**a**) NH_4-N (**b**) and SO_4-S (**c**) at 12 °C for unplanned columns, and columns planted with *Carex* and *Schoenoplectus*. Continuous line: simulation results, dots: measurement means, intervals: standard deviation (adapted from [28]).

It was further assumed that particles are transported in the liquid phase until they are attached. Once attached, the particles undergo hydrolysis and CR (readily biodegradable organic matter as defined in the CWM1 biokinetic model [14]) in the liquid phase that is produced. An additional assumption was that particles with different sizes have the same density and degradability.

The model was tested using measurements of particle concentrations in different depths of the 50-cm main layer of the VF pilot-scale wetland as described before. Particle measurements at the inlet were used to calculate fractionation of COD influent. Measured and simulated soluble particulate COD concentrations for the 10–30 μm particle class before and after loading of the VF wetland along the filter depth are shown in Figure 8. "Measured" particulate COD was calculated from particle measurements. In general, the first test showed a good agreement between measured and simulated particulate COD. However, there is still a great need for further testing of the particle transport model and for finding an optimal parameter set for various conditions present in treatment wetlands.

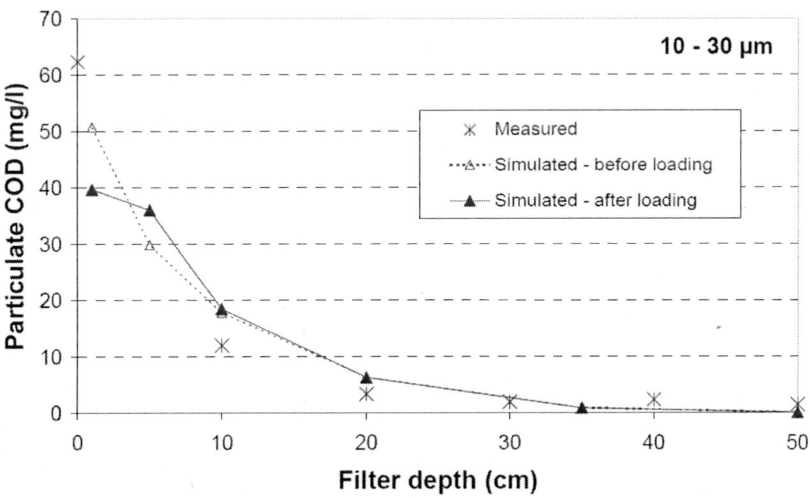

Figure 8. Comparison of measured and simulated particulate COD for the 10–30 μm particle class (adapted from [49]).

Several clogging models have been developed for BIO_PORE [7]. Samsó and García [50] described the relation between bacterial communities and accumulated solids, leading to clogging in a HF wetland. The so-called "Cartridge Theory" describes how the active bacterial zone moves in the filter. In this model, the effect of the inert organic matter accumulated in pores on the hydraulic conductivity is not considered. A description of limited growth of bacteria was implemented by adding two parameters. A negative feedback term is provided by the first parameter to prevent unlimited bacteria growth in areas where substrate concentrations are high (which is especially relevant for the influent zones of HF wetlands). The second parameter decreases the growth rate of bacteria when pore space is limited. Bacteria density and accumulated inert solids also had an effect on simulated effluent concentrations [50] and on effluent pollutant concentrations; again, only without considering the effect of organic matter accumulation on water flow. Additionally, the effect of a pore size reduction on the hydraulic conductivity is described using the biological clogging model proposed by Mostafa and Van Geel [51]. Using this approach, a realistic behaviour of HF wetlands, including overland flow and re-infiltration of water in the unused filter material could be simulated [52,53]. In the work of Rajabzadeh et al. [54], a computational fluid dynamics model was used to simulate biofilm development and the effect on local porosity in a saturated vertical downflow wetland. There was good agreement between simulation results and measured data for predicting clogging due to biomass growth were good [54].

3.2.6. Wetland Models as Design Tools

Further developments of existing models are needed to make them a useful and applicable tool for a TW design [3]. A simplified computer-based TW design tool, based on process-based numerical models, should be developed that

- could be used when having knowledge about a TW design, but without requiring special knowledge about numerical modelling,
- allows the design of TWs for different boundary conditions (such as different climatic conditions, wastewater characterization, filter material, etc.), and
- makes the description of the dynamic behaviour of the designed TW possible, thus allowing the higher robustness of wetland treatment systems, for example, for fluctuating inflows and peak loads.

Several developments have been made to develop such a simplified design tool. RSF_Sim [55] supports the design of wetlands for a combined sewer overflow treatment (in Germany named retention soil filters, i.e., RSFs). This simplified, yet robust and reliable, model was developed for design purposes based on experiences with the HYDRUS Wetland module. Only few parameters are required in RSF_Sim which makes it very simple to use. RSF_Sim was developed specifically for German conditions. RSF_Sim was also adapted for a combined sewer overflow treatment for French conditions considering e.g., different design rules and effluent requirements [56,57].

A similar approach was taken for developing a simple model to support the design of French VF wetlands [58]. Numerical simulations using the HYDRUS Wetland Module were first performed, varying a number of design parameters. The design parameters included the depth of the main layer of the first stage filter, the hydraulic loading rate, and the inlet COD concentration and temperature. Effluent concentrations of the first stage filter were then verified using measured data and a surrogate model was developed to estimate COD and NH4-N removal in relation to the design parameters. Although the first results were promising, it was concluded that more work needs to be done to develop a robust design model [58].

4. CONCLUSIONS

Interest in modelling processes occurring in treatment wetlands has increased over the last 15 years. Several mechanistic models to describe processes in subsurface flow treatment wetlands have been developed. The HYDRUS Wetland Module, up to now, is the only one that is commercially available and thus available for wider use.

Experiences with applying the HYDRUS Wetland Module and other existing simulation tools can be summarised as follows:

- Process-based models are a powerful tool for understanding the processes in treatment wetlands in more detail.
- All sub-models required to describe the different processes in treatment wetlands are important.
  - Good calibration of the water flow model is a pre-requisite for achieving a good match between measured and simulated pollutant concentrations. Due to the intermittent loading, more data are required for VF wetlands compared to HF wetlands. If the water flow model is calibrated, good results can be obtained in most applications when using the standard parameter sets of the biokinetic CW2D and CWM1 biokinetic models [7].
  - A good description of solute transport in the porous media is of equal importance for good simulation results for pollutant degradation and removal. To be able to model realistic solute transport behaviour, it is advised to carry tracer experiments.

Adsorption and desorption is an important process for several parameters (such as ammonia and phosphate), especially when reactive filter media are used.

- Influent fractionation (i.e., fractionation of influent COD and the N and P contents of different COD fractions) has a high impact on simulation results and thus is an essential part of calibrating reactive transport models. This is especially true when simulating wetlands treating other than domestic wastewater. Parameters of the biokinetic model should be changed only after a good calibration of the water flow and transport models has been achieved and after the effect of the influent fractionation on simulation results has been considered.
- Simulating the influence of wetland plants is possible by considering uptake and release via the plant roots linked to evapotranspiration. Modelling the effect of plants on the treatment performance has been shown to be most important for HF wetland and batch-operated systems and less important for VF wetlands.
- Models describing clogging are important to describe the long-term behaviour of treatment wetlands. Up to now, clogging models are the least established and tested among the sub-models mentioned. Further research is needed to get better reliability in clogging model predictions.

- One of the main obstacles for the wider use of available simulation tools is that they are rather complicated and difficult to run. Simplified, yet robust and reliable models for the design of treatment wetlands, need to be developed (such as RSF_Sim, developed to support the design of treatment wetlands treating combined sewer overflow [55]).

REFERENCES

1. Kadlec, R.H.; Wallace, S. *Treatment Wetlands*, 2nd ed.; CRC Press: Boca Raton, FL, USA, 2009.
2. García, J.; Rousseau, D.P.L.; Morató, J.; Lesage, E.; Matamoros, V.; Bayona, J.M. Contaminant Removal Processes in Subsurface-Flow Constructed Wetlands: A Review. *Crit. Rev. Environ. Sci. Technol.* **2010**, *40*, 561–661.
3. Langergraber, G. Numerical modelling: A tool for better constructed wetland design? *Water Sci. Technol.* **2011**, *64*, 14–21.
4. Langergraber, G.; Rousseau, D.; García, J.; Mena, J. CWM1—A general model to describe biokinetic processes in subsurface flow constructed wetlands. *Water Sci. Technol.* **2009**, *59*, 1687–1697.

5. Langergraber, G. Modeling of processes in subsurface flow constructed wetlands—A review. *Vadose Zone J.* **2008**, *7*, 830–842.
6. Langergraber, G.; Morvannou, A. Modelling of Treatment Wetlands. *Sustain. Sanit. Pract.* **2014**, *18*, 31–36.
7. Langergraber, G.; Šimůnek, J. Reactive Transport Modeling of Subsurface Flow Constructed Wetlands Using the HYDRUS Wetland Module. *Vadose Zone J.* **2012**, *11*, 460.
8. Samsó, R.; Garcia, J. BIO PORE, a mathematical model to simulate biofilm growth and water quality improvement in porous media: Application and calibration for constructed wetlands. *Ecol. Eng.* **2013**, *54*, 116–127.
9. Langergraber, G.; Giraldi, D.; Mena, J.; Meyer, D.; Peña, M.; Toscano, A.; Brovelli, A.; Korkusuz, E.A. Recent developments in numerical modelling of subsurface flow constructed wetlands. *Sci. Total Environ.* **2009**, *407*, 3931–3943.
10. Kumar, J.L.G.; Zhao, Y.Q. A review on numerous modeling approaches for effective, economical and ecological treatment wetlands. *J. Environ. Manag.* **2011**, *92*, 400–406.
11. Meyer, D.; Chazarenc, F.; Claveau-Mallet, D.; Dittmer, U.; Forquet, N.; Molle, P.; Morvannou, A.; Pálfy, T.; Petitjean, A.; Rizzo, A.; et al. Modelling constructed wetlands: Scopes and aims—A comparative review. *Ecol. Eng.* **2015**, *80*, 205–213.
12. Samsó, R.; Meyer, D.; García, J. Subsurface Flow Constructed Wetland Models: Review and Prospects. In *The Role of Natural and Constructed Wetlands in Nutrient Cycling and Retention on the Landscape*; Vymazal, J., Ed.; Springer: Cham, Switzerland, 2015; pp. 149–174.
13. Šimůnek, J.; Šejna, M.; van Genuchten, M.T. The HYDRUS Software Package for Simulating the Two- and Three-Dimensional Movement of Water, Heat, and Multiple Solutes in Variably-Saturated Media. In *Technical Manual*, version 2.0; PC-Progress: Prague, Czech Republic, 2011; p. 254.
14. Langergraber, G.; Šimůnek, J. Modeling variably-saturated water flow and multi-component reactive transport in constructed wetlands. *Vadose Zone J.* **2005**, *4*, 924–938.
15. Šimůnek, J.; van Genuchten, M.T.; Šejna, M. Development and applications of the HYDRUS and STANMOD software packages and related codes. *Vadose Zone J.* **2008**, *7*, 587–600.
16. Langergraber, G. Simulation of subsurface flow constructed wetlands—Results and further research needs. *Water Sci. Technol.* **2003**, *48*, 157–166.
17. Langergraber, G. Simulation of the treatment performance of outdoor subsurface flow constructed wetlands in temperate climates. *Sci. Total Environ.* **2007**, *380*, 210–219. Langergraber, G.; Tietz, A.; Haberl, R. Comparison of measured and simulated distribution of microbial biomass in subsurface vertical flow constructed wetlands. *Water Sci. Technol.* **2007**, *56*, 233–240.
18. Hochfeldt, V. Numerical Simulation of a Full-Scale Two-Stage Constructed Wetland System. Master's Thesis, University of Applied Sciences Ostwestfahlen-Lippe, Höxter, Germany, 2016.
19. Pucher, B.; Ruiz, H.; Paing, J.; Chazarenc, F.; Molle, P.; Langergraber, G. Using simulation of a one stage vertical flow constructed wetland to determine filter depth of a zeolite layer to improve the treatment performance. *Water Sci. Technol.* **2016**. submitted.
20. Dittmer, U.; Meyer, D.; Langergraber, G. Simulation of a Subsurface Vertical Flow Constructed Wetland for CSO Treatment. *Water Sci. Technol.* **2005**, *51*, 225–232.
21. Henrichs, M.; Langergraber, G.; Uhl, M. Modelling of organic matter degradation in constructed wetlands for treatment of combined sewer overflow. *Sci. Total Environ.* **2007**, *380*, 196–209.
22. Henrichs, M.; Welker, A.; Uhl, M. Modelling of biofilters for ammonium reduction in combined sewer overflow. *Water Sci. Technol.* **2009**, *60*, 825–831.
23. Meyer, D. Modellierung und Simulation von Retentionsbodenfiltern zur weitergehenden Mischwasserbehandlung (Modelling and Simulation of Constructed Wetlands for Treatment of Combined Sewer Overflow). Ph.D. Thesis, Technical University of Kaiserslautern, Kaiserslautern, Germany, 2011. Available online: https://kluedo.ub.uni-kl.de/frontdoor/index/index/docId/2843 (accessed on 14 August 2013). (In German)
24. Pálfy, T.G.; Molle, P.; Langergraber, G.; Meyer, D. Simulation of constructed wetlands treating combined sewer overflow using HYDRUS/CW2D. *Ecol. Eng.* **2016**, *87*, 340–347.

25. Toscano, A.; Langergraber, G.; Consoli, S.; Cirelli, G.L. Modelling pollutant removal in a pilot-scale two-stage subsurface flow constructed wetlands. *Ecol. Eng.* **2009**, *35*, 281–289.
26. Pálfy, T.G.; Gribovszki, Z.; Langergraber, G. Design-support and performance estimation using HYDRUS/CW2D: A horizontal flow constructed wetland for polishing SBR effluent. *Water Sci. Technol.* **2015**, *71*, 965–970.
27. Pálfy, T.G.; Langergraber, G. The verification of the Constructed Wetland Model No.1 implementation in HYDRUS using column experiment data. *Ecol. Eng.* **2014**, *68*, 105–115.
28. Rizzo, A.; Langergraber, G.; Galváo, A.; Boano, F.; Revelli, R.; Ridolfi, L. Modelling the response of horizontal flow constructed wetlands to unsteady organic loads with HYDRUS-CWM1. *Ecol. Eng.* **2014**, *68*, 209–213.
29. Rizzo, A.; Langergraber, G. Novel insights on the response of horizontal flow constructed wetland to sudden changes in influent loads from a modelling investigation. *Ecol. Eng.* **2016**, *93*, 242–249.
30. Karlsson, S.C.; Langergraber, G.; Pell, M.; Dalahmeh, S.; Vinnerås, B.; Jönsson, H. Simulation and verification of hydraulic properties and organic matter degradation in sand filters for greywater treatment. *Water Sci. Technol.* **2015**, *71*, 426–433.
31. Smethurst, P.J.; Petrone, K.; Langergraber, G.; Baillie, C.; Worledge, D. Nitrate dynamics in a rural headwater catchment: Measurements and modeling. *Hydrol. Process.* **2014**, *28*, 1820–1834.
32. Pugliese, L.; Bruun, J.; Kjaergaard, C.; Hoffmann, C.C.; Langergraber, G. Nonequilibrium model for solute transport in full size subsurface flow constructed wetland. *Ecol. Eng.* **2016**. submitted.
33. Šimůnek, J.; Jacques, D.; Langergraber, G.; Bradford, S.A.; Šejna, M.; van Genuchten, M.T. Numerical Modeling of Contaminant Transport Using HYDRUS and its Specialized Modules. *J. Indian Inst. Sci.* **2013**, *93*, 265–284.
34. Van Genuchten, M.T. A closed form equation for predicting the hydraulic conductivity of unsaturated soils. *Soil Sci. Soc. Am. J.* **1980**, *44*, 892–898.
35. Tietz, A.; Langergraber, G.; Sleytr, K.; Kirschner, A.; Haberl, R. Characterization of microbial biocoenosis in vertical subsurface flow constructed wetlands. *Sci. Total Environ.* **2007**, *380*, 163–172.
36. Langergraber, G. Erratum: Water Science and Technology 56, 2007, 233–240: Comparison of measured and simulated distribution of microbial biomass in subsurface vertical flow constructed wetlands. *Water Sci. Technol.* **2015**, *71*, 157–158.
37. Langergraber, G.; Prandtstetten, C.; Pressl, A.; Haberl, R.; Rohrhofer, R. Removal efficiency of subsurface vertical flow constructed wetlands for different organic loads. *Water Sci. Technol.* **2007**, *56*, 75–84.
38. Morvannou, A.; Forquet, N.; Vanclooster, M.; Molle, P. Which hydraulic model to use in vertical flow constructed wetlands? In Proceedings of the 13th IWA Specialized Group Conference on "Wetland Systems for Water Pollution Control", Perth, Australia, 25–29 November 2012; Conference Papers Volume 2. Dallas, S., Ed.; Murdoch University: Perth, Australia, 2012; pp. 145–153.
39. Molle, P.; Liénard, A.; Boutin, C.; Merlin, G.; Iwema, A. How to treat raw sewage with constructed wetlands: An overview of the French systems. *Water Sci. Technol.* **2005**, *51*, 11–21.
40. Troesch, S.; Esser, D. Constructed Wetlands for the Treatment of Raw Wastewater: The French Experience. *Sustain. Sanit. Pract.* **2012**, *12*, 9–15.
41. Langergraber, G. Development of a Simulation Tool for Subsurface Flow Constructed Wetlands. Ph.D. Dissertation, BOKU University, Vienna, Austria, 2001.
42. Morvannou, A.; Choubert, J.-M.; Vanclooster, M.; Molle, P. Solid respirometry to characterize nitrification kinetics: A better insight for modelling nitrogen conversion in vertical flow constructed wetlands. *Water Res.* **2011**, *45*, 4995–5004.
43. Puchner, B.; Langergraber, G. Simulation of two-stage vertical flow sand and zeolite filters treating domestic wastewater using the HYDRUS wetland module. In Proceedings of the 6th International Symposium on "Wetland Pollution Dynamics and Control" and Annual Conference of the Constructed Wetland Association, York, UK, 13–18 September 2015;

Dotro, G., Gagnon, V., Eds.; Book of Abstracts. Cranfield University: Cranfield, UK, 2015; pp. 198–199.

44. Dal Santo, S.; Canga, E.; Pressl, A.; Borin, M.; Langergraber, G. Investigation of nitrogen removal in a two-stage subsurface vertical flow constructed wetland system using natural zeolite. In Proceedings of the 12th IWA Specialized Group Conference on "Wetland Systems for Water Pollution Control", San Servolo, Venice, Italy, 3–8 October 2010; Masi, F., Nivala, J., Eds.; IRIDRA Srl: Florence, Italy, 2010; pp. 263–270.
45. Pálfy, T. Verification of the Implementation of CWM1 in the HYDRUS Wetland Module. Master's Thesis, BOKU University, Vienna, Austria, 2013.
46. Langergraber, G. The role of plant uptake on the removal of organic matter and nutrients in subsurface flow constructed wetlands—A simulation study. *Water Sci. Technol.* **2005**, *51*, 213–223.
47. Bradford, S.A.; Šimůnek, J.; Bettahar, M.; van Genuchten, M.T.; Yates, S.R. Modeling Colloid Attachment, Straining, and Exclusion in Saturated Porous Media. *Environ. Sci. Technol.* **2003**, *37*, 2242–2250.
48. Langergraber, G.; Šimůnek, J. Simulating particle transport in subsurface flow constructed wetlands with CW2D/HYDRUS. In Proceedings of the 3rd International Symposium on "Wetland Pollutant Dynamics and Control—WETPOL 2009", Barcelona, Spain, 20–24 September 2009; Bayona, J.M., García, J., Eds.; Book of Abstracts. Universitat Politècnica de Catalunya: Barcelna, Spain, 2009; pp. 135–136.
49. Samsó, R.; García, J. The Cartridge Theory: A description of the functioning of horizontal subsurface flow constructed wetlands for wastewater treatment, based on modelling results. *Sci. Total Environ.* **2014**, *473–474*, 651–658.
50. Mostafa, M.; Van Geel, P.J. Conceptual models and simulations for biological clogging in unsaturated soils. *Vadose Zone J.* **2007**, *6*, 175–185.
51. Samsó, R.; Blázquez, J.; Agulló, N.; Grau, J.; Torres, R.; García, J. Effect of bacteria density and accumulated inert solids on the effluent pollutant concentrations predicted by the constructed wetlands model BIO_PORE. *Ecol. Eng.***2015**, *80*, 172–180.
52. Samsó, R.; García, J.; Molle, P.; Forquet, N. Modelling bioclogging in variably saturated porous media and the interactions between surface/subsurface flows: Application to Constructed Wetlands. *J. Environ. Manag.* **2015**, *165*, 271–279.
53. Rajabzadeh, A.R.; Legge, R.L.; Weber, K.P. Multiphysics modelling of flow dynamics, biofilm development and wastewater treatment in a subsurface vertical flow constructed wetland mesocosms. *Ecol. Eng.* **2014**, *74*, 107–116.
54. Meyer, D.; Dittmer, U. RSF_Sim—A simulation tool to support the design of constructed wetlands for combined sewer overflow treatment. *Ecol. Eng.* **2015**, *80*, 198–204.
55. Pálfy, T.G.; Meyer, D.; Molle, P. Orage: Simulation of planted detentive filters treating urban storm water. In Proceedings of the 6th International Symposium on Wetland Pollution Dynamics and Annual Conference of the Constructed Wetland Association, York, UK, 13–18 September 2015; Dotro, G., Gagnon, V., Eds.; Book of Abstracts. Cranfield University: Cranfield, UK, 2015; pp. 216–217.
56. Pálfy, T.G.; Gourdon, R.; Meyer, D.; Troesch, S.; Molle, P. A dynamic design tool for CWs treating combined sewer overflow. In Proceedings of the 15th IWA Specialized Group Conference on "Wetland systems for Water Pollution Control", Gdansk, Poland, 4–9 September 2016; Gajewska, M., Matej-Lukowicz, K., Wojciechowska, E., Eds.; Conference Proceedings Volume 2. Gdańsk University of Technology: Gdańsk, Poland, 2016; pp. 771–790.
57. Morvannou, A.; Froquet, N.; Troesch, S.; Molle, P. Impact of design and operating conditions on vertical flow constructed wetland performances: Modelling and global assessment confrontation. In Proceedings of the 6th International Symposium on Wetland Pollution Dynamics and Annual Conference of the Constructed Wetland Association, York, UK, 13–18 September 2015; Dotro, G., Gagnon, V., Eds.; Book of Abstracts. Cranfield University: Cranfield, UK, 2015; pp. 196–197.

CHAPTER 10

Main Driving Factors for Seacoast Vegetation in the Southern and Eastern Baltic

Piotr Hulisz[1], Agnieszka[2], Piernik Jasmin Mantilla-Contreras[3], Tiina Elvisto[4]

[1]*Faculty of Earth Sciences, Nicolaus Copernicus University, ul Lwowska 1, 87-100 Toruń, Poland*

[2]*Faculty of Biology and Environment Protection, Nicolaus Copernicus University, ul Lwowska 1, 87-100 Toruń, Poland*

[3]*Institute for Biology and Chemistry, University of Hildesheim,Universitätsplatz 1, 31141 Hildesheim, Germany*

[4]*School of Natural Sciences and Health, Tallinn University, 25 Narva Rd, 10120 Tallinn, Estonia*

ABSTRACT

This study aimed to determine the role of environmental factors in the pattern of vegetation related to marsh soils, which are rare in the Baltic coastal zone. It was assumed that the pattern of vegetation directly affected by seawater depends on seawater salinity, and in general reflects the decreasing salinity gradient of the Baltic Sea. The research of vegetation and soils (0–25 cm) has been undertaken on 11 selected sites along the Baltic seacoast in Germany, Poland and Estonia. According to TWINSPAN classification, five plant communities were distinguished: the *Puccinellia maritima-Spergularia marina* community, *Juncus gerardi* community, *Agrostis stolonifera-Phragmites australis-Scirpus lacustris* ssp.

tabernaemontani community, *Phragmites australis-Calystegia sepium* community and *Elymus repens* community. A canonical correspondence analysis showed that the most important environmental factors influencing vegetation differentiation were: electrical conductivity of the saturation paste extract, hay/pasture management, redox potential, soil moisture and actual pH. In conclusion, the spatial distribution of the investigated vegetation and soil properties along the Baltic Sea shore were inconsistent with the spatial salinity gradient of the open surface seawater, but was significantly dependent on local conditions. This was confirmed by the highest soil salinity and most abundant occurrence of *Salicornia europaea* in Estonia.

Keywords: CCA, Coastal meadows, Coastal marsh, soils, Gradient analysis, Halophytes, Soil salinity

1. INTRODUCTION

The Baltic Sea is a brackish, semi-enclosed, almost non-tidal, inland water body in Northern Europe. Due to limited water exchange with the North Sea and its receiving a large freshwater runoff, there is a strong, west-to-east salinity gradient across the open Baltic surface seawater (0–10 m) (20 ‰ in the Kattegat; 8.0 ‰ in the Bay of Greifswald, 7.0 ‰ in the Gulf of Gdańsk, about 5.5 ‰ in northern Estonia, in the Gulf of Bothnia and less than 3 ‰ in the eastern part of the Gulf of Finland). Fluctuations in seawater salinity are low during the year, with the maximum usually recorded in the spring and summer periods due to inflow from melting ice, river water and precipitation (Majewski and Lauer 1994).

The coastal vegetation reflects plant species' tolerance to salinity and waterlogging, which is visible in their zonation (Chapman 1960; Adam 1990). Therefore, besides salinity, the dominant factor for Baltic coastal vegetation is water level fluctuations, which are dependent on the season, and determined by wind direction and force, air pressure and discharge from rivers (Dijkema 1990), as well as microtidal activity (Adam 1990). Another very important factor seems to be rises in sea level, which may result in the loss of coastal habitats (Nicholls et al. 1999). Some authors (*i.a.,* Morris et al. 2002; Moeslund et al. 2011) emphasise that the Baltic coastal areas are highly susceptible to this global change because the tidal influence is negligible. On the other hand, for some northern and eastern Baltic regions the sea-level rise can be counteracted by ongoing postglacial land uplift (Kall et al. 2014).

The Baltic coastal marsh vegetation can be generalised as salt grasslands (managed habitats) and brackish reed beds (natural vegetation). Salt grasslands comprise areas that have developed due to sedimentation, peat formation or land rise, and are irregularly flooded by saline or brackish water (Wanner 2009). However, some of them on the Baltic coast are affected not only by seawater but also by brine associated with fossil salt deposits uplifted to the surface, or relict seawater (Grube 2000; Bosiacka et al. 2011). They are characterised by relatively high salinity and a high number of halophytic plant species (Piotrowska 1974; Bosiacka et al. 2011). Moreover, salt grasslands are found at a higher level in the zonation, and more inland on so-called salt patches, which developed as a result of chloride and sulphate accumulation in the topsoils during dry periods (Dijkema 1990).

In general, Baltic salt grasslands are considered to be valuable and endangered ecosystems (Dijkema 1990; Rannap et al. 2007; Wanner 2009; Bosiacka et al. 2016). They are designated as wetlands of international importance according to the Ramsar Convention on Wetlands (1971), as habitats of European importance according to the EU habitat directive (European Commission 1992) and as special protection areas for birds according to the EU birds directive (European Commission 1979).

The Baltic Sea coast is dominated, especially in Poland, by sandy beaches, dunes or cliffs. Although small, there are many salt marsh sites with brackish or salt marsh soils, which have developed under the direct influence of open Baltic water or brackish water in river deltas and lagoons (Hulisz 2013). They have formed on both marine and marine-alluvial sediments. The properties of these soils are mostly dependent on local water properties and flooding frequency. Such salt marsh sites have been described in Germany (Bernhardt and Koch 2003), Sweden (van der Graaf et al. 2007), Estonia (Ratas et al. 2006; Hulisz et al. 2011) and Poland (Pracz 1989; Hulisz 2013; Hulisz 2016). The spatial differentiation of vegetation directly related to marsh soil properties, had not been well described until now, whereas the general characteristics of the Baltic coastal grasslands have (Rebassoo 1975; Dijkema 1990; Fock et al. 2002; Czyż 2006; Wanner 2009; Ludewig et al. 2015).

Therefore, the aim of our study was to determine the role of environmental factors in the pattern of vegetation related to brackish/salt marsh soils along the southern and eastern Baltic seashore. It was assumed that this pattern depends on water salinity, and in general reflects the decreasing salinity gradient across the Baltic Sea. Moreover, we tested if field measurements of actual soil properties reflect vegetation differentiation and could be included in simple monitoring of these valuable habitats.

1.1 Study Area

The research was carried out in four study areas along the coastal zone of the southern and eastern Baltic Sea Coast: Bay of Greifswald, Germany (G); Szczecin Lagoon, Poland (PS); Gulf of Gdańsk, Poland (PG) and the western and northern coast of Estonia (E) - Fig. 1.

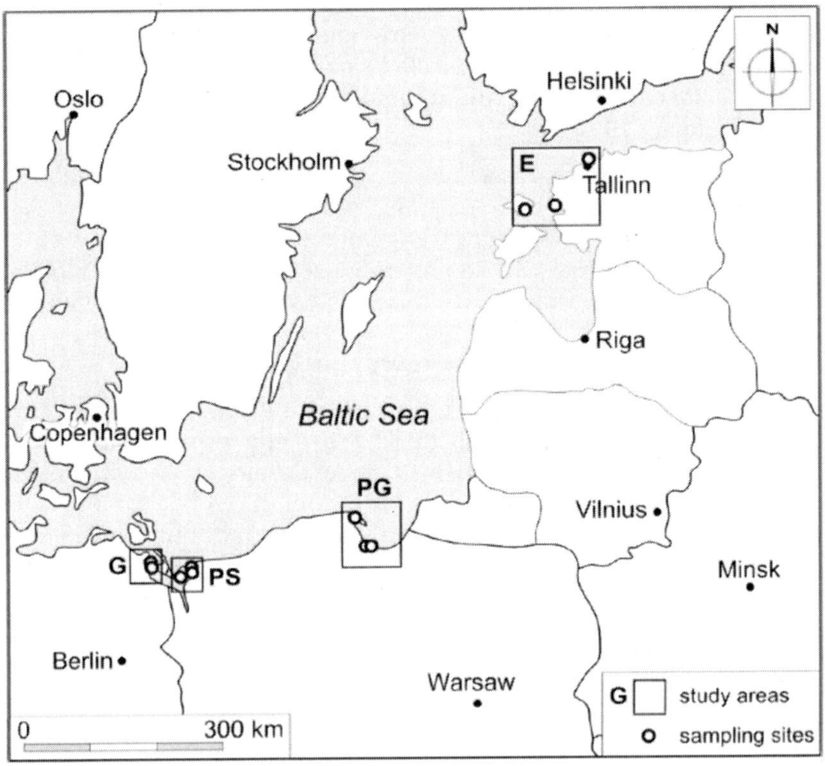

Figure. 1 Location of the study areas and the sampling sites: G – Germany; PS – Poland, Szczecin Lagoon; PG – Poland, Gulf of Gdańsk; E – Estonia

The Bay of Greifswald is a lagoon located in north-eastern Germany, and locally called "Greifswalder Bodden". There are flat plains of glacial origin built from sand and tills, and local depressions filled with organic deposits (Niedermeyer et al. 1995). The study was conducted at two sites seasonally flooded with seawater: Karrendorfer Wiesen and Kooser Wiesen. They are managed, and grazed by livestock (cattle). In contrast to the Karrendorfer Wiesen, revitalised after removal of flood embankments, the Kooser Wiesen have never been diked and have been preserved as nearly-natural, undisturbed salt grasslands (Lampe and Janke 2004).

The Szczecin Lagoon is a large, brackish water body (salinity 0.6–0.9 %) in Germany and Poland. It constitutes the Oder River estuary system, and is affected by seawater intrusions (backwater) through the Peenestrom, Świna and Dziwna straits. The first site is situated within the reverse delta of the Świna River (Karsiborska Kępa Island), whereas the two others (Chrząszczewo Island, Wrzosowo) are located on the banks of the Dziwna strait. The PS study sites were dominated by peat, mud and silty alluvial soils, and used as a pasture in some places. Karsiborska Kępa Island (315 ha) was diked and turned into a polder during the first half of the twentieth century. Chrząszczewo Island (10 km^2) is partially protected against flooding by embankments.

The Gulf of Gdańsk is located in northern Poland. The Władysławowo site is situated in the coastal zone of the Puck Lagoon on the Hel Peninsula, the long sandy spit that constitutes a border with the open Baltic waters. The other sites were located within the Vistula River delta, and therefore strongly affected by backwater. Soils of the PG study area were developed from stratified, organic-matter-rich, marine and alluvial deposits, which reflect the changing conditions of sedimentation often caused by human impact. This is particularly evident in the Vistula River delta, where the land was drained and the river banks were strengthened (Hulisz 2013). At present the Władysławowo site is used as pasture, while the others are wasteland.

Three study sites are on the Estonian sea coast (E). Two of them were located in the north-west (Kassari Island) and eastern (Põgari-Sassi) sections of the Väinameri Strait. They are on the shores of shallow lagoons, with low land relief and mineral soils formed of bipartite marine sediment deposits overlaying weakly-permeable, carbonate-rich glacial sediments (Hulisz et al. 2011; Hulisz 2013). These sites are traditionally grazed meadows. The third study site (Paljassaare) is in the eastern part of the Saartevahe Haak, a small bay at the tip of the Paljassaare peninsula in the Tallinn area on the Gulf of Finland. Overgrown with reed beds, it is a sandy beach zone with a marine sand layer, resulting from accumulating wave activity, which sits on till layers (Elvisto 2010). This site is not managed.

Most study sites have a protected status (e.g. a halophyte reserve in Władysławowo, Käina Bay-Kassari Landscape Protection Area and Matsalu National Park), and were included in the Natura 2000 network. Some are also a refuge for rare bird species (e.g. Karsiborska Kępa Island, the Vistula River delta and Paljassaare) listed in the EU birds directive (European Commission 1979).

In the Bay of Greifswald area, similarly to the Polish coast, the climate is warm temperate, fully humid with warm summer (Cfb). The climate of Estonian coast is significantly colder, and classified as snow, fully humid with moderately warm summer (Dfb) (Kottek et al. 2006). The changes in seawater level in the study areas are wind-induced and partly depend on the varying seasonal atmospheric pressure. According to Majewski and Lauer (1994), extreme deviations from the mean sea level are usually recorded during the autumn and winter months. They range respectively from +0.95 to -1.0 m (G), from +1.1 to -0.7 m (PS), from +1.5 to -1.0 m (PG) and from +1.2 to -0.9 m (E).

2. METHODS

2.1 Field Work

In total, 39 phytosociological relevés were collected in patches directly affected by seawater: 6 in Germany (G1–G2 – Karendorfer Wiesen, G3–G6 – Kooser Wiesen), 23 in Poland (PS1–PS11 – Karsiborska Kępa Island, PS12–PS13 – Chrząszczewo Island, PS14–PS15 – Wrzosowo, PG1 – Sobieszewo Island, PG2–PG7 – Władysławowo, PG8 – Brave Vistula River outlet), and 10 in Estonia (E1–E5 – Kassari Island, E6–E8 – Põgari-Sassi, E9–E10 – Paljassaare) – Fig. 1. Relevés were taken along a gradient from the waterline, i.e. from the pioneer zone to the meadow or reed beds – usually 2–3 points. The Braun-Blanquet method was used (Braun-Blanquet 1964; Westhoff and van der Maarel 1978). The size of each relevé was ca. 10 m^2 to represent local soil conditions. Nomenclature follows Flora Europaea (Tutin et al. 1964–1980). In each plot a soil sample (0–25 cm) was taken. Three random samples were taken by sampler in each plot and mixed to create one representative for the relevé. The following soil properties related to current environmental conditions were measured in the field: redox potential (E_h) and reaction (pH_a) by potentiometric method, electrical conductivity (EC_a) and moisture content (% *v*/v) of bulk soil (M) by Time Domain Reflectrometry (TDR). In addition to soil material, 28 Baltic Sea water samples, representative for each of the four study areas, were taken and analysed. As far as possible, each plot was classified within categories of management type: (i) not used lands (salt meadows excluded from exploitation) and (ii) hay/pasture lands (managed salt meadows).

2.2 Laboratory Analysis

In the air-dry soil samples, the organic matter (OM) content was determined by loss on ignition at 550 °C. Saturated paste extracts were prepared to evaluate soil salinity level. Reaction (pH_e), electrical conductivity (EC_e), content of chlorides (Cl^-), sodium (Na^+), calcium (Ca^{2+}) and magnesium (Mg^{2+}) ions were determined (van Reeuwijk 2006). In addition, reference measurements (pH, EC and ionic composition) were taken in the seawater (Hermanowicz et al. 1999). Based on the obtained results, the following calculations were made (Jackson 1958; van Reeuwijk 2006): sodium adsorption ratio (SAR), salt percentage in saturation extract (P_{ew}) and salt percentage in **soil (P_{es}).**

2.3 Data Analysis

The influence of location (G, PS, PG and E study areas) on soil properties was tested by Kruskal-Wallis one-way analysis of variance by ranks with post-hoc mean rank multiple comparison (Dunn test) (Zar 1999). Box-and-whisker plots were drawn using STATISTICA 9.0 software (Statsoft Inc.).

Before the analysis of species data, the cover/abundance values obtained with the Braun-Blanquet scale were transformed into ordinal scale values according to van der Maarel (1979).

Plant communities were separated from the set of phytosociological relevés by the Two Way Indicator Species Analysis (TWINSPAN, Hill 1979) with the use of the TWINSPAN package (Hill and Šmilauer 2005). They were defined as groups of similar species composition and abundance.

The pattern of vegetation distribution in the gradient of measured environmental variables was analysed by ordination methods. Canonical Correspondence Analysis was applied (CCA) because Detrended Correspondence Analysis (DCA) results detected a strong unimodal structure in the vegetation data (Jongman et al. 1995).

The relative importance of each environmental variable and its statistical significance in vegetation differentiation was assessed by the forward selection of the explanatory variables and Monte Carlo permutation test. The independent and dependent effect of each factor was taken into account. The Canoco 5.0 package was used for calculations (ter Braak and Šmilauer 2012).

For comparison of soil parameters between distinguished plant communities, the same nonparametric tests were used as in the case of research area comparisons (Kruskal-Wallis one-way analysis of variance by ranks with post-hoc Dunn test) (Zar 1999).

3. RESULTS

3.1 Water and Soil Properties

The coastal seawater was characterised by typical brackish features (pH 7.6–9.0; EC 8.0–11.1 dS·m^{-1}; salinity 4.7–7.0 ‰) - Table 1. Although the samples were taken in different geographical locations, a relatively low spatial variability of seawater properties was observed. However, the highest pH, EC values and Na^{+}, Cl^{-}, SO_4^{2-} ion content were recorded in the coastal waters in northern Poland (PG) and Estonia (E). Two hydrochemical types of water are distinguished: Cl-Na (PS, PG, E) and Cl-Na-Mg (G). The latter contained up to six times more magnesium than calcium.

Table 1 Ranges of selected properties of the Baltic coastal seawater in the studied areas (2009–2011; *n* = 28)

Parameter		G	PS	PG	E
pH	Min	8.1	7.9	7.6	8.3
	Max	8.6	8.4	8.7	9.0
EC [dS·m^{-1}]	Min	8.2	9.2	8.0	9.5
	Max	10.8	10.6	11.1	11.1
Salinity [‰]	Min	5.0	5.1	4.7	5.8
	Max	5.5	5.5	6.8	7.0
	Max	5.5	5.5	6.8	7.0
HCO_3^- [g·dm^{-3}]	Min	0.14	0.11	0.14	0.08
	Max	0.23	0.13	0.17	0.14
SO_4^{2-} [g·dm^{-3}]	Min	0.38	0.26	0.41	0.43
	Max	0.47	0.40	0.56	0.55
Cl^- [g·dm^{-3}]	Min	2.79	2.84	2.58	3.23
	Max	3.07	3.06	3.76	3.88
Na^+ [g·dm^{-3}]	Min	1.21	1.46	1.02	1.80
	Max	1.72	1.83	2.25	2.43
Ca^{2+} [g·dm^{-3}]	Min	0.11	0.08	0.07	0.09
	Max	0.12	0.09	0.10	0.18
Mg^{2+} [g·dm^{-3}]	Min	0.21	0.17	0.17	0.19
	Max	0.39	0.22	0.23	0.21

The characteristics of the investigated soils is shown in Fig. 2. They were formed on stratified marine-alluvial and marine deposits dominated by sands with varying admixtures of finer fractions. Many of these soils had thick surface horizons (up to 30 cm), rich in both autochthonous and allochthonous organic matter.

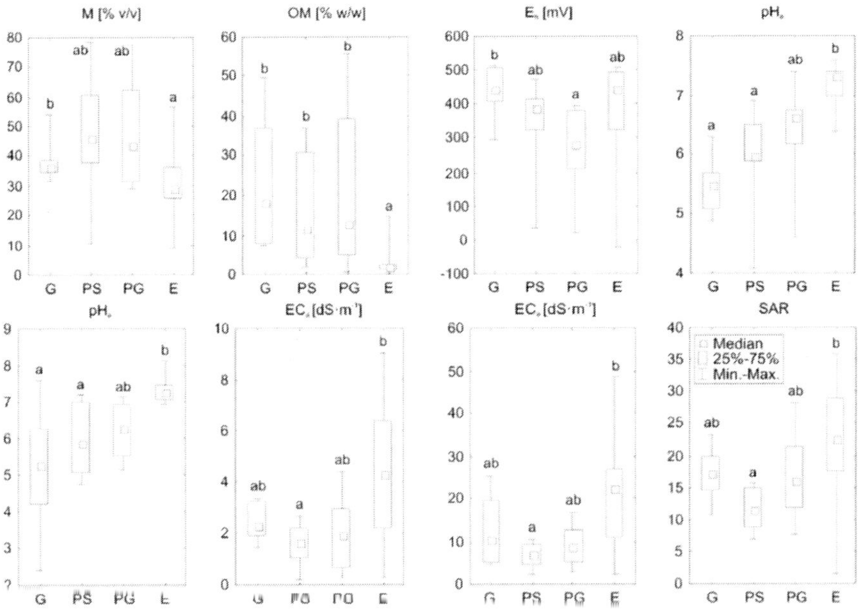

Figure. 2 Comparison of soil properties (0–25 cm) in four study areas. Kruskal-Wallis test was applied with Dunn test as post-hoc nonparametric multiple comparisons. Treatments with no difference are marked by the same letter (**a** or **b**). M – actual soil moisture, OM – organic matter content, E_h – redox potential, pH_e – pH of the soil saturated paste extract, pH_a – actual pH measured in the field, EC_a – electrical conductivity of the bulk soil, EC_e – electrical conductivity of the soil saturated paste extract, SAR – sodium adsorption ratio. Study areas abbreviations as in Fig. 1

The Kruskal-Wallis test with post-hoc mean rank multiple comparison revealed that most of the analysed soil parameters (with the exception of Mg^{2+} and P_{es}) significantly differed between the study areas. The salt percentage (P_{sw}), and major constituents of seawater such as Cl^-, Na^+, Ca^{2+} ions were excluded from the diagram because these parameters demonstrated a strong correlation with such salinity indices as EC_e and SAR.

The actual moisture (M) and organic matter (OM) content in all analysed soil samples was very heterogeneous (M 9–78 %, OM 0.3–56 %). The first parameter was significantly higher (p = 0.043) in soils from study area G than in study area E, while the second was significantly lower in E than in other regions (p = 0.010, 0.014 and 0.033, respectively). The studied soils had a redox potential (E_h) in the range of -25 to 517 mV, which suggests the presence of both anaerobic conditions and strong oxidation of the soil environment. The E_h values recorded in soils from the PG study area were significantly lower (0.043) than in soils from G (Fig. 2).

The values of pH_a varied from 4.1 to 7.6, and pH_e from 2.2 to 8.1. The strong acid reaction (pH_e) was measured in iron-sulphide-rich and carbonate-poor soil samples (G and PS), which were oxidised under laboratory conditions. This may suggest the presence of potential acid sulphate soil material. The pH_a and pH_e parameters showed a similar pattern, and were significantly lower in regions G (p = 0.000 and 0.008) and PS (p = 0.002 and 0.004), as compared to region E.

Electrical conductivity (EC_a 2.4–9.0 $dS \cdot m^{-1}$; EC_e 12–48 $dS \cdot m^{-1}$) and sodium adsorption ratio (SAR 16–36) reached significantly higher maximum values in study area E than in study area PS (p = 0.008, 0.028 and 0.012, respectively) – Fig. 2.

3.2 Vegetation of Seashore Zone

In total, 93 species were noted in the relevés. The most frequent species from the whole study area were: *Agrostis stolonifera* (present in 64 % of relevés), *Atriplex hastata* var. *salina* (54 %), *Spergularia marina* (44 %), *Phragmites australis* (41 %) and *Juncus gerardi* (38 %). Among halophytes, the most frequent besides *A. prostrata*, *S. marina* and *J. gerardi* were: *Plantago maritima* (33 %), *Triglochin maritimum* (26 %) and *Bolboschoenus maritimus* (21 %).

According to TWINSPAN classification, five groups of phytosociological relevés were distinguished: the *Puccinellia maritima-Spergularia marina* community, *Juncus gerardi* com., *Agrostis stolonifera-Phragmites australis-Scirpus lacustris* ssp. *tabernaemontani* com., *Phragmites australis-Calystegia sepium* com. and *Elymus repens* com. (Fig. 3).

The *Puccinellia maritima-Spergularia marina* com. Was typical for the first vegetation zone close to the waterline, and was noted on the Estonian (Kassari Island, Põgari-Sassi) and German (Kooser Wiesen) coasts. It was characterised by a low number of species and relatively low total species cover. However, on the Estonian coast *Salicornia europaea* and *Suaeda maritima* frequently accompanied two main species, i.e. *P. maritima* and *S. marina*.

Sample name	P	PPPPPPPPPP	PPPPPPEE	PEEGGPPPGPPG	EEEGGEEE
	S	GGSSSSSSSS	SGGSSG91	S7542GGG6SS1	61335824
	1	4818181167	365211.0	1....723.54.	....
	1	..5.0.34..	....2...	....	....
Vegetation cluster	Er	Ph-Cs	As-Ph-St	Jg	Pm-Sm
Juncus effusus	5	----2-----	--------	2-----------	--------
Phragmites australis	-	**-799999958**	**3--75723**	--3---------	--------
Calamagrostis epigejos	5	----2---2-	--------	------------	--------
Calystegia sepium	-	**-3233-3-35**	2-------	------------	--------
Carex gracilis	-	--2------3	--------	------------	--------
Cirsium arvense	-	-3---2----	--------	------------	--------
Epilobium tetragonum	-	----32----	--------	------------	--------
Filipendula ulmaria	-	----3---57	2-------	------------	--------
Holcus lanatus	-	----2---3-	--------	------------	--------
Lycopus europaeus	-	----2--2--	-------1	------------	--------
Phalaris arundinacea	-	--2-25----	--------	------------	--------
Rumex hydrolapathum	-	--2----2--	--------	------------	--------
Stachys palustris	-	--2-----22	--------	------------	--------
Symphytum officinale	-	---2252--2	--------	------------	--------
Urtica dioica	-	---23-3---	-------1	------------	--------
Deschampsia cespitosa	-	--------73	--------	------------	--------
Epilobium hirsutum	-	--------2-	-------2	------------	--------
Lysimachia vulgaris	-	-3------32	--------	------------	--------
Mentha arvensis	-	--------33	2-------	------------	--------
Sonchus arvensis	-	2-------22	--------	------------	--------
Hydrocotyle vulgaris	-	--------2-	3-------	------------	--------
Rumex crispus	-	2------1--	2-------	------------	--------
Eleocharis uniglumis	-	----------	---57---	------------	--------
Juncus articulatus	-	----------	3--8----	------------	--------
Ranunculus repens	-	--------2-	2---2---	------------	--------
Scirpus lacustris ssp. tabernaemontani	-	2---------	**-5-8-522**	------------	--------
Taraxacum officinale	-	----------	2---2---	------------	--------
Juncus ranarius	-	----------	3------7	------------	--------
Lotus tenuis	-	----------	5-------	----2-------	--------
Bolboschoenus maritimus	-	5-5-------	-52---2-	-------52---	----2---
Festuca arundinacea	-	98--------	--------	------3-2--7	--------
Potentilla anserina	-	233-----22	255-2---	-3---5-22---	--------
Trifolium repens	-	----------	5-------	2---2-------	--------
Agrostis stolonifera	-	--5-2--252	**35938837**	8577875-525-	-57-----
Elymus repens	9	----------	--2-5---	32--2---5--5	--------
Festuca rubra	-	-----2--78	795-----	55-535-79---	--------
Poa pratensis	-	--------5-	--------	-------5----	--------
Alopecurus geniculatus	-	----------	----7---	3--------53-	--------
Atriplex hastata var. salina	-	2--2222323	3-------	222222-22357	----2---
Leontodon autumnalis	-	----------	2-2-----	122---------	--------
Poa subcaerulea	-	----------	3-------	5-----------	--------
Aster tripolium	-	----------	-----5--	-2---128----	--------
Ranunculus sceleratus	-	----------	-------2	---------23-	--------
Trifolium fragiferum	-	----------	2-------	-2--22------	--------
Polygonum aviculare	-	----------	--------	---2-------3	--------
Ranunculus trichophyllus	-	----------	--------	---------32-	--------
Odontites littoralis	-	----------	--------	-21---------	--------
Puccinellia distans	-	----------	--------	------55----	--------
Triglochin maritima	-	----------	---3----	**2225727-----**	-22-----
Glaux maritima	-	----------	33------	-772-253----	32327---
Juncus gerardi	-	----------	--------	**-7575-87352-**	-372223-
Plantago intermedia	-	----------	--------	------22----	----2---
Plantago maritima	-	----------	5------2	-8533222----	273--2--

Figure. 3 Results of TWINSPAN classification of recorded relevés on German (G), Polish (PS and PG) and Estonian (E) coasts. Vegetation cluster

abbreviations: Er – *Elymus repens* community, Ph-Cs – *Phragmites australis-Calystegia sepium* com., As-Ph-St – *Agrostis stolonifera-Phragmites australis-Scirpus lacustris* ssp. *tabernaemontani* com., Jg – *Juncus gerardi* com., Pm-Sm – *Puccinellia maritima-Spergularia marina* com. The cover of species transformed from Braun-Blanquet scale into ordinal scale. Rare species with one occurrence are omitted

Patches of *Juncus gerardi* com. were present in all investigated sites as salt meadows. They were characterised by relatively high cover of *J. gerardi* and the presence of other halophytic species, such as: *Glaux maritima, Triglochin maritimum, Puccinellia maritima, Spergularia marina, Plantago maritima* and *Aster tripolium*.

The next vegetation cluster, i.e. *Agrostis stolonifera-Phragmites australis-Scirpus lacustris* ssp. *tabernaemontani* com., covered vegetation patches of wet subhalophilous meadows with relatively high abundance of grasses, such as *Agrostis stolonifera* and *Festuca rubra,* and reed species, such as *Scirpus lacustris* ssp. *tabernaemontani, Bolboschoenus maritimus* and *Phragmites australis*. Single halophytes were noted in the relevés (Fig. 3) on the Polish and Estonian (Paljassaare) coasts.

Phragmites australis-Calystegia sepium com. was typical for the Polish coast, and was present in all investigated stands next to pioneer vegetation zones directly affected by waves. It was characterised by dominance of *Phragmites australis* accompanied by *Calystegia sepium, Symphytum officinale, Phalaris arundinacea, Filipendula ulmaria* and halophilous *Atriplex hastata* var. *salina*.

Elymus repens com. with dominant *Elymus repens, Calamagrostis epigejos* and *Juncus effusus* was represented by only one relevé, taken behind the small dike on the Karsiborska Kępa Island (Szczecin Lagoon).

3.3 Factors Responsible for Vegetation Differentiation

Results of CCA analysis revealed that among the measured environmental variables, only OM, pH_e and P_{es} were not significant in the species distribution pattern (Table 2). Because many factors were correlated with each other, conditional effects were analysed. It was discovered that salinity (expressed as EC_e), management type (hay/pasture vs. not used), redox potential, actual moisture and pH_a were the most important (Table 2). These significant factors all together accounted for ca 24 % of the total species variance. Salinity accounted for the highest amount of variance, i.e. 6.9 %, then management type (5.3 %), followed by E_h (4.6 %), M (3.6 %) and pH_a (3.4 %).

Table 2 Results of forward selection, Monte Carlo permutation test and percent of species variance explained by each environmental factor. Significant factors are in bold. Simple term effect – independent effect of each factor, conditional effect – dependent effect of stepwise selected factors

Simple term effects:				Conditional term effects:			
	% variance	pseudo-F	p		% variance	pseudo-F	p
EC_e	**6.9**	**2.7**	**0.002**	**EC_e**	**6.9**	**2.7**	**0.002**
P_{ew}	**6.9**	**2.7**	**0.002**	**hay/pasture**	**5.3**	**2.2**	**0.002**
Cl^-	**6.9**	**2.7**	**0.002**	**E_h**	**4.6**	**1.9**	**0.002**
hay/pasture	**6.9**	**2.7**	**0.002**	**M**	**3.6**	**1.5**	**0.022**
not used	**6.9**	**2.7**	**0.002**	**pH_a**	**3.4**	**1.5**	**0.020**
Na^+	**6.7**	**2.7**	**0.002**	Ca^{2+}	2.8	1.2	0.146
SAR	**6.6**	**2.6**	**0.002**	EC_a	2.7	1.2	0.164
EC_a	**5.8**	**2.3**	**0.002**	OM	2.7	1.2	0.198
Ca^{2+}	**5.5**	**2.2**	**0.002**	pH_e	2.3	1.0	0.442
E_h	**5.3**	**2.1**	**0.002**	Na^+	2.1	0.9	0.576
Mg^{2+}	**4.7**	**1.8**	**0.028**	P_{es}	1.8	0.8	0.712
M	**3.8**	**1.5**	**0.022**	TDS	1.9	0.8	0.708
pH_a	**3.8**	**1.5**	**0.016**	Mg^{2+}	1.2	0.5	0.986
P_{es}	3.4	1.3	0.100				
OM	2.9	1.1	0.302				
pH_e	2.5	1.0	0.524				

The most saline places were occupied by *Puccinellia maritima-Spergularia marina* com. (Fig. 4). Next in the salinity gradient were *Juncus gerardi* com., *Agrostis stolonifera-Phragmites australis-Scirpus lacustris* ssp. *tabernaemontani* com. and *Phragmites australis-Calystegia sepium* com. *Elymus repens* com. was not separated on the ordination diagram and was located among patches of *Ag-Ph-St* com.

The management type gradient was of the same direction as the salinity gradient. All patches classified as *Pm-Sm* com. and *Jg* com. were used as hay meadows or pastures, compared with ca. 40 % of *Ag-Ph-St* com. and only 20 % of *Ph-Cs* com. (Table 3). Simple term effects comparison demonstrated that management type was significantly different for patches of *Ph-Cs* com. and *Pm-Sm* com. and *Jg* com.

Soil properties (mean ± SD) and management type in the distinguished vegetation clusters together with Kruskal-Wallis statistical comparisons with Dunne post-hoc test have been listed in Table 3.

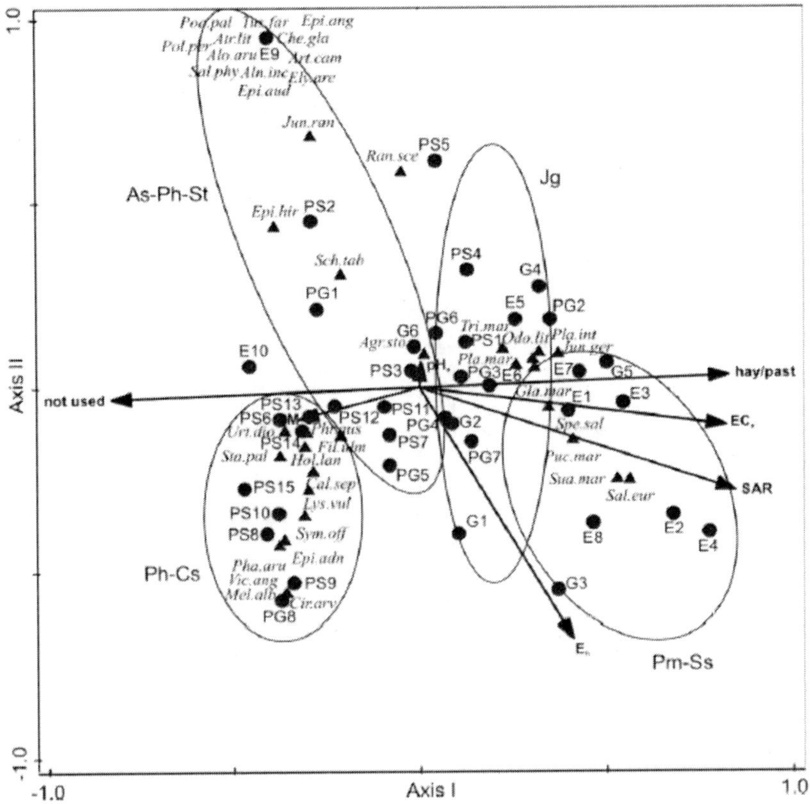

Figure. 4 CCA ordination diagram of the Baltic seashore vegetation showing correlation between typical species (of over 10 % of fit into ordination space) and significant environmental variables ($p < 0.05$). EC_e – electrical conductivity of the soil saturated paste extract. SAR – sodium adsorption ratio, E_h – redox potential, M – actual soil moisture, pH_a– actual pH, hay/pasture – hay or pasture lands, not used – abandoned places. Species abbreviations cover three letters of genus name and three letters of species name. Full species names and names of circled vegetation clusters as in Fig. 3

4. DISCUSSION

There were only small differences in the coastal water salinity within the analysed fragments of the coastal zone in the territory of Germany, Poland and Estonia – Table 1.

Table 3 Soil properties (mean ± SD) and management type in the distinguished vegetation clusters. The Kruskal-Wallis test with Dunne post-hoc multiple comparison between groups were applied. Significantly different means are marked by different letters. Community abbreviations as in Fig. 3. ns – not significant

	Er	Ph-Cs	As-Ph-St	Jg	Pm-Sm	p
EC_a [$dS \cdot m^{-1}$]	0.2	1.5 ± 0.7	1.3 ± 0.7	3.0 ± 1.2	4.6 ± 2.8	ns
EC_e [$dS \cdot m^{-1}$]	10.1	5.6^a ± 2.5	6.5^a ± 4.2	10.1^{ab} ± 4.7	27.5^b ± 10.5	0.0001
P_{es} [%]	0.4	0.3 ± 0.3	0.5 ± 0.9	0.7 ± 0.5	0.8 ± 0.7	ns
Na^+ [$mg \cdot dm^{-3}$]	1422	861^a ± 518	914^a ± 656	1634^{ab} ± 692	4731^b ± 2680	0.0001
Mg^{2+} [$mg \cdot dm^{-3}$]	264	158^a ± 92	254^a ± 279	281^a ± 170	1425^b ± 1542	0.0005
Ca^{2+} [$mg \cdot dm^{-3}$]	335	164^a ± 99	209^a ± 131	221^a ± 130	516^b ± 154	0.0028
SAR	14	12^{ac} ± 4.6	10^a ± 4.2	18^{bc} ± 4.9	25^b ± 6.8	0.0002
pH_a	4.7	6.4^{ab} ± 0.4	6.5^{ab} ± 0.6	5.7^a ± 1.0	7.0^b ± 0.6	0.02
pH_e	4.9	6.5 ± 0.8	6.4 ± 0.9	5.9 ± 1.0	6.4 ± 2.0	ns
E_h [mV]	375	373^{ab} ± 60	176^a ± 167	381^{ab} ± 114	458^b ± 66	0.0026
OM [%]	11.0	14.5 ± 14	14.0 ± 18	18.5 ± 15	12.5 ± 20	ns
M [%]	16.0	53.1 ± 18	43.2 ± 16	41.3 ± 17	31.9 ± 14	ns
Hay/pasture [% of patches]	–	20^a	40^{ab}	100^b	100^b	0.0001

Therefore, the obtained results can be considered as inconsistent with the spatial gradient of salinity of the Baltic Sea water (Majewski and Lauer 1994). This can probably be explained by specific, local environmental conditions in the coastal zones, especially the limited exchange of waters, the occurrence of the shallow lagoons and the riverine inflow (Hulisz 2013).

The significant impact of local factors on the soil environment in the area of the southern and eastern Baltic Sea coast was strongly confirmed by the results of the Kruskal-Wallis test with post-hoc mean rank multiple comparison (Fig. 2). The obtained data proved the statistical significance of the selected soil parameters in the differentiation of the study areas, which did not correspond to their location in the salinity gradient of the Baltic Sea open surface waters (nor to coastal water properties – Table 1). Characteristics related to salinity (EC_a, EC_e and SAR) were one of the most important criteria differentiating the soils at sites located in the four Baltic Sea regions in Germany (G), Poland (PS, PG) and Estonia (E). In addition, the unique combination of environmental factors in the Estonian coastal zone may contribute to very high soil salinity ($EC_e > 30$ dS·m^{-1}) and sodicity (SAR >20). The coastal flat plains in this region are developed from marine sediments overlaying glacial tills, which are a barrier to the vertical infiltration of saline water (Kalm and Kadastik 2001). In some cases, their long-term stagnation on poorly permeable layers and intensive evaporation, mostly during the summer season, can cause a very high increase in the salt concentration (EC up to 76 dS·m^{-1}) (Hulisz et al. 2011). Taking into account the brackish character of the Baltic sea waters, that phenomenon should be considered as very rare and untypical.

Single relevés of vegetation patches in the direct ordination results (Fig. 4) were not exactly located according to the averaged soil properties of investigated Baltic regions. That is to say, halophyte occurrence and the distribution of distinguished plant communities more closely reflected soil properties within the regions (on a local scale). As stated by Hulisz (2013), some soils directly affected by seawater may present a very specific spatial distribution pattern, where salinity increases with distance from the waterline (up to 100 m). This can be explained by small-scale morphological diversity and other soil properties. A similar spatial heterogenity of local environmental conditions reflected by halophilous vegetation can be also found in Polish inland areas (Piernik et al. 2015).

The parameters responsible for vegetation pattern were EC_e, E_h, M and pH_a – Fig. 4. Soil salinity has already been reported as the principle decisive factor for vegetation pattern from other salt marsh areas in the world (Chapman 1960; Adam 1990; Ortiz et al. 1995; Wanner 2009). The second important factor besides salinity was the management type of the studied sites. Some research has demonstrated, however, that management type can be even more important than salinity in influencing vegetation patterns (Amiaud et al. 1998; Jutila 2001; Piernik 2012). Moreover, the vegetation pattern can be affected in a different way by different intensities of cutting and grazing (Bakker 1989; Kiehl et al. 1996). We investigated relatively small areas directly affected by salt/brackish water and, there, it was difficult to determine if the area is permanently mown or grazed (sometimes both). To obtain and assess the general pattern of investigated vegetation, we decided to distinguish only two categories of management, as is mentioned in the section *Methods*. In the case of our research, salinity and intensity of land use were in the same direction (Fig. 4), which is to say that the most saline places in the investigated gradient were managed. The third factor – redox potential – was related to the actual water table and its low values indicate anoxic conditions that can limit some species' occurrence. In tidal marshes, species have to face such conditions much more frequently than at the Baltic Sea coast (Adam 1990), but even on the Baltic Sea this factor is important. The last variables, i.e. actual moisture and pH, are known to be important for halophytic species distribution (Piernik 2005; Wanner 2009; Bosiacka et al. 2011). What is quite surprising is that we did not detect any influence of organic matter content on vegetation distribution. This factor has been reported by many authors as important for halophytic vegetation because organic matter and nitrogen content can increase species' salt tolerance (Kiehl et al. 1997; Liu et al. 2004). At the same time, organic matter is responsible for many other properties of saline soils (Hulisz et al. 2013; Hulisz 2016). The obtained results suggest that, in the investigated seashore zone, which was directly related to the influence of water, organic matter did not affect general species pattern.

We related the general vegetation pattern to the abiotic condition. However, there are many studies, especially in coastal tidal marshes, that stress the following biotic factors as important: competition (Snow and Vince 1984; Bertness 1991; Pennings and Callaway 1992; Ungar 1998; Egan and Ungar 2001), impact of parasites (Pennings and Callaway 1996) and herbivores (Miller et al. 1996), and the presence of mycorrhizae (Neto et al. 2006). These influence the general pattern and can modify the vegetation pattern on a local scale.

The distribution of species along the salinity gradient as we have described (Fig. 4) reflected a pattern known already from coastal and inland saline areas in Europe, with *Salicornia europaea* in the most saline places, followed by *Suaeda maritima*, *Atriplex pedunculata*, *Puccinellia maritima* and *Spergularia marina*, then *Juncus gerardi*, *Plantago maritima*, *Aster tripolium*, *Atriplex hastata* var. *salina*, and reed beds species *Scirpus lacustris* ssp. *tabernaemontani*, *Eleocharis uniglumis*, *Bolboschoenus maritimus* and *Phragmites australis* (Chapman 1960; Adam 1990; Nienartowicz and Wilkoń-Michalska 1993; Piernik 2005). The gradient of land-use intensity demonstrates the presence of dominant *Phragmites australis* in unmanaged places, and species typical of salt meadows (dominated by *Juncus gerardi*) in patches used as pastures or hay meadows. *Phragmites australis* is a salt-resistant species that can grow in extremely saline sites (Mauchamp and Mésleard 2001; Piernik 2012). Therefore, we relate its presence to management rather than salinity, although in our study, salinity and management were of the same direction (as explained above). A similar pattern to that of the Baltic Sea coast was described by Jutila (2001) and Wanner (2009).

Vegetation patches in the investigated sites were classified into five plant communities (Fig. 3). In managed places of the highest salinity (EC_e, TDS, Mg^{2+}, Ca^{2+}) on the German and Estonian Baltic Sea coast, *Puccinelia maritima-Spergullaria marina* community was present (Table 3). This community has been described as typical for lower marshes on the Baltic coast (Dijkema 1990) and is present in lower marshes of the Wadden Sea (Bakker 1989). Second in the gradient of salinity was *Juncus gerardi* community, present on managed coastal meadows in all investigated areas. This vegetation type typically had significantly lower pH_a than in patches of the previous community. Low pH_a values can likely be explained by high levels of soil aeration and weak buffer capacity (lack of carbonates) favouring oxidation of sulphides (Hulisz 2013). This community is typical for middle marshes (Bakker 1989; Dijkema 1990).

The next two plant communities, i.e. *Agrostis stolonifera-Phragmites australis-Scirpus lacustris* ssp. *tabernaemontani* com. and *Phragmites australis-Calystegia sepium* com. Were found in places of significantly lower salinity compared to *Pm-Sm* community. In general, they did not differ significantly in salinity level compared with *Jg* com. However, they cover mostly unmanaged patches. Therefore, they can be thought of as early and late successional stages, as described by Wanner (2009).

Ag-Ph-St com. in this case represented the early successional stage. Dijkema (1990) identified *Agrostis stolonifera-Juncus gerardi* community within lower marshes on the Baltic Northeaster Archipelagos. Therefore, this community could represent lower marsh vegetation, because it was characterised by a relatively lower salinity level and worse oxygen condition for vegetation.

Ph-Cs community was typical for unmanaged, backwater-affected areas. Dijkema (1990) placed this type of vegetation as typical for emerging flats on the Baltic coast. The presence of *Atriplex hastata* var. *salina* indicated soil salinity in the vegetation patches (Piernik 2003), whereas *Festuca arundinacea*, *Potentilla anserina*, *Agrostis stolonifera* and *Festuca rubra* could be evidence of the potential grassland origin of these places. Therefore, they could represent the late successional stage, dominated by *Phragmites australis* after abandonment (Wanner 2009).

Only one relevé represented *Elymus repens* com. The communities with abundant *Elymus* species are more typical for upper coastal marshes (Bakker 1989) and for inland saline areas (Piernik 2012).

In conclusion, the spatial distribution of the investigated vegetation and soil properties along the southern and eastern Baltic seashore were inconsistent with the spatial salinity gradient of the open surface waters of the Baltic Sea. However, the obtained data proved the statistical significance of the selected soil parameters in the differentiation of the four study areas, which did not correspond to their geographical location. This is in contrast to our initial assumptions. Instead, vegetation pattern was dependent on the local environmental conditions. The main driving environmental factors for vegetation differentiation were: EC_e, management type, redox potential, actual moisture and pH_a. Results confirmed the key role of management for presence and structure of halophilous vegetation. Comparing the use of field and laboratory measurements for monitoring purposes, it seems appropriate to include those measured-in-field soil parameters which were significant for vegetation (M, E_h, pH_a), together with electrical conductivity of saturated paste extract (EC_e) as a salinity measure.

Our results are important for understanding vegetation-environment relationships on non-tidal salt marshes and for monitoring of these habitats. They are important for projects concerning landscape management strategy in coastal zones.

ACKNOWLEDGMENTS

The authors would like to thank Prof. Renata Bednarek, Prof. Urszula Pokojska, Prof. Luise Giani and Prof. Andrzej Nienartowicz for helpful comments on this work, to Dr. Mirosław T. Karasiewicz for his contribution in the fieldwork and Ms. Ewelina Urbańska, Mr. Adam Michalski and Mr. Michał Dąbrowski for their assistance with laboratory analysis. This study was financed by the Polish Ministry of Science and Higher Education (grant N305 231135, 2008-2011).

REFERENCES

1. Adam P (1990) Saltmarsh ecology. Cambridge University Press, Cambridge
2. Amiaud B, Bouzille JB, Tournade F, Bonis A (1998) Spatial patterns of soils salinities in old embanked marshlands in western France. Wetlands 18:482–494
3. Bakker JP (1989) Nature management by grazing and cutting. Geobotanica 14. Kluwer Academic Publishers, Dordrecht

4. Bernhardt KG, Koch M (2003) Restoration of a salt marsh system: temporal change of plant species diversity and composition. Basic applied. Ecology 4:441–451
5. Bertness MD (1991) Interspecific interactions among high marsh perennials in a New England alt marsh. Ecology 72:125–137
6. Bosiacka B, Podlasiński M, Pieńkowski P (2011) Salt marshes determined by ascending brine in northern Poland: land-use changes and vegetation-environment relations. Phytocoenologia 41:201–213
7. Bosiacka B, Kull T, Więcław H, Marciniuk P, Podlasiński M (2016) Habitat requirements of marsh dandelions (Taraxacum sect. Palustria) in polish and Estonian coastal grasslands. Polish Journal of Ecology 64:481–498
8. Braun-Blanquet J (1964) Pflanzensoziologie, 3rd edn. Springer, Wien
9. Chapman VJ (1960) Salt marshes and salt deserts of the world. Leonard Hill Books Limited, London
10. Czyż H (ed) (2006) Salt grasslands and coastal meadows. AR Szczecin, Szczecin
11. Dijkema KS (1990) Salt and brackish marshes around the Baltic Sea and adjacent parts of the North Sea: their vegetation and management. Biological Conservation 51:191–209
12. Egan TP, Ungar IA (2001) Competition between *Salicornia europaea* and *Atriplex prostrata* (*Chenopodiaceae*) along an experimental salinity gradient. Wetlands Ecology and Management 9:457–461
13. Elvisto T (2010) Eutrophication level of *Phragmites australis* habitats at a shallow coastal lake, Paljassaare peninsula, Tallinn, Estonia. Estonian. Journal of Ecology 59:83–98
14. European Commission (1979) Council Directive 79/409/EEC on the conservation of wild birds
15. European Commission (1992) Council Directive 92/43/EEC of 21 May 1992 on the conservation of natural habitats and of wild fauna and flora. Official Journal L 206, 22/07/1992 P 0007–0050
16. Fock T, Hergarden K, Repasi D (eds) (2002) Salt grasslands and coastal meadows in the Baltic region. Fachhochschule Neubrandenburg, Neubrandenburg
17. Grube A (2000) Widespread geogenic salt water occurrence in North Germany. Proceedings of the 16th Salt Water Intrusion Meeting, Międzyzdroje-Wolin: 55–61
18. Hermanowicz W, Dojlido J, Dożańska W, Koziorowski B, Zerbe J (1999) Fizyczno-chemiczne analizy wody i ścieków. Wyd. Arkady, Warszawa
19. Hill MO (1979) TWINSPAN: A FORTRAN program for arranging Multivariate Data in order two-way table by classification of the individual and attributes. Cornell University, Ithaca, New York
20. Hill MO, Šmilauer P (2005) Twinspan for Windows 2.3. Centre for Ecology and Hydrology, University of South Bohemia, Hutington, Česke Budějovice
21. Hulisz P (2013) Geneza, właściwości i pozycja systematyczna marszy brakicznych w strefie oddziaływania wód Bałtyku. Nicolaus Copernicus University Press, Toruń
22. Hulisz P (2016) Coastal marsh soils in Poland: characteristics and problems of classification. Soil Science Annual 67:37–44
23. Hulisz P, Elvisto T, Karasiewicz MT, Piernik A (2011) Abiotic factors influencing the occurrence of *Salicornia europaea* in West Estonia. Ecological Questions 14:57–60
24. Hulisz P, Gonet SS, Giani L, Markiewicz M (2013) Chronosequential alterations in soil organic matter during initial development of coastal salt marsh soils at the southern North Sea. Zeitschrift für Geomorphologie 57:515–529
25. Jackson ML (1958) Soil chemical analysis. Constable Ldt, London
26. Jongman RHG, ter Braak CJF, van Tongeren DFR (eds) (1995) Data analysis in community and landscape ecology. Pudoc, Wageningen
27. Jutila H (2001) How does grazing by cattle modify the vegetation of coastal grasslands along the Baltic Sea? Annales Botanici Fennici 38:181–200
28. Kall T, Oja T, Tänavsuu K (2014) Postglacial land uplift in Estonia based on four precise levelings. Technophysics 610:25–38
29. Kalm V, Kadastik E (2001) Waterlain glacial diamicton along the Palivere ice-marginal zone on the west Estonian archipelago, eastern Baltic Sea. Proceedings of the Estonian. Academy of Sciences 50:114–127

30. Kiehl K, Eischeid I, Gettner S, Walter J (1996) Impact of different sheep grazing intensities on salt marsh vegetation in northern Germany. Journal of Vegetation Science 7:99–106
31. Kiehl K, Esselnik P, Bakker JP (1997) Nutrient limitation and plant species composition in temperate salt marshes. Oceanologia 111:325–330
32. Kottek M, Grieser J, Beck C, Rudolf B, Rubel F (2006) World map of the Köppen-Geiger climate classification updated. Meteorologische Zeitschrift 15:259–263
33. Lampe R, Janke W (2004) The Holocene sea level rise in the southern Baltic as reflected in coastal peat sequences. Polish Geological Institute Special Papers 11:19–30
34. Liu X, Yang Y, Li W, Li C, Duan D, Tadano T (2004) Interactive effects of sodium chloride and nitrogen on growth and ion accumulation of a halophyte. Communications in Soil Science and Plant Analysis 35:2111–2123
35. Ludewig K, Wanner A, Jensen K (2015) Recolonization and facilitation in Baltic salt marsh vegetation. Annales Botanici Fennici 52:181–191
36. Majewski A, Lauer Z (eds) (1994) Atlas Morza Bałtyckiego. IMGW, Warszawa
37. Mauchamp A, Mésleard F (2001) Salt tolerance in *Phragmites australis* populations from coastal Mediterranean marshes. Aquatic Botany 70:39–52
38. Miller DL, Smeins FE, Webb JW (1996) Mid-Texas coastal marsh change (1939–1991) as influenced by lesser snow goose herbivory. Journal of Coastal Research 12:462–476
39. Moeslund JE, Arge L, Bøcher PK, Nygaard B, Svenning JC (2011) Geographically comprehensive assessment of salt-meadow vegetation-elevation relations using LiDAR. Wetlands 31:471–482
40. Morris JT, Sundareshwar PV, Nietch CT, Kjerfve B, Cahoon DR (2002) Responses of coastal wetlands to rising sea level. Ecology 83:2869–2877
41. Neto D, Carvalho LM, Cruz C, Martinez-Lucao MA (2006) How do mycorrhizas affect C and N relationships in flooded *Aster tripolium* plants. Plant and Soil 279:51–63
42. Nicholls RJ, Hoozemans FMJ, Marchand M (1999) Increasing flood risk and wetland losses due to global sea-level rise: regional and global analyses. Global Environmental Change 9, Suppl. 1:69–87
43. Niedermeyer RO, Scholle T, Flemming BW, Hertweck G (1995) Holocene sediments and basin development of a Pomeranian coastal lagoon (Greifswalder Bodden; Southern Baltic). Prace PIG 149:117–121
44. Nienartowicz A, Wilkoń-Michalska J (1993) The application of numerical analysis to comparison of ecological amplitudes of halophytic species. Variability and Evolution 2(3):103–112
45. Ortiz R, Alvarez Rogel J, Alcaraz F (1995) Soil-vegetation relationships in two coastal salt marshes in southeastern Spain. Arid Soil Research and Rehabilitation 9:481–493
46. Pennings SC, Callaway RM (1992) Salt marsh plant zonation: the relative importance of competition and physical factors. Ecology 73:681–690
47. Pennings SC, Callaway RM (1996) Impact of a parasitic plant on the structure and dynamics of salt marsh vegetation. Ecology 77:1410–1419
48. Piernik A (2003) Inland halophilous vegetation as indicator of soil salinity. Basic applied. Ecology 4:525–536
49. Piernik A (2005) Vegetation-environment relations on inland saline habitats in Central Poland. Phytocoenologia 35:19–37
50. Piernik A (2012) Ecological pattern of inland salt marsh vegetation in Central Europe. Nicolaus Copernicus University Press, Toruń
51. Piernik A, Hulisz P, Rokicka A (2015) Micropattern of halophytic vegetation on technogenic soils affected by the soda industry. Soil Science & Plant Nutrition 61:98–112
52. Piotrowska H (1974) Nadmorskie zespoły solniskowe w Polsce i problemy ich ochrony. Ochrona Przyrody 39:7–63
53. Pracz J (1989) Właściwości gleb tworzących się przy udziale słonej wody gruntowej w polskiej strefie przybałtyckiej. SGGW-AR, Warszawa
54. Rannap R, Lõhmus A, Jakobson K (2007) Consequences of coastal meadow degradation: the case of the natterjack toad (*Bufo calamita*) in Estonia. Wetlands 27:390–398

55. Ratas U, Puurmann E, Pappel P (2006) Soils and water conditions in land uplift coastal areas of the Baltic Sea, Estonia. In: Czyż H (ed) Salt grassland and coastal meadows Wyd AR, Szczecin, pp 91–96
56. Rebassoo HE (1975) Sea-shore plant communities of the Estonian islands. Academy of Sciences of the Estonian SSR, Tartu
57. Snow AA, Vince SW (1984) Plant zonation in Alaskan salt-marsh. II. An experimental study of the role of edaphic conditions. Journal of Ecology 72:669–684
58. ter Braak CJF, Šmilauer P (2012) CANOCO Reference manual and User's guide: Software Ordination (version 5.0). Biometrics, Wageningen and České Budějowice
59. Tutin TG, Heywood VH, Moore DM, Valentine DH, Walters SM, Weeb DA (1964–1980) Flora Europaea, vol 1–5. Cambridge University Press, Cambridge
60. Ungar IA (1998) Are biotic factors significant in influencing the distribution of halophytes in saline habitats? The Botanical Review 64:176–199
61. van der Graaf AJ, Stahl J, Veeneklaas RM (2007) Vegetation characteristics of a brackish marsh on Gotland and foraging choices of migrating and brood rearing geese. Annales Botanici Fennici 44:33–41
62. van der Maarel E (1979) Transformation of cover-abundance values in phytosociology and its effect on community similarity. Vegetatio 39:97–114
63. van Reeuwijk LP (2006) Procedures for soil analysis. 7th ed. Technical Report 9. ISRIC – World Soil Information, Wageningen
64. Wanner A (2009) Management, biodiversity and restoration potential of salt grassland vegetation of the Baltic Sea: Analyses along a complex ecological gradient. PhD thesis, University of Hamburg, Hamburg
65. Westhoff V, van der Maarel E (1978) The Braun-Blanquet approach. In: Whittaker RH (ed). Classification of plant communities Junk, Thee Hague, pp 287–339
66. Zar JH (1999) Biostatistical analysis. Prentice-Hall, Englewood Cliffs

CHAPTER 11

Early Snowmelt Enhances the Carbon Sequestration of Hummock-Forming Sphagnum Mosses on Boreal Wetlands

Niko Silvan[1*], Kari Jokinen[2]

[1]*Natural Resources Institute Finland, Natural Resources and Bioproduction, Parkano Unit, Parkano, Finland*

[2]*Natural Resources Institute Finland, Natural Resources and Bioproduction, Viikki Unit, Helsinki, Finland*

ABSTRACT

Sphagnum mosses are globally important owing to their considerable peat-forming ability and their potential impact on global climatic cycles acting as a long-term net carbon sink. However, changes in climatic conditions due to global warming may affect the relations between Sphagnum mosses and vascular plants but also the competition among Sphagnum, and thus alter the accumulation of carbon on boreal wetlands. Sphagnum mosses are a plant genus with a favorable ability to grow in low solar irradiance and temperature conditions compared to vascular plants. This may be increasingly beneficial in increased wintertime temperatures and predated snowmelt conditions. To understand particularly the importance of early spring photosynthetic activity and thus the role of the length of growing season on carbon balance, we analyzed the CO_2 exchange of Sphagnum mosses with closed chamber technique in two categories of microtopographical habitats, hummocks and lawns, during four seasons 2010-2013 on a raised bog in Central Finland. During CO_2 exchange measurements, instantaneous net ecosystem exchange (NEE) and ecosystem respiration (RE) were measured. Our results show that the mean measured seasonal NEE, i.e. the instantaneous net carbon sequestration, of hummocks was generally only slightly higher than the NEE of lawns, but the mean measured seasonal RE of hummocks was clearly and significantly higher than the RE of lawns in every study year. A reason for the observed still higher seasonal carbon sequestration of hummocks than that of lawns besides the slightly higher rate of carbon accumulation was the longer duration of physiologically active growing season. Therefore, hummock-forming Sphagnum mosses exposed firstly from snow cover showed to get the extra time for photosynthesis and thus extra benefit compared to other mire plants. This may be further enhanced by the expansion of hummock-forming Sphagnum moss dominated raised bogs towards northern aapa-mire region due to the global warming.

Keywords: Sphagnum Mosses, Boreal Wetlands, Mire Microtopography, Carbon Dynamics, Global Warming

1. INTRODUCTION

Sphagnum moss species are globally important owing to their considerable peat-forming ability and their potential impact on global climatic cycles [1] [2] . Sphagnum is the most important peat forming plant genus and there may be more carbon stored in Sphagnum than in any other genus [3] . The genus Sphagnum includes 250 - 450 species, but rather fewer than 50 species contribute to peat formation on a large scale [4] . These peat-producing species are most abundant on boreal wetlands above latitude 45N [5] . Thus, Sphagnum forms a major component of boreal wetlands, which cover 2% of the total land area [6] . These wetlands have acted as a long-term net carbon sink and take up 0.07 Pg carbon per year [1] [7] , equivalent to 5% of the total net carbon flux from atmosphere to land [6] . Sphagnum mosses solely have thus been responsible for the sequestration of large quantities of carbon and play an important role in the global carbon cycle [8] .

Changes in climatic conditions due to the global warming may affect the relations between Sphagnum mosses and vascular plants [9] . Changes in climatic conditions may also affect competition among Sphagnum, both through direct moisture changes and through the effects on Sphagnum density and growth rate [10] [11] . These may have important consequences for the long-term stability of the mire vegetation community [12] . On boreal wetlands, vascular plant performance is further controlled by the presence of Sphagnum, which often dominates the bottom layer and the peat deposits in which the plants grow [3] . Sphagnum mosses strongly reduce the availability of nutrients to vascular plants on nutrient-poor mires by efficiently accumulating nutrients from atmospheric deposition and reducing decay and mineralization rates [9] [13] - [15] . At the same time, however, they force vascular plants to keep pace with their progressively increasing height, in order not to become buried [9] [10] [16] . Therefore global warming may impact on the functioning of mire ecosystems via alterations in species composition, and thus alter also the accumulation of carbon on boreal wetlands [8] [12] [17] .

According to the six-year study of Aurela et al. [18] on a subarctic fen in northern Finland, the interannual variation of the CO_2 balances originates almost completely from the variations during the snow-free period. In their study, the snowmelt timing was proven to be the most important single determinant of the annual carbon balance. In contrast to a commonly-held view, the hydrometeorological conditions during the growing season had only a minor effect on the annual balance, emphasizing the importance of the start of the growing season [18] . Sphagnum mosses have a favorable ability to grow in low solar irradiance and temperature conditions compared to vascular plants [19] [20] . In the study of Harley et al. [19] , the photosynthesis of Sphagnum angustifolium saturated already in the photosynthetic photon flux density (PPFD) under 500 $\mu mol \cdot m^{-2} \cdot s^{-1}$. Photosynthesis started at ca. ±0°C, when photosynthetic rates at 5°C were already ca. 50% of maximum and the saturation temperature was at ca. 10°C [19] . These requirements would normally be met already in late March at least in southern and middle boreal region assuming that no snow cover occurs.

Boreal raised bogs show a typical pattern of microtopographical habitats, ranging from relatively dry but regularly (in spring after snowmelt) inundated lawns to dry high hummocks (without inundation period), with a different set of Sphagnum species occupying each microhabitat [21] . Therefore the decrease in snow cover in early spring might be more beneficial for hummock-forming Sphagnum species than for lawn species. This could enhance the expansion of hummock dominated raised bogs towards northern aapa-mire region due to the global warming.

We hypothesized therefore 1) that productivity of lawn species (e.g. S. fallax and S. balticum) will not increase with decreasing snow cover, whereas biomass production of hummock species (e.g. S. fuscum and S. magellanicum) will increase, and 2) that global climate warming with shortened snow cover period may thus promote rather than threaten the carbon sequestration ability of boreal wetlands by favoring the growth of hummock-forming Sphagnum.

2. MATERIALS AND METHODS

2.1. Study Site

The study was carried out during 2010-2013 on a pristine raised bog at Kauhaneva, central Finland (62°13'N, 22°26'E). The mire site type of the study site was low sedge S. papillosum fen with large Sphagnum hummocks [22] . The study site lies in the transitional zone between southern-boreal and middle-boreal coniferous forest zones, on the edge of a large raised bog area with the characteristics of both poor minerotrophy (lawns) and ombrotrophy (hummocks). The long term (1983-2013) annual mean temperature of the site was 3.8°C, the annual mean precipitation 612 mm and the accumulative temperature sum (>+5°C) 1156 degree-days.

The surface of the study site consisted of a typical pattern of boreal raised bogs: relatively dry but regularly inundated lawns to dry hummocks. The height of the studied hummocks was ca. 0.4 m compared to the surrounding lawn area. Soil of the site consisted of weakly decomposed Sphagnum-Carex peat with the depth of ca. 1 meter. Dominant mosses on the hummock-plots were S. fuscum and S. magellanicum, and on the lawn-plots S. balticum and S. fallax.

2.2. Vegetation Monitoring

A leaf area index (LAI) estimation was employed in order to relate CO_2 fluxes to the spatial and temporal variation in vegetation. For vascular plants, species-specific LAIs in each sample plot were estimated in early August during 2010-2013 as a product of the number of leaves and the average leaf size [23] . The number of green leaves of each species was estimated by counting the number of the leaves in permanent sub-sample plots within each sample plot that accounted for 10% of the total area of the plot.

2.3. CO_2 Exchange Measurements

Six CO_2 exchange measurement plots were established at the site. Three of the measurement plots were established on the hummock surfaces and the other three on the lawn surfaces. CO_2 exchange measurement occasions with closed chamber technique [24] were carried out during the whole snow-free period (March/April-October) 2010-2013. The timing of snowmelt was monitored at the site, and CO_2 exchange measurements were started in each spring immediately after the snow cover was melt at least from all of the hummock plots.

Prior to measurements, an aluminium collar (60 × 60 cm) was permanently inserted into 30 cm depth in each plot. The collar was surrounded by a water groove that allowed chamber placement and air-tight sealing of the measurement system. Instantaneous net ecosystem CO_2 exchange (NEE) in each plot was measured with a transparent plastic chamber and a portable infra-red gas analyzer (EGM-4, PP Systems, UK). Measurements over 90 - 180 s were carried out in the full light and under an artificial shade reducing the ambient light by 40% - 60%.

During the measurements, the CO_2 concentration in the chamber headspace, photo synthetically active radiation (PAR) under the chamber roof, and the chamber temperature were recorded at 15-s intervals. After the measurements in light, the chamber was covered with an opaque hood and the ecosystem respiration (RE) in the dark was measured. Water table level (WTL) in a perforated tube next to each plot, and peat temperatures at 5 cm below the moss surface were measured simultaneously with the CO_2 exchange measurements to relate the fluxes to the prevailing environmental conditions. NEE and RE were calculated as the linear change in CO_2 concentration as a function of time by fitting a linear regression line.

2.4. Statistical Analyses

Since we had a limited amount of data on the CO_2 exchange, we could not reliably model fluxes on which to base simulations of seasonal estimates of NEE and RE. Instead, we concentrated on testing the significance of the effect of microtopographic habitat on instantaneous NEE and RE, and interpolating the seasonal mean carbon balances for both microtopographical habitats over the growing seasons. Since observations were made on the same measurement plots over a period of time (four growing seasons 2010-2013), the observed responses were therefore correlated. As opposed to two-way ANOVA, linear mixed models have been designed to handle correlated data with unequal variances from normal distributions [25] .

Thus, the analyses of instantaneous flux rates were based on fixed effects models with restricted maximum likelihood estimation method using the linear mixed procedure in SPSS Statistics ver. 22 (IBM Corp., USA). Test results were considered significant when $p < 0.05$.

3. RESULTS

3.1. Environmental Conditions during the Study Period

The thermal growing season started in late April or early May and lasted until the end of October during study years 2010-2013 (Figure 1). The mean seasonal (April-October) temperatures at Kauhaneva were slightly higher in all study years 2010-2013 except 2012 than the long term average of 1984-2014 (10.5, 11.4, 9.4, 10.6 and 9.7°C, respectively) (Figure 1). However, the monthly temperatures of early season (April-July) in 2010 and 2011 were clearly higher than the long term average of 1984-2014 (Figure 1). Peat temperatures at 5 cm below the moss surface measured simultaneously with the CO_2 exchange measurements were typically close to 0°C during the first measurements after snowmelt and rose to their seasonal maximum in July at ca. +20°C (Figure 2). Peat temperatures in hummocks were generally slightly higher than in lawns (Figure 2). The highest photosynthetically active radiations (PAR) under the chamber roof measured simultaneously with the CO_2 exchange measurements varied from ca. 200 $\mu mol \cdot m^{-2} \cdot s^{-1}$ during late season in October to 1600 $\mu mol \cdot m^{-2} \cdot s^{-1}$ during peak season in July (Figure 3). Oppositely, measured PAR values were during early season in April only ca. 20% lower than during peak season in July (Figure 3).

The timing of snowmelt varied rather slightly during the study period, occurring early April in all study years except 2012. Although the season of 2012 was the coldest during our study period 2010-2013, the snowmelt was, however, the earliest taking place in the middle of March in 2012. Short-term (ca. one month) inundation occurred immediately after snowmelt especially in spring 2010 on the lawn plots (Figure 2). The snow and/or inundation free period lasted for seven months covering the whole growing season for the hummocks, whereas only ca. six months for the lawns having the inundation after the snowmelt and delayed snowmelt compared to hummocks. In 2012, the difference between hummocks and lawns in the snow and/or inundation free period was almost one and a half months.

The cumulative precipitations during April-October were rather close to the long-term average of 1984-2014 (408 mm) in all study years 2010-2013 except in 2012 (678 mm) (Figure 1). Especially the monthly precipitations in July and October 2012 were clearly higher than the monthly long term averages of 1984-2014 (Figure 1). Water table levels (WTL) measured in a perforated tube next to each CO_2 exchange measurement plot varied from ca. −10 cm to −40 cm below the soil surface for hummocks, and from ca. +10 cm above the soil surface (inundation) in spring to −15 cm below the soil surface for lawns (Figure 2). In early spring also solid soil frost occurred during the first CO_2exchange measurement occasions in 2011-2013, and therefore WTLs cannot be measured (Figure 2).

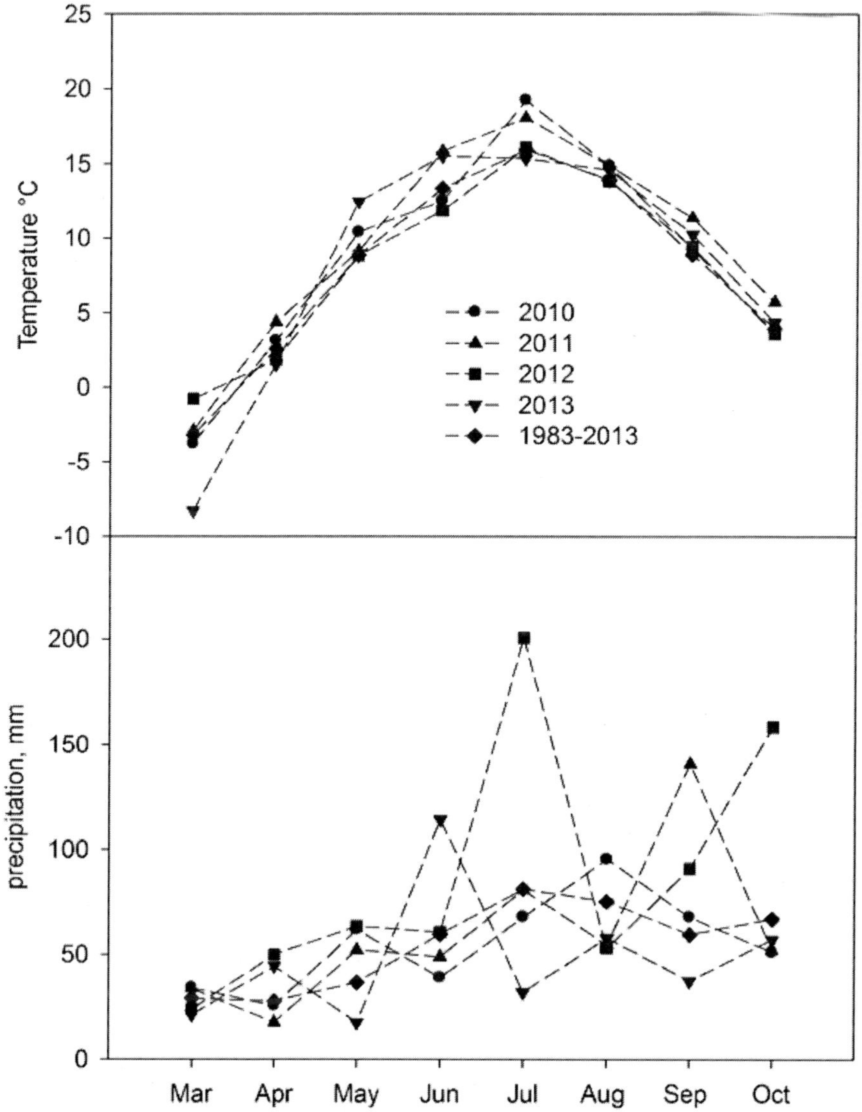

Figure 1. Mean air temperature and monthly precipitation at Kauhaneva from March to October during 2010-2013 and as 30-year average 1983-2013.

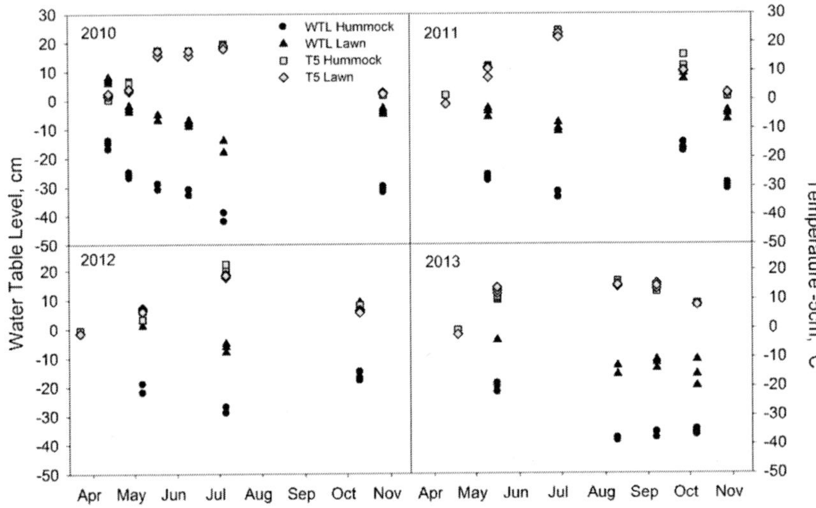

Figure 2. Measured water table level (WTL) and soil temperature (T5) values at 5 cm depth below soil surface at Kauhaneva during 2010-2013. WTL and T5 values based on manual measurements made in conjunction with CO_2 exchange measurements. Three WTL measurements from March-April 2011-2013 are missing because of the solid soil frost.

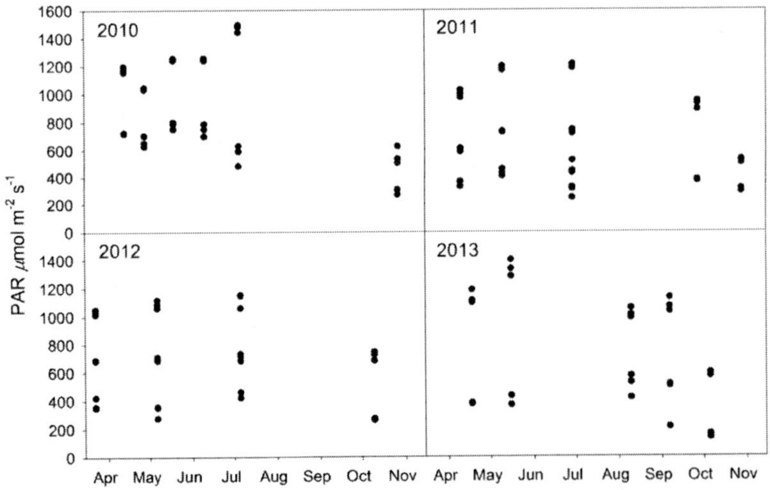

Figure 3. Measured photosynthetically active radiations (PAR) at Kauhaneva during 2010-2013. PAR values are based on manual measurements made in conjunction with CO_2 exchange measurements. Part of the measurements was done under an artificial shade reducing the ambient light by 40% - 60% resulting the rather large variation in PARs.

The maximum leaf area index (LAI) of vascular plants was generally very low at Kauhaneva, in average ca. 0.14 $m^2 \cdot m^{-2}$ for hummocks and ca. 0.17 $m^2 \cdot m^{-2}$ for lawns, respectively. The variation in LAIs was very small between the study years and the plots. Dominant vascular plants on the hummocks were Empetrum nigrum, Andromeda polifolia and Rubus chamaemorus, on the lawns Eriophorum vaginatum, Andromeda polifolia and Vaccinium oxycoccos, respectively.

3.2. CO_2 Exchange

The mean measured seasonal NEE of hummocks during 2010-2013 was 329, 218, 201 and 162 mg CO_2 $m^{-2} \cdot h^{-1}$, respectively, and for lawns 220, 182, 240 and 156 mg CO_2 $m^{-2} \cdot h^{-1}$, respectively (Figure 4). All mean seasonal NEE values showed a positive seasonal net balance, i.e. all measured CO_2 exchange plots acted as a sink for CO_2 during 2010-2013, but both microtopographical habitats showed relatively large differences in NEE between the years (Figure 4). The mean measured seasonal NEE of hummocks was higher than the NEE of lawns in every year during 2010-2013, except in 2012 (Figure 4). However, the difference between the NEE of hummocks and lawns was statistically significant only in 2010 (Table 1). Assuming that the growing season at Kauhaneva over the four-year study period 2010-2013 is considered to last ca. seven months for hummocks and ca. six months for lawns, the seasonal carbon balance was 310 g CO_2-C m^{-2} for hummocks and 233 g CO_2-C m^{-2} for lawns, respectively.

The mean measured seasonal RE in the dark for hummocks during 2010-2013 was 723, 537, 402 and 566 mg CO_2 $m^{-2} \cdot h^{-1}$, respectively, and for lawns 417, 259, 282 and 270 mg CO_2 $m^{-2} \cdot h^{-1}$, respectively (Figure 4). The mean measured seasonal RE of hummocks was clearly higher than the RE of lawns in every year during 2010-2013 (Figure 4). The difference between the RE of hummocks and lawns was significant in every year during 2010-2013 (Table 1).

4. DISCUSSION

In our four-year study the inter-annual variability in climatic conditions was considerable and partly unexpected as reported in other multi-year studies (e.g. [26]) as well. For instance, during the study period there was a very warm growing season in 2010, but a cold season in 2012 too. Additionally, although the season of 2012 was the coldest during our study period 2010-2013, the snowmelt was, however, the earliest occurring already in the middle of March.

In general, instantaneous NEE and RE in our study roughly corresponded to those measured in other studies on pristine mires [26] - [29] . Likewise to the above studies, annual variation in NEE and RE results was considerable in our study, highlighting the importance of multi-year empirical studies. The mean measured seasonal NEE, i.e. the net carbon sequestration, of hummocks during 2010-2013 was generally slightly higher than the NEE of lawns, but the differences between the NEE of microtopographical habitats were rather small and not significant, except in 2010. However, the mean measured seasonal RE of hummocks was clearly higher than the RE of lawns in every year during 2010-2013, and the difference between the RE of hummocks and lawns was significant in every year during 2010-2013. A reason for the more significant differences in RE than NEE, may be the higher sensitivity of RE on alterations in WTL, in particular during early spring and late autumn when the differences in WTL between hummocks and lawns were at greatest. Also Riutta et al. [26] and Ballantyne et al. [30] reported the significantly increased RE with lowered WTL on boreal poor fens.

The other important reason for the observed higher seasonal carbon sequestration of hummocks than that of lawns besides the slightly higher rate of carbon accumulation was the longer snow- and inundation-free period on hummocks, i.e. the duration of physiologically active growing season. There is strong evidence that the interannual variation of the carbon sequestration on boreal mires originates from the variations in the snow-free period [18] or from the variations in the inundation [21] . Our study shows that longer snow-free period due to early snowmelt enhances more the carbon sequestration of microtopographically higher habitats (hummocks) than the surroundings (lawns). This is mainly because of snow cover and too high WTL (inundation) for the effective carbon sequestration in early spring on the microtopographically lower lawns.

In our study, the photosynthesis of mire vegetation did not occur under snow cover or under inundation, although the photosynthetic activity of boreal dwarf shrubs under snow cover has shown to take place [31] [32] . At Kauhaneva, the LAI of vascular plants was very low, and the major part of photosynthetic activity was due to Sphagnum mosses that are not likely capable for photosynthesis under snow. However, Sphagnum mosses have the ability to grow under lower solar irradiance and temperature conditions than the vascular plants [19] [20] . Accordingly, Sphagnum mosses grow in low temperature conditions immediately after snowmelt and/or after the removal of inundation in early spring, or in low solar irradiance conditions in late autumn. Since the increase of greenhouse gases in the atmosphere is expected to instigate warming in high northern latitudes especially during winter and early spring [33] - [35] , the snowmelt will take place earlier in the future, and this will probably prolong the effective growing season especially during early spring in rather high solar irradiance conditions [36] [37] . The ability of plants to benefit from intermittent warm periods during winter and early spring, without succumbing to freezing damage during subsequent periods of frost, depends on a delicate balance between activity and protection, which may easily tip over in conditions where the annual winter- and springtime temperatures are rising [31] [32] .

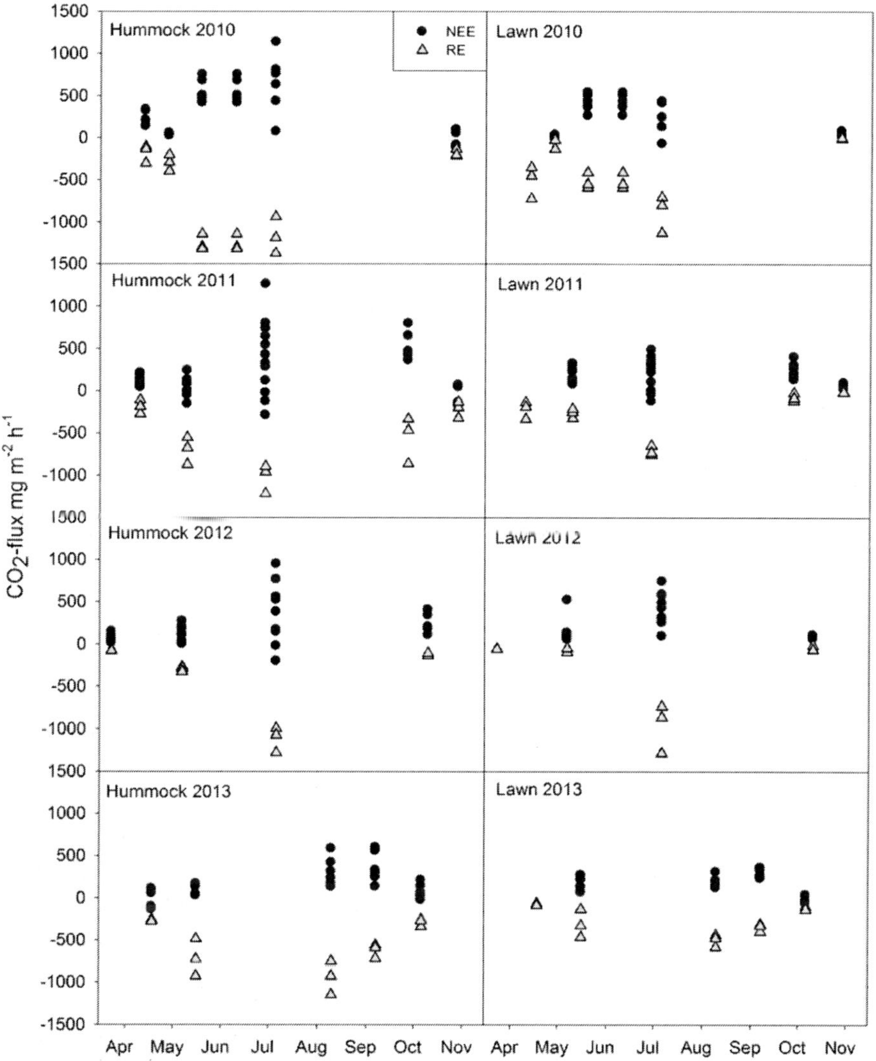

Figure 4. Measured instantaneous CO_2 exchange (NEE and RE) of the microtopographical habitats (hummocks and lawns) during 2010-2013.

Plants having a capability for the effective winter- and springtime photosynthesis, like Sphagnum mosses, may significantly benefit from increased wintertime temperatures and earlier snowmelt.

Table 1. The significances (F- and their p-values) of differences between the NEE/RE of the hummocks and lawns (microform effect) during 2010-2013. Test results were considered significant when p < 0.05, and marked by the asterisk.

Microform effect	2010	2011	2012	2013
NEE	$F = 5.24; p < 0.05$*	$F = 0.42; p = 0.522$	$F = 1.88; p = 0.177$	$F = 1.70; p = 0.199$
RE	$F = 9.91; p < 0.05$*	$F = 18.28; p < 0.05$*	$F = 10.62; p < 0.05$*	$F = 38.39; p < 0.05$*

Sphagnum mosses are known as a mire plant group that is the most sensitive to environmental conditions, especially to soil moisture, as a result of an optimum WTL close to the surface and narrowest tolerance to WTL variation [26] [38] [39] . Furthermore, unfavorable environmental conditions may diminish directly the amount of phytomass, i.e. in terms of the abundance and coverage, of a certain Sphagnum group [26] .

Also our study demonstrated that microtopographically scattered, functional plant groups on a bog differed in their responses to snow cover and WTL and, consequently, changes in snow cover and WTL may modify their contribution to the ecosystem CO_2 dynamics in the future. In particular, the hummock-forming Sphagnum mosses that exposed firstly from snow cover showed to gain additional time for photosynthesis and thus further benefit compared to other mire plants.

This may further enhance the expansion of hummock-forming Sphagnum moss dominating the raised bogs towards northern aapa-mire region due to the global warming. Since Sphagnum is one of the main peat forming and carbon accumulating plant genus globally, the large-scale shifting from lawn-forming Sphagnum mosses towards hummock-forming Sphagnum mosses in northern boreal region may play a significant role in the global carbon balance.

The results of this study are therefore of general interest for assessing the annual photosynthetic cycle of this abundant mire plant genus throughout the boreal zone, emphasizing the role of early growing season.

REFERENCES

1. Gorham, E. (1991) Northern Peatlands: Role in the Carbon Cycle and Probable Responses to Climatic Warming. Ecological Applications, 1, 182-195. http://dx.doi.org/10.2307/1941811
2. Franzen, L.G. (1994) Are Wetlands the Key to the Ice-Age Cycle Enigma? Ambio, 23, 300-308.
3. Clymo, R.S. and Hayward, P.M. (1982) The Ecology of Sphagnum. In: Smith, A.J.E., Ed., Bryophyte Ecology, Chapman & Hall, New York, 229-289. http://dx.doi.org/10.1007/978-94-009-5891-3_8
4. Shaw, A.J. (2000) Phylogeny of the Sphagnopsida Based on Chloroplast and Nuclear DNA Sequences. Bryologist, 103, 277-306. http://dx.doi.org/10.1639/0007-2745(2000)103[0277:POTSBO]2.0.CO;2
5. Gunnarsson, U. (2005) Global Patterns of Sphagnum Productivity. Journal of Bryology, 27, 269-279. http://dx.doi.org/10.1179/174328205X70029
6. Houghton, J.T., Ding, Y. and Griggs, D.J. (2001) Climate Change 2001: The Scientific Basis Contribution of Working Group I to the Third Assessment Report of the Intergovernmental Panel on Climate Change. Cambridge University Press, Cambridge.
7. Clymo, R.S., Turunen, J. and Tolonen, K. (1998) Carbon Accumulation in Peatland. Oikos, 81, 368-388. http://dx.doi.org/10.2307/3547057
8. Dorrepaal, E., Aerts, R., Cornelissen, J.H.C., Callaghan, T.V. and van Logtestijn, R.S.P. (2003) Summer Warming and Increased Winter Snow Cover Affect Sphagnum fuscum Growth, Structure and Production in a Sub-Arctic Bog. Global Change Biology, 10, 93-104. http://dx.doi.org/10.1111/j.1365-2486.2003.00718.x
9. Malmer, N., Svensson, B.M. and Wallen, B. (1994) Interactions between Sphagnum Mosses and Field Layer Vascular Plants in the Development of Peat-Forming Systems. Folia Geobotanica & Phytotaxonomica, 29, 483-496. http://dx.doi.org/10.1007/BF02883146
10. Rydin, H. (1997) Competition among Bryophytes. Advances in Bryology, 6, 135-168.
11. Sonesson, M., Carlsson, B.A., Callaghan, T.V., Halling, S., Bjorn, L.O., Bertgren, M. and Johanson, U. (2002) Growth of Two Peat-Forming Mosses in Subarctic Mires: Species Interactions and Effects of Simulated Climate Change. Oikos, 99, 151-160. http://dx.doi.org/10.1034/j.1600-0706.2002.990115.x
12. Dorrepaal, E., Aerts, R., Cornelissen, J.H.C., van Logtestijn, R.S.P. and Callaghan, T.V. (2006) Sphagnum Modifies Climate-Change Impacts on Subarctic Vascular Bog Plants. Functional Ecology, 20, 31-41. http://dx.doi.org/10.1111/j.1365-2435.2006.01076.x
13. Lee, J.A. and Woodin, S.J. (1988) Vegetation Structure and the Interception of Acid Deposition by Ombrotrophic Mires. In: Verhoeven, J.T.A., Heil, G.W. and Werger, M.J.A., Eds., Vegetation Structure in Relation to Carbon and Nutrient Economy, Academic Publishing, The Hague, 137-147.
14. Van Breemen, N. (1995) How Sphagnum Bogs Down Other Plants. Trends in Ecology and Evolution, 10, 270-275. http://dx.doi.org/10.1016/0169-5347(95)90007-1
15. Li, Y.H. and Vitt, D.H. (1997) Patterns of Retention and Utilization of Aerially Deposited Nitrogen in Boreal Peatlands. Ecoscience, 4, 106-116.
16. Backéus, I. (1985) Aboveground Production and Growth Dynamics of Vascular Bog Plants in Central Sweden. Acta Phytogeographica Suecica, 74, 1-98.
17. Moore, P.D. (2002) The Future of Cool Temperate Bogs. Environmental Conservation, 29, 3-20. http://dx.doi.org/10.1017/s0376892902000024
18. Aurela, M., Laurila, T. and Tuovinen, J.-P. (2004) The Timing of Snowmelt Controls the Annual CO2 Balance in a Subarctic Fen. Geophysical Research Letters, 31, Article ID: L16119. http://dx.doi.org/10.1029/2004GL020315
19. Harley, P.C., Tenhunen, J.D., Murray, K.J. and Beyers, J. (1989) Irradiance and Temperature Effects on Photosynthesis of Tussock Tundra Sphagnum Mosses from the Foothills of the Philip Smith Mountains, Alaska. Oecologia, 79, 251-259. http://dx.doi.org/ 10.1007/BF00388485

20. Balagurova, N., Drosdov, S. and Grabovik, S. (1996) Cold and Heat Resistance of Five Species of Sphagnum. Annales Botanici Fennici, 33, 33-37.
21. Robroek, B.J.M., Limpens, J., Breeuwer, A. and Schouten, M.G.C. (2007) Effects of Water Level and Temperature on Performance of Four Sphagnum Mosses. Plant Ecology, 190, 97-107. http://dx.doi.org/10.1007/s11258-006-9193-5
22. Laine, J., Vasander, H., Hotanen, J.-P., Nousiainen, H., Saarinen, M. and Penttila, T. (2012) Suotyypit ja turvekankaat —Opas kasvupaikkojen tunnistamiseen. Metsakustannus, 160 p. (In Finnish)
23. Wilson, D., Alm, J., Riutta, T., Laine, J., Byrne, K.A., Farrel, E.P. and Tuittila, E.-S. (2007) A High Resolution Green Area Index for Modelling the Seasonal Dynamics of CO2 Exchange in Peatland. Plant Ecology, 190, 37-51. http://dx.doi.org/10.1007/s11258-006-9189-1
24. Alm, J., Shurpali, N.J., Tuittila, E.-S., Laurila, T., Maljanen, M., Saarnio, S. and Minkkinen, K. (2007) Methods for Determining Emission Factors for the Use of Peat and Peatlands—Flux Measurements and Modeling. Boreal Environment Research, 12, 85-100.
25. Pan, W. and Connett, J.E. (2002) Selecting the Working Correlation Structure in Generalized Estimating Equations with Application to the Lung Health Study. Statistica Sinica, 12, 475-490.
26. Riutta, T., Laine, J. and Tuittila, E.-S. (2007) Sensitivity of CO2 Exchange of Fen Ecosystem Components to Water Level Variation. Ecosystems, 10, 718-733. http://dx.doi.org/10.1007/s10021-007-9046-7
27. Frolking, S., Roulet, N., Moore, T., Lafleur, P., Bubier, J. and Crill, P. (2002) Modeling Seasonal to Annual Carbon Balance of Mer Bleue Bog, Ontario, Canada. Global Biogeochemical Cycles, 16, 1-21. http://dx.doi.org/10.1029/2001GB001457
28. Aurela, M., Lohila, A., Tuovinen, J.-P., Hatakka, J., Riutta, T. and Laurila, T. (2009) Carbon Dioxide Exchange on a Northern Boreal Fen. Boreal Environment Research, 14, 699-710.
29. Chivers, M.R., Turetsky, M.R., Waddington, J.M., Harden, J.W. and McGuire, A.D. (2009) Effects of Experimental Water Table and Temperature Manipulations on Ecosystem CO2 Fluxes in an Alaskan Rich Fen. Ecosystems, 12, 1329-1342.
30. Ballantyne, D.M., Hribljan, J.A., Pypker, T.B. and Chimner, R.A. (2013) Long-Term Water Table Manipulations Alter Peatland Gaseous Carbon Fluxes in Northern Michican. Wetlands Ecology and Management, 22, 35-47. http://dx.doi.org/10.1007/s11273-013-9320-8
31. Starr, G. and Oberbauer, S.F. (2003) Photosynthesis of Arctic Evergreens under Snow: Implications for Tundra Ecosystem Carbon Balance. Ecology, 84, 1415-1420. http://dx.doi.org/10.1890/02-3154
32. Lundell, R., Saarinen, T., Astrom, H. and Hanninen, H. (2008) The Boreal Dwarf Shrub Vaccinium vitis-idaea Retains Its Capacity for Photosynthesis through the Winter. Botany, 86, 491-500. http://dx.doi.org/10.1139/B08-022
33. Kattenberg, A., Giorgi, F., Grassl, H., Meehl, G.A., Mitchell, J.F.B., Stouffer, R.J., Tokioka, T., Weaver, A.J. and Wigley, T.M.L. (1996) Climate Models—Projections of Future Climate. In: Houghton, J.T., Filho, L.G.M., Callander, B.A., Harris, N., Kattenberg, A. and Maskell, K., Eds., Climate Change 1995: The Science of Climate Change, Cambridge University Press, Cambridge, 285-357.
34. IPCC (2007) Climate Change 2007. Fourth Assessment Report. Intergovernmental Panel on Climate Change, Cambridge University Press, Cambridge.
35. Moss, R.H., Edmonds, J.A., Hibbard, K.A., Manning, M.R., Rose, S.K., van Vuuren, D.P., Carter, T.R., Emori, S., Kainuma, M., Kram, T., Meehl, G.A., Mitchell, J.F.B., Nakicenovic, N., Riahi, K., Smith, S.J., Stouffer, R.J., Thomson, A.M., Weyant, J.P. and Wilbanks, T.J. (2010) The Next Generation of Scenarios for Climate Change Research and Assessment. Nature, 463, 747-756. http://dx.doi.org/10.1038/nature08823
36. Walther, G.-R., Post, E., Convey, P., Menzel, A., Parmesan, C., Beebee, T.C.J., Fromentin, J.-M., Hoegh-Guldberg, O. and Bairlein, F. (2002) Ecological Responses to Recent Climate Change. Nature, 416, 389-395. http://dx.doi.org/10.1038/416389a
37. Raisanen, J., Hansson, U., Ullerstig, A., Doscher, R., Graham, L.P., Jones, C., Meier, M., Samuelsson, P. and Willen, U. (2003) GCM Driven Simulations of Recent and Future

Climate with the Rossby Centre Coupled Atmosphere— Baltic Sea Regional Climate Model RCAO. Reports Meteorology and Climatology, SMHI, Norrkoping, 1-61.

38. Silvola, J. and Aaltonen, H. (1984) Water Content and Photosynthesis in the Peat Mosses Sphagnum fuscum and Sphagnum angustifolium. Annales Botanici Fennici, 21, 1-6.

39. Schipperges, B. and Rydin, H. (1998) Response of Photosynthesis of Sphagnum Species from Contrasting Microhabitats to Tissue Water Content and Repeated Desiccation. New Phytologist, 140, 677-684. http://dx.doi.org/10.1046/j.1469-8137.1998.00311.x

CHAPTER 12

Design and Hydrologic Performance of a Tile Drainage Treatment Wetland in Minnesota, USA

Christian Lenhart [1], Brad Gordon [2], Joshua Gamble [3], Dean Current[4], Nikol Ross [5], Lydia Herring [6], John Nieber [7] and Heidi Peterson [5]

[1]*Ecological Engineering Group, Department of Bioproducts and Biosystems Engineering, University of Minnesota, 303 BAE Hall, 1390 Eckles Ave., St. Paul, Minneapolis, MN 55108, USA*

[2]*Water Resources Science Program, University of Minnesota, 311 BAE Hall, 1390 Eckles Ave., St. Paul, Minneapolis, MN 55108, USA*

[3]*USDA-ARS Soil and Water Management Research Unit, 439 Borlaug Hall, 1991 Upper Buford Circle, St. Paul, Minneapolis, MN 55108, USA*

[4]*Center for Integrated Natural Resources Management (CINRAM), Department of Forest Resources, 100 Green Hall, St. Paul, Minneapolis, MN 55108, USA*

[5]*Minnesota Department of Agriculture, 625 Robert Street North, St. Paul, Minneapolis, MN 55155, USA*

[6]*Tetra Tech, Inc., 661 Andersen Drive, Pittsburgh, PA 15220, USA*

[7]*Ecological Engineering Group, Department of Bioproducts and Biosystems Engineering, University of Minnesota, 203 BAE Hall, 1390 Eckles Ave., St. Paul, Minneapolis, MN 55108, USA*

ABSTRACT

Treatment wetlands are increasingly needed to remove nitrate from agricultural drainage water to protect downstream waters, such as the Gulf of Mexico. This project sought to develop a new edge-of-field treatment wetland, designed to remove nitrate-nitrogen and enhance phosphorus removal by plant harvest and to monitor its effectiveness. A 0.10 ha wetland was designed and installed to treat subsurface drainage flow from farmland in southwestern Minnesota, USA, in 2013, and monitored for three years by recording flow, nitrate-nitrogen, total phosphorus (TP) and soluble orthophosphorus (OP) input to and output from the wetland. Prior to construction, a level-pool routing, mass balance approach with DRAINMOD flow inputs was used to predict nitrate removal efficiency. Nitrate load removal averaged 68% over three years, nearly matching model predictions. However, most denitrification occurred in the sub-soil of the wetland rather than in surface flow as predicted. Phosphorus removal was approximately 76% over three years, and phosphorus removed by plant uptake exceeded inflow mass in the third year. The edge-of-field design has potential as a cost-effective method to treat field outflows because agricultural landowners can adopt this treatment system with minimal loss of productive farmland. The wet-prairie vegetation and shallow depth also provide the opportunity to remove additional phosphorus via vegetative harvest.

Keywords: treatment wetland; nitrate; subsurface drainage water; plant harvest

1. INTRODUCTION

Although it is widely recognized that wetlands provide nutrient removal and water storage benefits to downstream waters [1], many of the native wetlands in the Midwestern U.S. have been drained for agriculture and development, reducing those functions in the present-day landscape. Prior to the adoption of the Wetland Conservation Act in 1992, over half of the original wetlands in southern Minnesota were drained through the installation of agricultural tile lines and surface ditches. During this same time, Minnesota's first Conservation Reserve Enhancement Program (CREP) initiated, with over 50% of the enrolled land area (>20,000 ha) dedicated to wetland restorations in southern Minnesota. As the program evolved, 10–15 year contracts were granted rather than permanent easements, which lead to increased landowner adoption. As these contracts began to expire between 2007 and 2010, much of the land area was not renewed due to increased commodity prices. Since 2007, Minnesota has sustained the largest loss of emergent wetland area in the Midwest [2]. The reduced water holding and filtration capacity has resulted in water quality issues such that nitrate pollution in the Midwest has been implicated as a major contributor to the Gulf of Mexico hypoxia [3,4]. Recent flood events have also been connected to the reduced landscape water holding capacity resulting from the declining wetland area [5,6,7].

To remediate these changes and specifically address water quality concerns, additional restored or created wetlands are needed in the Midwestern agricultural landscape. Past studies have shown Midwestern wetlands are effective at removing nitrates, however, phosphorus removal efficiency is variable depending on whether the load is attributed to the dissolved or particulate fraction [8]. Past studies of water quality treatment wetlands indicate that large restored Midwestern wetlands are highly effective at peak flow reduction, water storage, and nitrate removal but to a lesser extent for phosphorus removal [9,10]. Nitrate-nitrogen (NO_3-N) export was reduced by 85% during the primary loading period of April to June. Particulate phosphorus and sediment from inflowing field and gully erosion were effectively settled out in the wetlands, however, soluble orthophosphorus (OP) export, (reaching a maximum concentrations of 1.3 mg/L) and wind suspension of organic matter from the wetlands, resulted in no significant removal of total phosphorus (TP) and total suspended solids. Although the wetlands were a net remover of TP, high levels of soluble OP were exported out of the wetland in mid- to late-summer reaching levels of 1.3 mg/L.

Although nitrate has been the focus of Midwestern water quality treatment wetlands, phosphorus is increasingly being considered as well. Research has indicated that TP and soluble OP as a measured load may be less than 2% of the applied phosphorus, however, when compared to the amount responsible for harmful algal blooms, the losses cannot go unaddressed. Long-term phosphorus removal depends upon plant uptake and physical settling [11]. Designing a treatment wetland to enable plant harvesting is one example of how phosphorus could be permanently removed from the system to prevent re-mobilization. Native perennial species that are adapted to wet soil conditions, or along the fringe of wetlands, have the capacity to aid in removing excess phosphorus from wetlands and thus provide an opportunity for addressing phosphorus in drainage water and wetlands used to treat those waters.

While many wetlands have been restored for ecological reasons, there is a growing need to develop wetlands to specifically capture subsurface drainage flow [12]. Subsurface drainage systems short-circuit riparian buffers or grass filters increasing NO_3-N loadings to surface waters and reduce the opportunity for denitrification to occur in the soil of the riparian buffer [13,14]. To address this nutrient loss, tile-drainage systems can be routed to discharge into a constructed treatment wetland for treatment prior to discharging into drainage ditches or streams.

Similar to natural wetlands, constructed wetlands can provide many of the benefits of restored wetlands, including nutrient removal, flood storage, the decoupling of storm rainfall and surface runoff, and cycling of nutrients [15]. Habitat for fish and wildlife, recreation and aesthetic benefits may be provided but often to a lesser extent than natural wetlands.

Given the increasing value of farmland in the Midwest, there is a pressing need for wetland systems that are compatible with current agricultural systems to improve landowner adoption. Innovative designs, strategies, and management must be explored in order to maximize the nutrient reduction efficiency of these treatment wetlands while minimizing the construction footprint on land that would need to be taken out of agricultural production. One way to accomplish this goal is by creating edge-of-field systems. These designs include fitting wetlands into riparian zones and land adjacent to drainage ditches to minimize the land area and associated costs of easements or land purchase. Crumpton et al. [8] developed a treatment wetland design that is built in-line with public drainage ditches and was highly effective at removing nitrate. The disadvantages were high initial costs for land easements, the legal and social difficulties of modifying public ditch systems, and phosphorus removal was not considered. In 2012, a treatment wetland was proposed on private farmland in southern Minnesota to test some of the ideas discussed above.

1.1 Study Purpose

The project was developed to design, construct and test a small-scale, treatment wetland system compatible with tile-drained row crop agriculture in southern Minnesota, USA. The location was cited at the edge of an agricultural field, to minimize the land area removed from agricultural production, and to adequately receive tile water prior to creek outflow. This is referred to as "edge-of-field" treatment. The project was intended to serve as a research and demonstration site for the wetland and other innovative farm management practices on the site for landowners and government agency staff.

From a research perspective, the primary objectives were to measure the flow of water into and out of the wetland and determine the nitrogen and phosphorus removal efficiency in the treatment wetland. A secondary objective was to quantify phosphorus uptake by plants in the wetland system.

2. MATERIALS AND METHODS

2.1. Study Area and Site Description

The research site is located within the Des Moines Lobe glacial till plain, which covers much of southern Minnesota and northern Iowa of the United States. These glacial till deposits are fine-textured fertile soils on a flat to gently rolling glacial till plain; when adequately drained this area becomes productive farmland. The region historically supported abundant pothole wetlands [16] as well as wet and mesic prairie. As the land was converted to agricultural uses, most of the wetlands were drained via ditches and sub-surface drainage pipes. Today the area is predominantly in corn (*Zea mays* L.) and soybean (*Glycine max* [L.] Merr.) with 80% of the Elm Creek basin in row crops.

The soil types found in the vicinity of the wetland include Clarion-Storden loam on 6%–12% slopes on adjacent uplands. Coland clay loam and Spillville loam cover much of the floodplain at the study site and along the length of Elm Creek. Coland is a hydric soil and would naturally support wetland vegetation and hydrology. It has approximately one meter of clay loam on top of a sandy layer underneath. In terms of water supply, the area receives an annual average of 81 cm/year of precipitation. The wetland itself lies within the floodplain of Elm Creek, a tributary of the Blue Earth River in the upper Mississippi basin (Figure 1). The wetland is exposed to floods from Elm Creek occasionally, calculated to be every 2–5 years. Along with surface water inputs, there is evidence of groundwater flow in sand layers underlying the heavy loams on top of the floodplain at a depth of 2–3 m.

The owner of the project site typically grows row crops in the 10.1 ha drainage area contributing to the wetland. Planning for the wetland construction began in 2012, with construction in the winter and spring of 2013. The wetland project was completed in conjunction with a new sub-surface drainage system and a bioreactor installed near the wetland for additional drainage water treatment, particularly for nitrate.

2.2. Wetland Design

The wetland was designed as a surface wetland system to capture and treat nitrogen and phosphorus in subsurface tile drainage flow from 10 ha. It was intended to be a model for small edge-of-field wetland systems that are compatible with current cropping systems by limiting the land area used. The design was based on guidance described by Kadlec and Wallace [17] and current conservation practice standards for

Constructed Wetlands from the Natural Resource Conservation Service (NRCS), specifically the CP39 design guidance [18]. In addition, guidance for wetland-to-watershed ratio (1%–2% for optimal benefit) developed in Iowa was used to size the treatment wetland. Given its small size of 0.10 ha, retention time was maximized by using a sinuous path through the wetland to prevent short-circuiting of flow water. The wetland was broken up into three equal-area cells to extend the flow length and to facilitate comparisons of different plant treatments.

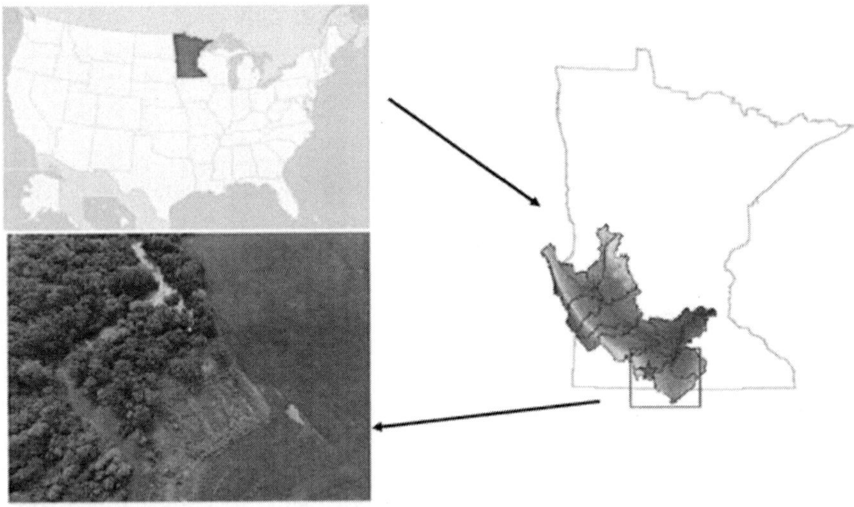

Figure 1. Project location showing, upper left: location of Minnesota within the United States, right: location of the Minnesota River basin and treatment wetland site within Minnesota, and lower left showing location of the wetland in relationship to farmland and Elm Creek. Lower left photo by David Hansen.

The wetland was designed to receive only subsurface tile drainage water to eliminate problems with sedimentation, erosion and/or flooding problems in the treatment cell area. Berms were constructed around the entire wetland to prevent inflow of surface water runoff and gully flow into the wetland.

2.3. Wetland Modeling

The flow through the wetland was modeled assuming the wetland water surface remains as a level pool, thus allowing the classical level-pool routing method to be used to route flows through the wetland. The transport of chemicals through the wetland was modeled assuming that the wetland behaves as a completely mixed reactor [19]. Since the outflow from the wetland was controlled by a weir, it was assumed that the unique stage–discharge relation represented by that weir could be used in the flow routing procedure.

The water balance of the wetland (modeled as a level pool) may be written in the following form [17]:

$$\frac{dV}{dt} = Q_{in} + (P - ET - I) * A - Q_{out} \tag{1}$$

where, V = water volume stored in wetland (m^3); A = surface area of wetland (m^2); P = precipitation (m/day); Q_{in} = inflow (m^3/day); Q_{out} = outflow (m^3/day); Q_c = runoff from surrounding area (m^3/day); ET = evaporation/transpiration (m/day); R = irrigation addition (m/day); I = groundwater loss or gain to wetland (m/day).

For the completely mixed reactor assumption, the transport of a chemical (e.g., nitrate) through the wetland can be expressed by the mass balance equation,

$$\frac{d(VC)}{dt} = Q_{in}C_{in} - Q_{out}C - IAC - kAC \tag{2}$$

where, C_{in} = concentration of chemical entering the wetland (mg/L); C = concentration of chemical in the wetland (mg/L); k = areal removal rate of the chemical (m/day).

Equations (1) and (2) are solved by a finite difference scheme. An essential input for the water balance equation is the relation between water stage and discharge. This was assumed to be given by the weir at the outlet of the wetland, and was assumed to be adequately represented by the equation for a rectangular weir, Equation (3). The parameter (b1 = 9.95) was determined using Equation (1). The weir width was assumed to be 2.5 m for the model and the discharge coefficient was assumed to be 3.98, a typical value for a rectangular broad crested weir. In comparison, the as-built outlet weir width placed in the constructed treatment wetland was 20 cm in width.

$$Q_o = \underbrace{C_E * W_w}_{\text{b1}}(H_o - H_w)^{1.5} \tag{3}$$

where Q_o = outflow rate (m^3/day); C_E = weir discharge coefficient (m^3/day); W_w = width of weir (m); H_o = water surface elevation at wetland outlet (m); H_w = weir crest elevation (m).

All of the parameters used to calculate wetland discharge from the water control structure and nutrient removal efficiency are listed in Table 1.

Table 1. Level-pool routing, mass balance routine input parameters and descriptors *.

Parameter	Symbol	Value	Units
Head increment	dh	0.1	m
$Q = b1 \times h^{1.5}$	b1	9.95 **	1
Weir crest	hc	0.25	m
Wetland porosity	por	0.7	1
Time step	dt	1	days
Wetland length (per cell)	L	36.6	m
Wetland width (per cell)	W	9.14	m
Wetland side slopes	SS	3	1
Concentration–discharge (Q–C) relation		0.44	mg/L/m^3/day
Kinetic reaction	k	0.04	m/day
# cells		1	1
Wetland area per cell		334.45	m^2
Total wetland area		1012	m^2

Notes: * Assuming no bank loss; ** Value of 9.95 obtained from table of weir equations [21].

Many of the input variables for the level pool routing, mass balance model were obtained through the DRAINMOD model [20]. The others are listed in Table 1.

- V = found with mass balance routine;
- A = treatment surface area of individual cell at each depth;
- P = DRAINMOD precipitation values applied to wetland cell area only;
- Q_{in} = Subsurface Drainage flow from DRAINMOD;
- Q_{out} = outflow in cell 1 is inflow to cell 2, outflow in cell 2 is inflow to cell 3;
- Q_c = 0 (surface water is directed elsewhere for this application);
- ET = DRAINMOD values for ET applied to wetland cell area only;
- R = 0 (no irrigation);
- I = assuming small leakages, with unsaturated conditions beneath the wetland;

- *DP* = vertical seepage from DRAINMOD applied to wetland cell area only.

2.4. Nutrient Removal Predictions

Using the level pool routing, mass balance routine to model nitrate reduction, successive nutrient removal performance of the wetland cells can be predicted. Input flow-concentrations (Q–C) were based on a value of 0.44 mg/L/ m^3/day prior to wetland monitoring. This estimate was found by dividing a mean nitrate concentration of 15.3 mg/L for a study done in the Elm Creek watershed [9] by the average daily flow (~35 m^3/day). The mean nitrate concentration was obtained by averaging five tile nitrate concentrations from a nearby water quality treatment wetland and one grab sample from the study site. The kinetic reaction rate coefficient was set based on the median annual value determined in a compilation study that utilized total nitrogen data from 141 wetlands. The median annual rate constant was 12.6 m/year, making the daily rate constant ~0.04 m/day [16].

2.4.1. Data Collection to Support Construction Design

Soil borings were advanced to get detailed data on stratigraphy and particle size data for soils down to a depth of 1.8 m. A temporary monitoring well was installed to obtain depth to water table information. Flood frequency estimates of the floodplain where the wetland is located were calculated using local stream gauge information, cross-sectional surveys and regional regression equations specific to Minnesota.

2.4.2. Vegetation: Seed Mix Selection and Plant Harvesting

A unique component of this project was the use of plant harvest to maximize phosphorus removal. Three vegetative treatments were seeded in the wetland along with an oat cover crop. Vegetative treatments included a low diversity wet prairie mix (12 species; Table 2), a medium diversity wet prairie mix (20 species), and a high diversity wet prairie mix (32 species), which were seeded into Cells 1, 2, and 3, respectively. Following wetland construction, a firm seedbed was prepared in each wetland cell and seed mixes were broadcast by hand and raked lightly into the soil. Heavy precipitation from 21 June through 23 June 2013 resulted in flash flooding, which at its peak rose above the interior berms. The flooding washed much of the seed out of the first wetland cell. Cell 1 was reseeded in early August with a 23 species native mixture (Table 2). In fall 2013 and spring 2014, counts of plant populations and visual estimates of ground cover were conducted to assess stand establishment and stand vigor. Percent vegetative cover by functional group was visually estimated using six cover classes in two randomly selected 0.25-m^2 quadrats in each plot. The cover class midpoints of each observation were then averaged. Plants were also classified as to their wetness tolerance using the categories, FAC = facultative, FACW = facultative wetland, FACU = facultative upland, NA = not applicable, OBL = obligate.

Table 2. List of plant species planted in the treatment wetland in 2013. High, medium, and low diversity seed mixtures are indicated along with a 23-species mixture for reseeding. Wetland code refers to the wetness tolerance of plant species.

Scientific Name	Common Name	Wetland Code	Low	Medium	High	Reseed
Andropogon gerardii	Big Bluestem	FAC		x	x	x
Bromus ciliatus	Fringed Brome	FACW			x	x
Calamagrostis canadensis	Blue Joint Grass	OBL			x	x
Elymus canadensis	Canada Wild Rye	FACU	x	x	x	x
Glyceria grandis	Reed Manna Grass	OBL		x	x	
Glyceria striata	Fowl Manna Grass	OBL			x	
Panicum virgatum	Switchgrass	FAC	x	x	x	x
Poa palustris	Fowl Bluegrass	FACW	x	x	x	x
Sorghastrum nutans	Indiangrass	FACU			x	x
Spartina pectinata	Prairie Cord Grass	FACW		x	x	x
Carex pellita	Broad-leaved Woolly Sedge	OBL		x	x	x
Carex stipata	Awlfruit sedge	OBL				x
Carex stricta	Tussock Sedge	OBL			x	x
Carex vulpinoidea	Brown Fox Sedge	FACW	x	x	x	x
Scirpus atrovirens	Green Bulrush	OBL	x	x	x	
Scirpus cyperinus	Woolgrass	OBL	x	x	x	
Anemone canadensis	Canada Anemone	FACW			x	x
Asclepias incarnata	Swamp Milkweed	OBL	x	x	x	x
Bidens frondosa	Beggar's tick	FACW				x
Desmodium canadense	Showy Tick Trefoil	FACU	x	x	x	
Eupatorium maculatum	Joe Pye Weed	OBL	x	x	x	x
Eupatorium perfoliatum	Boneset	OBL			x	
Helenium autumnale	Sneezeweed	FACW		x	x	x
Helianthus grosseserratus	Sawtooth Sunflower	FACW	x	x	x	
Liatris pycnostachya	Prairie Blazingstar	FAC		x	x	
Lobelia siphilitica	Great Blue Lobelia	OBL			x	
Mimulus ringens	Monkey Flower	OBL		x	x	
Physostegia virginiana	Obedient plant	FACW				x
Pycnanthemum virginianum	Mountain Mint	FACW			x	
Rudbeckia laciniata	Wild golden glow	FACW				x
Symphyotrichum novae-angliae	New England aster	FACW				x
Symphyotrichum puniceus	Swamp Aster	OBL	x	x	x	
Symphyotrichum umbellatus	Flat-topped Aster	FACW			x	x
Verbena hastata	Blue Vervain	FACW	x	x	x	x
Vernonia fasciculata	Common Ironweed	FACW		x	x	
Veronicastrum virginicum	Culver's Root	FAC			x	
Zizia aurea	Golden Alexanders	FAC			x	x
Avena sativa	Oat (as cover crop)	NA	x	x	x	

Notes: FAC = facultative; FACW = facultative wetland; OBL = obligate; FACU = facultative upland; NA = not applicable.

2.4.3. Project Planning and Implementation

After the conceptual design was developed, a grading plan was created and the project was sent out for contractor bids in the fall of 2012. The contractor did all grading, earth moving and erosion control on the berms in spring 2013. A water inlet was constructed and armored with riprap to prevent scouring. Agri-Drain brand water control structures that utilize movable plastic slats for setting water elevation were installed in between each of the cells and at the wetland outlet.

Connection of the subsurface drainage pipe that required re-routing into the wetland was problematic due to the deep depth of the subsurface drain outlet as it descended the gradual valley slope towards Elm Creek. Additionally, frozen ground followed by wet conditions and unforeseen difficulties with excavating the deep sub-surface drainage pipe going down the hill delayed connecting the drainage flow to the wetland until late May 2013. At this point hydrologic monitoring began.

2.5. Wetland Monitoring and Nutrient Removal Assessment

The components of the water budget were calculated using multiple methods. Inputs for the wetland were tile discharge at the inlet and rainfall. Water exchange via overland flow into and out of the wetland through the outer berms was considered negligible. Groundwater discharge was not observed within the wetlands based on well data.

2.5.1. Rainfall Measurement

Rainfall was measured using a HOBO® tipping bucket rain gauge each year, a Davis Vantage Pro2 weather station tipping bucket rain gauge the second year, and a Teledyne ISCO 674 tipping bucket rain gauge the third year. The precipitation was multiplied by the 0.9-acre catchment area of each cell in the wetland to determine the total volume in each cell. Potential evapotranspiration (PET) was calculated using the Hamon PET equation [22,23] using temperatures collected from an EasyLog USB temperature logger and the HOBO rain gauge.

2.5.2. Water Volume

Inflow from tile drainage and outflow to the creek were measured using Teledyne ISCO 2150 and 4150 area velocity probes and modules in 2013 and 2014. In the second year, a Solinst brand pressure transducer, referred to as a Levelogger, was placed behind the flat weirs in the outflow Agridrain control structure. It was paired with a barometric pressure logger (Solinst brand barologger) to compensate the Levelogger with barometric pressure. By 2015, both the inflow and outflow were being measured by placing V-notch weirs in Agridrain control structures with Solinst Leveloggers measuring the height of water flowing over the weirs. Leveloggers were placed in the Agridrain control structures between the cells in the second year as well, and the flat boards in these structures were replaced by V-notch weirs in the third year.

When flat boards were used with Leveloggers, the following equations were used to calculate flow [24] as shown in Equations (4) and (5):

$$Q=0.020(L-0.437H)H^{1.48}\ if\ \mathrm{H}\leq 0.44L \tag{4}$$

$$Q=0.027LH1.2\ if\ H\geq 0.44L \tag{5}$$

where, Q = flow rate (L/s); L = width of the gate (cm); H = flow depth above the gate (cm). After a V-notch weir was installed in the Agridrain control structures in 2015, the following equation was used to calculate flow:

$$Q=0.9833\mathrm{x}^{2.0801} \tag{6}$$

Q is the flow rate (expressed in cubic feet/second and converted to cubic meters/day), and x is the height of water above the apex of the V-notch weir (feet). Infiltration of water into the soil was calculated algebraically once the other variables in the water budget were determined. From 23 March to 5 June 2014, tile drain inflow into the wetland was estimated using precipitation to runoff ratios in the region [10] due to equipment failure. An assumption was made, based on regional discharge data that 30% of rainfall infiltrated and flowed from the watershed through tile drainage from 23 March through 5 June 2014. Rainfall captured at the site was used to estimate tile flow from the 10.1 ha watershed into the wetland.

2.5.3. Nutrient Reduction

Grab samples were taken from the inlet and outlet of the wetland throughout the three years. Samples were analyzed for nitrate/nitrite-N, TP, and soluble OP at the Minnesota Valley Testing Laboratory (MVTL) in New Ulm, MN. Some samples were also analyzed using a Hach Nitratax sc, Ultraviolet Nitrate sensor, for nitrate/nitrite-N. Nutrient loads infiltrating into the soil were estimated by multiplying the concentrations in each cell by the infiltrated volume calculated above. In 2015, reductions in the shallow, subsurface flow were estimated using shallow wells and piezometers drilled throughout the wetland. The level in each well was measured throughout each year to determine the direction of subsurface flow. Furthermore, nutrient samples were submitted to MVTL from each well and piezometer. Water samples were also taken for stable isotopes of Oxygen and Hydrogen (^{18}O and ^{2}H) analysis. The following equation was used to calculate the contribution of groundwater in each sample:

$$\% \text{ GW} - \text{Contribution} = \left(\delta_{Rmix} - \frac{\delta_{SW}}{\delta_{GW}} - \delta_{SW} \right) \times 100 \qquad (7)$$

δ_{Rmix} is the isotope ratio of the sample where surface water and ground water are known to be mixing, δ_{SW} is the isotope ratio of the surface water sample, and δ_{GW} is the isotope ratio of the ground water. This analysis allowed for an estimate of the percent groundwater and percent surface water within the samples taken from the wells. Two piezometers sampled groundwater before it entered the wetland while the other wells and piezometers contained what would be a mixture of groundwater and infiltrated wetland surface water. The following equations were used to calculate nutrient reductions:

$$\text{Total volume flowing past wells} = \frac{\text{SW volume infiltrating}}{\%\text{SW contribution}} \qquad (8)$$

$$\text{GW volume flowing past wells} = (\text{Total volume flowing past}) - (\text{SW volume infiltrating}) \qquad (9)$$

$$\text{GW nutrient load} = (\text{GW volume flowing past wells}) \times ([\text{Nutrient}] \text{ at GW well}) \qquad (10)$$

$$\%\text{Reduction} = 100 - \frac{[\text{Nutrient}] \times \text{Total volume flowing past}}{\text{GW load} + \text{SW infiltration load}} \times 100 \qquad (11)$$

3. RESULTS

3.1. Wetland Construction, Design, and Goal Attainment

The wetland was installed almost consistent with the design dimensions (Figure 2), however, root channel cracks from frost-heave in the high clay content subsoil and potentially, grading of sand into the wetland area, increased the rate of infiltration into the sub-surface. The wetland did contain all inflows from the tile drainage network from 2013 to 2015, never over-topping the 1.8 m-high berms (Figure 3 shows the modeled stage-discharge relationship for the wetland). The wetland was flooded by overbank flow from Elm Creek each year, with several large floods occurring in June each year.

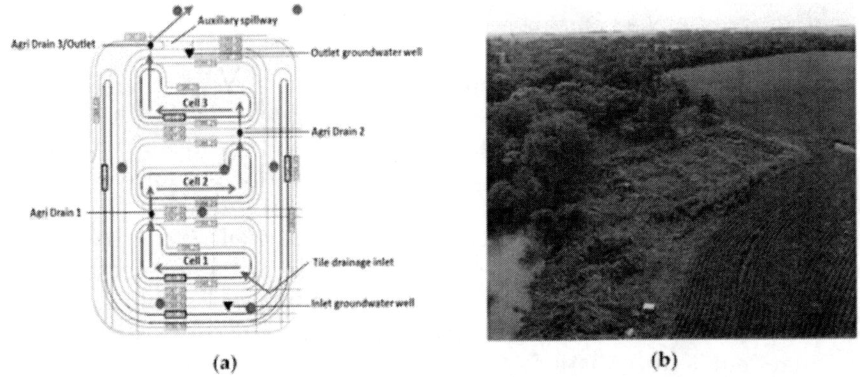

Figure 2. (**a**) Schematic of treatment wetland design; (**b**) Right: aerial view of treatment wetlands in July 2016. Subsurface tile flow from farmland located uphill drains into the wetland system and meanders through the three cells prior to discharging in the upper left corner to Elm Creek, shown in lower left of the right image. Photo by David Hansen.

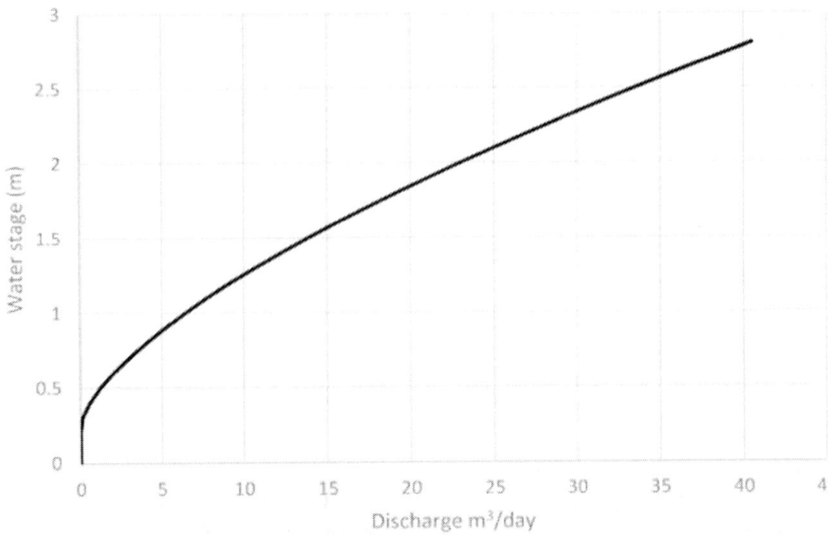

Figure 3. Modeled stage–discharge relationship for the treatment wetland. The wetland's maximum stage is 1.5 m, which is set by the berm height.

Wetland seed mixes were established with limited above-ground growth the first year (2013). By fall 2013, total ground cover was near 100%, but was dominated by weed species in each wetland cell. Weed cover ranged from 65% to 74% in all wetland cells, with the remainder being mostly seeded forbs. In Cell 1, ground cover of seeded species increased each year, from 31% in 2013, to 80% in 2014, and 81% in 2015. Likewise, in Cell 2, ground cover increased from 46% in 2013 to 87% in 2014, and 92% in 2015. Ground cover in Cell 3 showed a similar pattern, with 26% cover in 2013, 71% in 2014, and 98% in 2015. Weed cover declined each year, corresponding with increasing vigor and ground coverage of seeded species through time. Over years two–three, plant coverage became dominated by native FAC and FACW species including *Andropogon gerardii*, *Panicum virgatum* and *Spartina pectinata* in Cells 2 and 3. By 2015, the most abundant species in Cell 1 was FACW species *Helenium autumnale.* Aboveground plant biomass increased each year of the study, from 0.6 $Mg{\cdot}ha^{-1}$ in 2013 to 2.9 $Mg{\cdot}ha^{-1}$ in 2014, and 4.7 $Mg{\cdot}ha^{-1}$ in 2015.

3.2. Modeling Results

The level-pool routing and mass balance model with hydrologic inputs form DRAINMOD predicted drainage inflow to the wetland occurred 315 days, or 86% percent of the modeled time period (i.e., 1 year). Inflow was not constant but fluctuated rapidly with the occurrence of storm events in the watershed (Figure 4). The greatest volume of flow (54% of annual flow) was predicted to occur between the months of March to May with a rise in the fall during October and November. The total predicted inflow volume was 12,800 m^3/year.

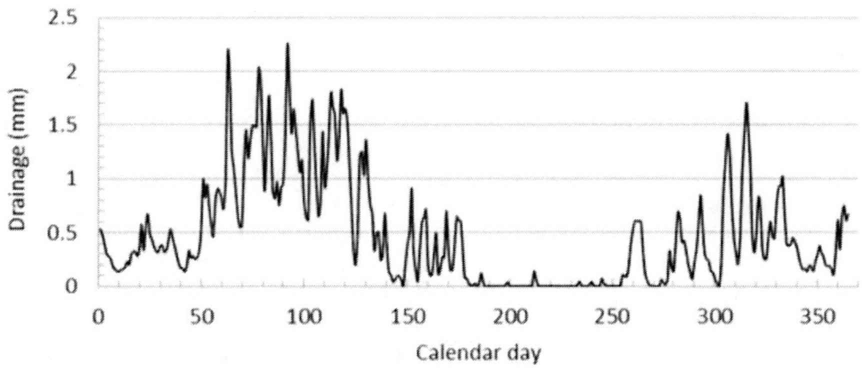

Figure 4. Predicted drainage flow in mm over the watershed per day over the model simulation period.

The wetland modeling results show attenuation of peak flow and delay of outflow moving through the three-cell system (Figure 5). Peak flow from the wetland was predicted to be reduced from 166 m^3/day to 71 m^3/day and delayed by 30 days for water exiting the wetland system.

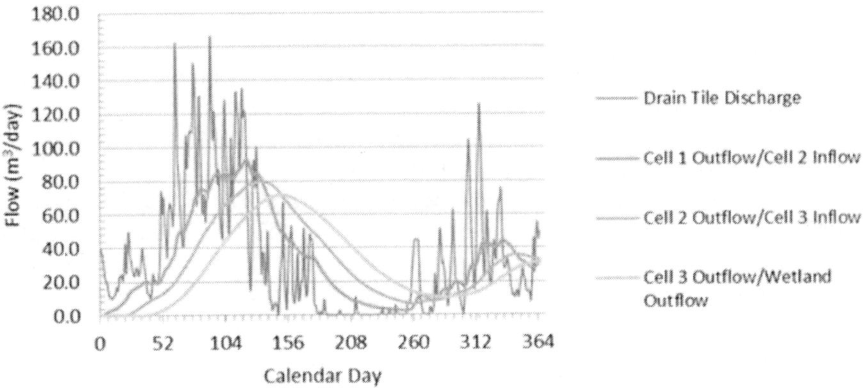

Figure 5. Predicted drainage inflow vs. pool outflow showing from each of the three cells within the treatment wetland.

Nitrate removal was predicted to be the most efficient in the months of July and August and least effective in the spring months from March to May (Figure 6). On an annual basis, nitrate removal by the wetland was predicted to be 67%. The nitrate load into the wetland was predicted to be strongly correlated with the volume of water entering the wetland as can be seen by comparing Figure 5 and Figure 6.

3.3. Hydrologic Monitoring Results

3.3.1. Inflow Data (Tile Flow and Rainfall)

In 2014, the wetland began flowing consistently on 27 April, and snow melt was observed in the wetland on 21 March 2014. In 2015, the wetland received some flow at the end of April, but it did not consistently flow until 14 May. The majority of flow into the wetland occurred as subsurface tile flow between the months of April to June with flow greatly reduced in late summer fall (Figure 7). There was more rain before the end of June 2015 than in the previous two years during that time. However, there was less total flow in 2015 during that time.

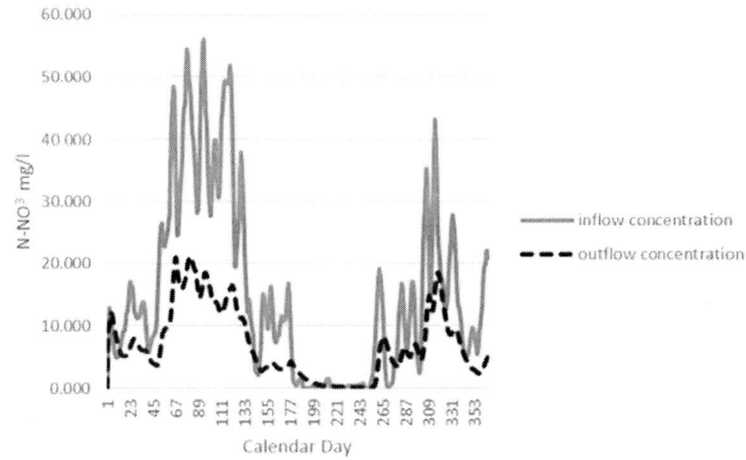

Figure 6. Modeled nitrate removal predictions. The inflowing concentration is shown in blue with the outflow as the black dashed line. Removal efficiency was predicted to be greatest during calendar days 180 to 250 (July and August) and least during calendar days 40–180 (months of February to May).

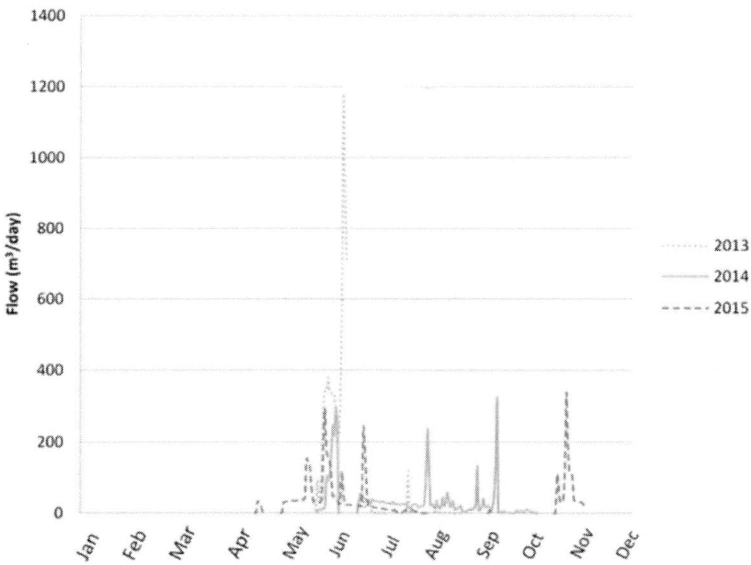

Figure 7. Daily tile flow volumes into the wetland in 2013–2015. Data are missing during the flood periods of 2013 and 2014 due to removal of equipment to avoid damage.

Flow into the wetland was highly variable and would cease part way through the growing season. In total, the tile drainage volume discharging into the wetland ranged from 5666 to 12,732 m^3/year, averaging 8546 (Table 3). Flow began in the wetland following snow melt and ended in the fall. There was no flow into the wetland during the winter, and flow ceased in July in two of the three summers until a large rain event would occur toward the end of the growing season.

Table 3. Volume of water flowing into the wetland from tile drainage each year of the study. The flow period is also listed by start and end dates each year.

Year	Tile Drain Inflow Volume (m^3)	Starting Date of Flow	Approximate Last Date of Flow
2013 *	7240 (5239–9240 **)	5 June	11 September
2014	12,732 (9215–16,277 **)	27 April	30 October
2015	5666 (5575–5669 **)	28 April	1 December

Notes: * Measurements began in late May of 2013, therefore early spring flow events are not included in the 2013 estimate; ** Range of values calculated based on instrument accuracies and uncertainty in inflow estimates when calculating from rainfall in 2014 rather than instrument measurements.

Flow into the wetland occurred 334 days over the three years of study or 30.5% of the monitoring period (Table 4). Of those days, the entire wetland was inundated on 13% of the days when water was flowing into the wetland. This means that the wetland was filled only 44 days in 3 years or 4% of the time. Cell 1, however, was filled 133 days over the 3 years or 40% of the days in which water flowed into the wetland.

Table 4. Approximate number of days of inundation observed in each wetland cell. This was defined as the number of days water was covering the entire cell and reached the control structure leading out of the cell.

Flow Statistic	Year			Total for Monitoring Period
	2013	2014	2015	
Flow into wetland	54	182	98	334
Inundation Cell 1	28	88	17	133
Inundation Cell 2	19	43	7	69
Inundation Cell 3	13	31	0	44
Flow out of wetland	13	27	0	40

According to data from the University of Minnesota Climatology Working Group, the wetland experienced wetter than average conditions between April and June in 2013 and April and May of 2014 (Table 5 and Table 6). June 2014 had greater than average rainfall. In 2015, April and June rainfall returned to average with a wet May 2015. The year 2015 had the most rainfall between March and November, but it had the least volume of flow into the wetland of the three years (Table 6).

Table 5. Cumulative rainfall amounts during the monitoring period from spring thaw to late fall during the years 2013–2015. Measurements of rainfall began after the ground thawed (April) and ended when equipment was removed and temperatures were consistently below freezing, (usually November).

Year	Rainfall (cm)	Direct Rainfall into Wetland Volume and Range (m^3)
2013	46.65	512.1 (506.9–517.2)
2014	46.53	510.7 (505.6–515.8)
2015	60.15	660.1 (653–667)

Table 6. Monthly rainfall totals during the years 2013–2015 with comparison to the Normal (30 year average) for Granada, Minnesota.

Month	2013 (cm)	2014 (cm)	2015 (cm)	Granada Normal (1981–2010) (cm)
January	0.91	1.73	1.37	1.96
February	2.82	2.03	1.24	1.78
March	4.95	2.90	1.17	4.39
April	12.45	5.23	5.97	7.87
May	11.76	11.33	12.80	9.40
June	13.74	15.60	8.23	11.46
July	4.04	1.75	10.31	10.97
August	7.98	7.26	9.68	10.29
September	2.01	2.90	6.91	7.98
October	7.19	2.46	3.94	5.97
November	1.24	2.59	6.30	4.24
December	1.70	2.31	4.93	2.82
Total Annual Precipitation	70.79	58.09	72.85	79.12

3.3.2. Hydrologic Regime and Discharge Data

The three wetland cells varied in their duration of inundation and saturation. Cell 1 contained water as long as the inlet was flowing, however, cells 2 and 3 would receive overflow from Cell 1 only during the spring following a series of multiple rain events or following large, single rain events. In 2013, only 17% of the water in the wetland flowed through the entire wetland and discharged into Elm Creek, with 22% in 2014. The majority of water infiltrated into the subsurface flow at a depth of 1 to 2 m below the ground surface (Figure 8).

Calculations of water loss from the wetland via evapotranspiration (ET) are listed in Table 7. The Hamon, Thornthwaite and Jensen–Haise equations for potential ET gave monthly average values ranging from 1.07 to 3.03 mm/day. Using the water level drawdown, transpiration via wetland plants was estimated to average 2.8 mm/day for 2015. Evaporation was limited by the small number of days that the wetland was inundated, with standing water occurring for 4%–12% of the monitoring period (Table 7). Calculated transpiration using the drawdown method was usually similar to PET estimates, although for brief periods during the growing season when the water table was below ground surface, transpiration was higher than PET ranging up to 5.5 mm/day. Since water was not present in the wetland for long time periods, the actual transpiration would have been less than PET, as expected.

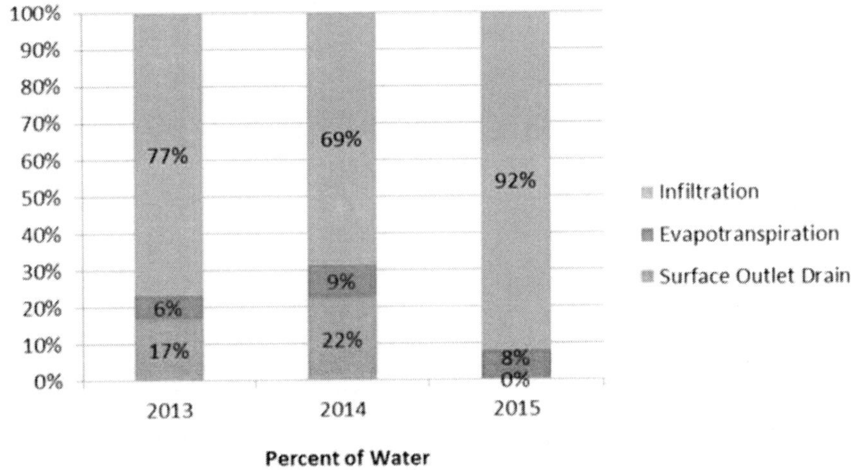

Figure 8. Outflow distribution of all water that entered the wetland each year. The total water volume includes direct precipitation over the wetland.

Table 7. Average potential evapotranspiration and transpiration per day during the monitoring period. Water level drawdown was estimated for a time period when water was in the shallow root zone (upper 1 m) with no ponding.

Year	Hamon PET (mm/Day)	Thornthwaite PET (mm/Day)	Jensen–Haise PET (mm/Day)	Transpiration: from Water Level Drawdown (mm/Day)
2013	2.35	2.84	NA	No data
2014	3.01	3.03	1.07	No data
2015	2.80	2.73	NA	2.80

3.3.3. Nutrient Removal Data

Nitrate removal over the three years of the study averaged approximately 68% of the load that entered the wetland through tile drainage. Percent reduction ranged from 60% to 93% in the three years including both surface and subsurface removals combined. More nitrate was removed through subsurface reductions than surface reductions due to the high volume of water that infiltrated into the soil. The nitrate load into the wetland was correlated with the volume of water entering the wetland with most loading occurring in the months of April to June.

Phosphorus removal was approximately 76% over the three-year monitoring period. Most phosphorus was removed by plant uptake since most of the phosphorus was in dissolved form traveling through the tiles. No direct surface flow from upstream farmland was received, thus minimizing the amount of sediment and particulate phosphorus often associated with it.

3.4. Comparison of Predicted vs. Observed Results

The duration and magnitude of tile inflow was less than predicted by the level pooling routing and DRAINMOD model. Total inflow was found to be 67% of that predicted, while the frequency of inflow was 35% of that predicted (Table 8). Nutrient removal percentages were almost identical to those predicted cumulatively from 2013 to 2015. Despite the differences in removal processes, the observed wetland removal rates were comparable to those modeled due to the high sub-surface flow nitrate removal rates.

Table 8. Summary comparison of predicted and observed results for hydrologic and water quality performance of treatment wetland.

Variable	Units	Predicted	Observed	% Observed/Predicted
Total annual inflow	m^3/year	12,800	8546 *	67%
Peak daily inflow	m^3/day	166	1172	700%
Frequency of inflow to wetland	# of days in year	315	111 *	35%
Duration of wetland inundation	% of days per year with ponded water in wetland	**	4%–12% ***	N/A
% nitrate removal	% (of total mass removal)	67%	68%	101%
% phosphorus removal	% (of total mass removal)	n/a ****	76%	N/A

Notes: * Average over 3 years; ** duration of ponding was not predicted by the model; *** represents the range of duration across the three wetland cells; the highest values were in the first cell with decreasing inundation moving downstream to the outlet; **** phosphorus removal was not modeled since the version of DRAINMOD we used does not have that capacity.

4. DISCUSSION

4.1. Wetland Construction, Implementation Issues and Lessons Learned

The wetland design functioned as intended in terms of storing the total volume of water discharging from the farmland drainage system. The meandering flow path also prolonged water residence time by extending the flow length through the wetland and by infiltration and subsequent subsurface flow towards the outlet. Much more infiltration occurred than anticipated given the high clay content subsoil, presumably due to soil cracks, macropores and sand lenses. Consequently, residence time was extended by moving water into the ground where it slowly infiltrates, in the range of days to weeks rather than hours. Nitrate removal rates were therefore higher in the sub-surface flow than in surface flow.

The vegetation was slow to establish with very little vegetative cover the first year but covered the entire basin by year two (2014). Most of the vegetation cover consisted of native wet prairie species that were facultative (FAC) or facultative wetland (FACW) species with a few obligates, reflecting the relatively short duration of flooding.

4.2. Modeled vs. Observed Results

4.2.1. Inflow Amounts and Peak

The volume of tile drainage inflow was about 50% less than predicted by the model overall to the intermittent nature of the tile flow (Table 4). This could be explained partly by the lack of late summer and winter flow in the wetland as the model predicted. Also, in 2015, a cover crop, cereal rye (*Secale cereale*) planted in the drainage area reduced inflow greatly from April until the end of May.

In contrast to flow volume, the peak was much greater than predicted. A daily maximum volume of seven times that predicted by the models was observed with 1172 m^3/day compared to 166 m^3/day predicted. This is, in part, due to the fact that Drainmod uses daily flow averages as the input and may not simulate extreme precipitation events as accurately. The peak daily flow observed in June 2013 followed an extremely wet precipitation period where the three preceding months exceed normal precipitation by over 10 cm.

4.2.2. Duration and Timing of Flows

The number of days with tile inflow was 33% of that predicted by the model. The timing of flows differed in that the model predicted flow throughout the year, except much of July and August, while the monitoring data showed that inflow occurred primarily from March to June with no winter flow. The period of late summer and fall had little or no inflow all three years due to crop transpiration in the watershed that significantly reduced tile flow. Less rainfall in the late summer months also contributed to falling water levels during that time period.

4.2.3. Flow Path in Wetland: Subsurface vs. Surface

Modeling assumed there would be prolonged surface ponding of water while monitoring data showed that most of the water infiltrated within hours to days of storm events.

The predominance of subsurface flow increased the residence time from hours to days or weeks. This was possible given the relatively small size of the contributing drainage area of 10 ha. Larger drainage basins would tend to create surface flow-dominated systems.

4.3. Nutrient Removal Effectiveness and Issues

Given the greater peak flows and shorter ponding period, surface water residence time was apparently much shorter than that predicted by the model, resulting in lower nitrate removal rates.

However, the lower inflow volume led to greater percentage nitrate removal in the subsurface flow. Intermittent flow leading to temporary soil drying may have reduced denitrifying bacteria activity. The mineral soils that were not high in organic matter may have limited bacterial activity as well as compaction by grading during construction.

Nitrate removal was approximately equal to the predicted rate. Given that a much greater inflow volume of water and total load of nitrate was predicted by the model, the wetland was actually less efficient than predicted in terms of the total load removed. Although not the focus of this paper, phosphorus removal was also very high due to plant uptake and harvest as described in the project report [25].

Phosphorus removal exceeded inflow from the tile drain in the third year indicating that the vegetation was extracting soil phosphorus and lowering residual levels.

4.4. Vegetation

The wetland vegetation that developed was a wet prairie plant community composition with primarily facultative to facultative wet grasses and forbs. In addition to providing enhanced evapotranspiration rates in 2014 and 2015, vegetation helped establish a source of organic carbon for denitrifying bacteria to utilize for reduction processes [26]. Vegetation establishment could have, in theory, been improved by water level drawdown in the wetland to promote growth in year 1. With surface water flooding from Elm Creek occurring June 2013, water level management was not possible, delaying full plant establishment until 2014–2015. When establishing future treatment wetland systems, if it is possible, vegetation establishment should be encouraged prior to diverting water into the wetland.

4.5. Cost and Landowner Adoption Issues

Construction cost was approximately $25,000/ha, which was comparable (on a per unit area of wetland installed) to other treatment wetlands. A three-cell design would be discouraged, which would significantly reduce this cost. The entire project area was only 0.2 ha; therefore, the land purchase or easement cost was greatly reduced. This could lead to a greater adoption rate, especially in states such as Minnesota with statutory buffer requirements. The design provided an edge-of-field construction location, adjacent or within standard buffer placement, to encourage landowner adoption by reducing the land area taken out of production. The cooperating farmer-landowner involved in this project, like other producers, was concerned about the reduction of production farmland that would be required for the construction of the treatment wetland. These smaller designs could be cited in low production areas or areas already requiring buffer adoption, making them an actionable practice for implementation by willing producers.

Some studies have shown that farmer-landowners are less willing to adopt innovative new management practices [27]. To address this barrier, field trips and walking demonstration tours were hosted onsite to promote the practice to regional landowners and agency staff. Although landowner surveys were not conducted, feedback obtained from meetings suggest that small, edge-of-field wetlands are more likely to be adopted than large wetlands that use up substantial amounts of farmland. Looking for areas on farms that are not in production and using them for establishing wetlands should improve adoption.

4.6. Future Research Needs

Typically restored or constructed wetlands in this region have lower removal rates for dissolved phosphorus than sediment, particulate phosphorus and nitrate [28]. There is a strong need to develop wetlands and other water-retention best management practices (BMPs) that are able to more effectively control eutrophication in downstream rivers and lakes that are especially sensitive to dissolved phosphorus.

The effects and combinations of practices placed in series from upstream to downstream are not well documented. In order to achieve nutrient reduction goals in Midwestern farmland, multiple BMPs in series will need to be utilized. In particular, practices that can reduce or store water are needed to reduce total loads. At the study site discussed in this paper, the use of cover crops in fall 2014 to spring 2015 greatly reduced flow into the treatment wetland, improving its performance. Combinations of in-field and edge-of-field practices show great promise for nutrient load reduction but have been little-studied to date.

5. CONCLUSIONS

A novel design for treating subsurface tile flow was developed for the region that is compatible with Midwestern tile-drained farm systems. While the wetland was effective at removing nitrate, the hydrologic inputs, hydraulics and flow-pathways were slightly different than predicted, particularly the inputs and pathways for subsurface drainage. The lessons learned from this project regarding wetland design, vegetation establishment, and the effects of hydrologic regime on nutrient removal performance will be useful in the development of regional watershed scale management and implementation plans.

Adoption of treatment wetlands and similar nutrient removal and water storage BMPs will be needed to achieve regional and state water quality goals, with the ultimate objective of reducing the input contribution to the Gulf of Mexico. Development of practical BMP designs to increase farmer-landowner adoption will encourage progress toward achieving the broader, national nutrient reduction goals [1].

ACKNOWLEDGMENTS

Funding was provided by the Minnesota Department of Agriculture through the Clean Water Legacy Fund. Ashley Brenke, Dustin Benes and

Richard Perrine of Martin SWCD provided assistance in water quality data collection. The Darwin Roberts family allowed the University of Minnesota to do research on their land for four years and continuing through 2018. Rural Advantage and University of Minnesota Extension helped coordinate public field trips to the site for public education. Numerous undergraduate and graduate students helped to collect soil and water data. Derek Lash of EOR, Inc. developed the grading plan for construction. The map of the Minnesota River basin in Figure 1 was used with permission of the Water Resources Center at Minnesota State University-Mankato. Finally, Scott Matteson of the Minnesota Department of Agriculture provided technical monitoring assistance.

REFERENCES

1. Mitsch, W.; Day, J. Restoration of wetlands in the Mississippi–Ohio–Missouri (MOM) River Basin: Experience and needed research. *Ecol. Eng.* **2006**, *26*, 55–69.
2. Dahl, T.E. *Status and Trends of Prairie Wetlands in the United States 1997 to 2009*; U.S. Department of the Interior, Fish and Wildlife Service, Ecological Services: Washington, DC, USA, 2014.
3. Rabalais, N.; Turner, R.; Justić, D.; Dortch, Q. Nutrient changes in the Mississippi River and system responses on the adjacent continental shelf. *Estuaries* **1996**, *19*, 386–407.
4. Goolsby, D.; Battaglin, W. *Nitrogen in the Mississippi Basin-Estimating Sources and Predicting Flux to the Gulf of Mexico*; USGS Fact Sheet; USGS: Reston, VA, USA, 2000.
5. Burkhart, M. The hydrologic footprint of annual crops. In *A Watershed Year: Anatomy of the Iowa Floods of 2008*; Mutel, C.F., Ed.; University of Iowa Press: Iowa City, IA, USA, 2010.
6. Javaheri, A.; Babbar-Sebens, M. On comparison of peak flow reductions, flood inundation maps, and velocity maps in evaluating effects of restored wetlands on channel flooding. *Ecol. Eng.* **2014**, *73*, 132–145.
7. Walters, K.M.; Babbar-Sebens, M. Using climate change scenarios to evaluate future effectiveness of potential wetlands in mitigating high flows in a Midwestern U.S. watershed. *Ecol. Eng.* **2016**, *89*, 80–102.
8. Crumpton, W.; Kovacic, D.A.; Hey, D.L.; Kostel, J.A. Potential of restored and constructed wetlands to reduce nutrient export from agricultural watersheds in the corn belt. In *Final Report: Gulf Hypoxia and Local Water Quality Concerns Workshop*; American Society of Agricultural and Biological Engineers: St. Joseph, MI, USA, 2008; pp. 29–42.
9. Lenhart, C. The Influence of Watershed Hydrology and Stream Geomorphology on Turbidity, Sediment and Nutrients in Tributaries of the Blue Earth River, Minnesota, USA. Ph.D. Thesis, University of Minnesota-Twin Cities, Minneapolis, MN, USA, 2008.
10. Fransen, G. Hydrologic, Nutrient, and Sediment Responses of Restored Perennial Vegetation/Wetland Complexes in Southern Minnesota. Master's Thesis, University of Minnesota-Twin Cities, Minneapolis, MN, USA, 2012.
11. Kovacic, D.A.; David, M.B.; Gentry, L.E.; Starks, K.M.; Cooke, R.A. Effectiveness of constructed wetlands in reducing nitrogen and phosphorus export from agricultural tile drainage. *J. Environ. Qual.* **2000**, *29*, 1262–1274.
12. Randall, G.W.; Mulla, D.J. Nitrate Nitrogen in Surface Waters as Influenced by Climatic Conditions and Agricultural Practices. *J. Environ. Qual.* **2001**, *30*, 337–344.
13. Mirek, S. Effect of Nitrogen Fertilizer Management and Seasonality of Tile Flow on Nitrate-Nitrogen Leaching through Tile Drains. Master's Thesis, University of Illinois at Urbana-Champaign, Urbana, IL, USA, 2001.

14. Kellman, L. A study of tile drain nitrate-δ15N values as a tool for assessing nitrate sources in an agricultural region. *Nutr. Cycl. Agroecosyst.* **2005**, *71*, 131–137.
15. Perciasepe, R. *National Strategy for the Development of Regional Nutrient Criteria*; EPA 822-R-98-002; United States Office of Water: Washington, DC, USA, 1998.
16. Galatowitsch, S.; Van der Valk, A. *Restoring Prairie Wetlands: An Ecological Approach*; Iowa State University Press: Ames, IA, USA, 1994.
17. Kadlec, R.H.; Wallace, S.D. *Treatment Wetlands*, 2nd ed.; CRC Press: Boca Raton, FL, USA, 2009.
18. USDA-Natural Resource Conservation Service (NRCS). CP39 FWP Constructed Wetland. In *FSA Handbook, Agricultural Resource Conservation Program for State and County Offices, Short Reference*; 2-CRP, Revision 5; Farm Service Agency, U.S. Department of Agriculture: Washington, DC, USA, 2010; Exhibit 11; pp. 235–241.
19. Reed, S.; Crites, R.; Middlebrooks, E. *Natural Systems for Waste Management and Treatment*; McGraw-Hill, Inc.: New York, NY, USA, 1995.
20. Skaggs, R.; Breve, M.; Gilliam, J. Hydrologic and water quality impacts of agricultural drainage. *Crit. Rev. Environ. Sci. Technol.* **1994**, *24*, 1–32.
21. Civil Engineering Portal. Weirs [Internet]. Available online: http://www.engineeringcivil.com/weirs.html (accessed on 1 April 2012).
22. Federer, C.A.; Vörösmarty, C.; Fekete, B. Intercomparison of Methods for Calculating Potential Evaporation in Regional and Global Water Balance Models. *Water Resour. Res.* **1996**, *32*, 2315–2321.
23. Lu, J.; Sun, G.; McNulty, S.; Amatya, D. A comparison of six potential evapotranspiration methods for regional use in the southeastern United States. *J. Am. Water Resour. Assoc.* **2005**, *41*, 621–633.
24. Chun, J.; Cooke, R. Technical note: Calibrating Agridrain water level control structures using generalized weir and orifice equations. *Appl. Eng. Agric.* **2008**, *24*, 595–602.
25. Current, D.; Lenhart, C.F.; Gordon, B.; Gamble, J.; Ross, N. *On-Farm Evaluation of Treatment Methods for Excess Nutrients in Agricultural Subsurface Tile Drainage*; Minnesota Department of Agriculture: St. Paul, MN, USA, 2016.
26. Reed, S.; Brown, D. Subsurface flow wetlands—A performance evaluation. *Water Environ. Res.* **1995**, *67*, 244–248.
27. Nassauer, J.I.; Dowdell, J.A.; Wang, Z.; McKahn, D.; Chilcott, B.; Kling, C.L.; Secchi, S. Iowa farmers' responses to transformative scenarios for Corn Belt agriculture. *J. Soil Water Conserv.* **2011**, *66*, 18A–24A.
28. Miller, T.P.; Peterson, J.R.; Lenhart, C.F.; Nomura, Y. *The Agricultural BMP Handbook for Minnesota*; Minnesota Department of Agriculture: St. Paul, MN, USA, 2012.

CHAPTER 13

Public Support for Wetland Restoration: What is the Link With Ecosystem Service Values?

Samantha S. K. Scholte[1], Maya Todorova[2], Astrid J. A. van Teeffelen[1], Peter H. Verburg[1]

1 Environmental Geography Group, VU University Amsterdam, De Boelelaan 1085, 1081 HVAmsterdam, The Netherlands

2 WWF Danube-Carpathian Programme, Ruse, Bulgaria

ABSTRACT

Fostering public support for wetland restoration is essential for long-term sustainable management and use of wetland areas. This paper explores the socio-cultural dimension of wetland restoration, by looking at the importance of wetland ecosystem services for different user groups. We try to better comprehend such values by evaluating the awareness people have of ecosystem services and the direct benefits people obtain from the ecosystems in their surroundings. In addition, we study how the values people assign to ecosystem services are related to wetland restoration attitudes. We identified four perspectives towards wetland restoration, which could partially be explained by corresponding values for ecosystem services: an eco-centric, a cultural, an economic and a negative perspective. To gather public support for wetland restoration it is important to take into account the different motivations people have to support such initiatives.

Keywords: Socio-ecological systems, Nature conservation, Ecological restoration, Attitudes, Awareness

1. INTRODUCTION

Wetlands are of key importance to biodiversity conservation and important providers of a range of goods and services such as water quality improvement, flood abatement and carbon sequestration (Zedler and Kercher 2005; de Groot et al. 2012). Yet, wetland areas suffer vastly from overuse of resources, eutrophication and pollution, disconnection from parent rivers by dike construction, water abstraction for industrial and domestic use, and land reclamation (Junk et al. 2012; van Asselen et al. 2013). In Europe, 50 % of the wetlands have been converted to urban and agricultural lands (Gumiero et al. 2013). International agreements and EU legislation aim to prevent the further degradation of wetland ecosystems, e.g. the Ramsar convention, the EU Water Framework Directive and the Habitats Directive. Although such regulations have contributed towards wetland conservation and restoration, wetland areas are still susceptible to drainage and reclamation, especially when major economic interests are at stake (Čížková et al. 2011; Verhoeven 2014). Gathering public support for wetland restoration is essential to ensure social commitment towards sustainable use and management of restored wetlands.

A challenge for gathering public support for wetland restoration is that restoration may undermine local income, as a result of limited access to the natural resources, or conversion of farmland to natural land cover (Naughton-Treves et al. 2005). Ecological restoration initiatives are often part of an international framework that aims to achieve regional and/or global conservation targets (such as the Aichi Biodiversity Targets and EU biodiversity targets), for which local people may pay the price (Adams et al. 2010; Kari and Korhonen-Kurki 2013). In the conservation literature this tension has been referred to as the 'parks versus people' debate, where the protection of global biodiversity through so-called fortress conservation is at one end of the spectrum and a focus on improving local livelihoods is at the other end (Southworth et al. 2006; Minteer and Miller 2011). Bringing these two groups together often proves difficult, as ecological restoration often involves competing rather than complementary goals (McShane et al. 2011). To protect nature restoration and conservation areas from resource depletion or destruction, restrictions on natural resource use are necessary. Because of such restrictions, people have been displaced or denied access to the resources, threatening their rights and livelihoods (Brockington and Wilkie 2015).

An additional challenge for wetland restoration is that people commonly have negative associations with wetland ecosystems. Wetlands are perceived as useless, strange, difficult to access and unattractive (Meindl 2000; Nassauer 2004). For this reason, the goods and services provided by wetlands are often not recognized by society. In addition, although the importance of public engagement for ecological restoration has long been recognized, the communication of restoration benefits from science to the wider public has had limited success (Groffman et al. 2010). To tackle this issue, an ecosystem service (ES) approach is increasingly advocated. The ES framework aims to inform the public about the benefits humans receive from nature, by portraying explicit links between a wide range of (wetland) ecosystem functions and human well-being (Costanza et al. 1997; Daily 1997; de Groot et al. 2002).

The dominant narrative of the ES framework (as shown in the ES cascade depicted in Fig. 1) has focused much on the supply of ESs (Raymond et al. 2013; Comberti et al. 2015), portraying how ecosystems, through different ecological functions, provide services and goods that people appreciate and rely on (de Groot et al. 2010; Haines-Young and Potschin 2010). As such, the ES framework has commonly been used as an awareness raising tool, with the underlying assumption that once people become aware of ESs, they will be more willing to support nature conservation. The importance people assign to ecosystems and their related goods and services, however, is not only influenced by knowledge, but also depends on how people interact with ecosystems (Daw et al. 2011; Scholte et al. 2015). Given the supply-based focus in ES research, there is limited understanding of how values for (wetland) ESs are shaped by the way people perceive, depend on and use the ecosystem (Asah et al. 2014). To contribute to this research gap, we investigate how the importance local beneficiaries assign to wetland ESs, i.e. values for ESs, can be understood by looking at their knowledge of wetland ecosystems, i.e. awareness of ESs, and the way people depend on and use wetland ecosystems (Fig. 1). In addition, we aim to understand how values for ESs are related to attitudes towards wetland restoration.

We explicitly distinguish between the awareness of ESs and the socio-cultural values for ESs, because these two do not necessarily coincide. A person may be aware of the flood regulating capacity of a wetland as a benefit to others, but not consider it important for his or her own livelihood. As such, we define awareness of ESs as the ability of an individual to *recognize* the ESs delivered by an ecosystem and we define value for ESs as the importance people *assign* to ESs. Our focus is on a nature conservation area at the Lower Danube in Bulgaria, where recently several wetlands have been restored. We explore the perceptions, values and activities of local farmers, fishermen and residents to gain a better understanding of why local users are motivated to support wetland restoration efforts. By exploring the different perspectives towards wetland restoration this paper provides insights that can be used to better manage wetland conservation and restoration areas.

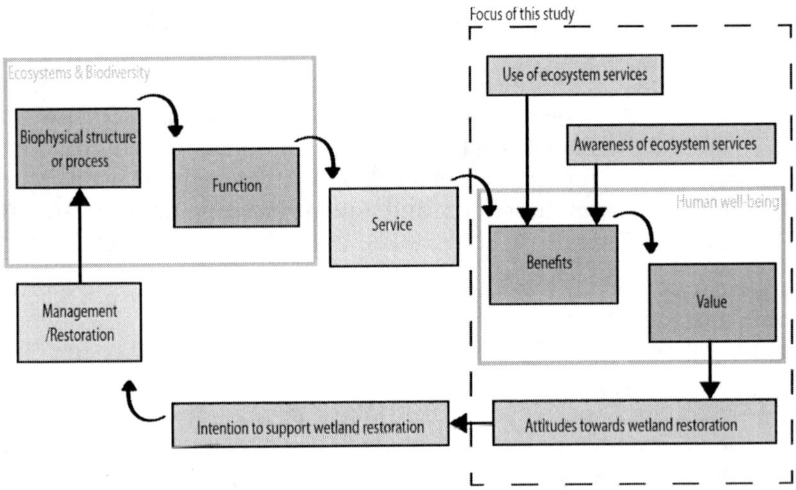

Figure 1 Conceptual diagram of the studied factors within the ES framework. Adapted from de Groot et al. (2010) and Haines-Young and Potschin (2010)

2. METHODS

2.1 Study area

Persina Nature Park (Persina) is the only nature conservation area along the Bulgarian part of the Danube River, covering an area of 21,762 ha. It was established in 2001 and is situated along the Lower Danube River in Bulgaria (Fig. 2). The Lower Danube comprises the last 860 km of the 3000 km long Danube River, which is one of the main watersheds in Europe. There are two municipalities within the boundaries of Persina: Belene (10,318 inhabitants) and Nikopol (9,305). The main economic activities in the area are agriculture and fishing. Most of the area is state-owned (60 %), while the rest of the area is owned by farmers. Small scale farms are most prominent, but are often owned and managed by big leaseholders.

Along the Lower Danube 75 % of the floodplains have been cut off from the river (Ebert et al. 2009), mostly in the 1960's to make room for agricultural lands, fishponds and hybrid poplar plantations (Schneider 2010). Recognizing the loss of important hydrological and ecological functions, such as the mediation of pollutants and the increased probability of a flood event (Hulea et al. 2009), the World Wide Fund for Nature (WWF) secured an agreement between the governments of Bulgaria, Romania, Ukraine and Moldova to establish a 'Lower Danube Green Corridor' (LDGC) in 2000. The aim of the LDGC project is to:

'make the Lower Danube a living River again, connected to its natural flooding areas and wetlands, reducing risks of major flooding in areas with human settlements and offering benefits both for local economies – fisheries, tourism – and for the ecosystems along the river' (WWF 2010).

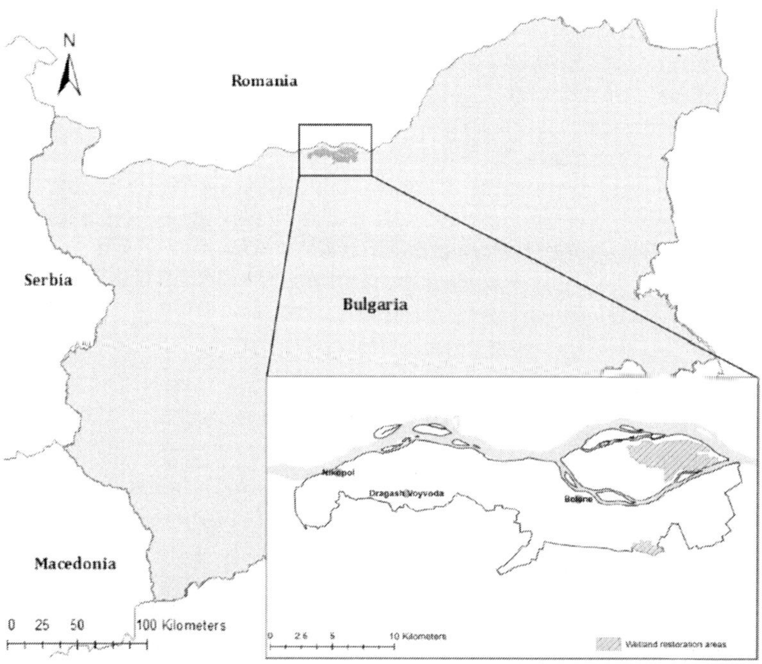

Figure. 2 Location of the study area

As part of the LDGC, Persina was established and designated as a protected area, falling under the NATURA 2000 network in Europe. The most important ecological asset of Persina is the Belene Island complex, which consists of 19 islands and was designated as wetlands of international importance under the EU RAMSAR Convention. The largest island of this complex, Persin Island, which is also the largest island along the Danube in Bulgaria, was one of the pilot sites for wetland restoration. The eastern part of the island has been restored into a natural wetland in 2004. An important aim for the WWF, who co-manages this restoration project, was to raise awareness about the benefits provided by wetlands and gather public support by local communities for wetland restoration. It is in this particular context that this study has taken place.

2.2 Wetland types

There are different types of wetlands, each type having their own distinct characteristics. Dobbie and Green (2013) demonstrated that people do not see wetlands as a homogeneous group, but distinguish between different types of wetlands by characteristics of water and vegetation: respondents in their study classified wetlands as 'grasslands', 'treed grasslands', 'wetlands with emergent vegetation','wetlands with open water', 'arid (water) scapes' or 'treed wetlands'. We address three specific wetland types that are prominent in our study region: (a) treed wetlands, i.e. riparian forests, (b) wetlands with open water, i.e. marshes, and (c) grasslands, i.e. meadows. These wetland types also correspond to classes that can be found under the Ramsar Classification System for Wetland Type, specifically 'Freshwater tree-dominated wetlands' (riparian forests, class Xf), 'Permanent freshwater marshes' (marshes, class Tp), and 'Seasonally intermittent freshwater marshes/pools, including seasonally flooded meadows' (meadows, class Ts) (Ramsar Convention Secretariat 2010).

2.3 Data collection and questionnaire design

We conducted structured face to face interviews amongst three different user groups in Persina: farmers, fishermen and local residents of Nikopol and Belene in April and May 2014. We recruited four local interviewers to interview the respondents using a pre-formulated questionnaire. We pre-tested the questionnaire in March 2014. We contacted farmers and fishermen with the support of local authorities, specifically the Agricultural Service and the Executive Agency for Fisheries and Aquaculture. We approached all farmers who applied for yearly funding at the Agricultural service between the period of April and May 2014 and requested their participation. The Executive Agency for Fisheries and Aquaculture provided us with contact details of all fishermen associations in the area. We approached local residents at public places, e.g. parks, restaurants and shops, in Belene and Nikopol. This lead to a total of 105 interviews. Three respondents did not complete the interview, leaving a total of 102 completed questionnaires. For an overview of the demographic characteristics of our sample: see Appendix (Table 5).

The questionnaire was divided into five sections, addressing 56 questions about: the use of ESs, awareness of ESs, socio-cultural values for ESs, attitudes towards wetland restoration and socio-demographics. For the questionnaire we identified a list of 14 ESs. We pre-selected ESs based on the Millennium Ecosystem Assessment (MEA) classification of ecosystem services (Millennium Ecosystem Assessment 2005) and adapted this list with input from stakeholders working at local environmental organizations, who expressed which ESs they thought were most relevant for wetland ecosystems in Persina. The final ESs included in this study were provisioning (food, materials, medicines and biomass for energy), regulating (climate regulation, maintaining water quality, ground water retention, soil erosion control and flood control) and cultural (environmental education, tourism, recreation, aesthetic values and existence values for biodiversity).

The first section of the questionnaire was aimed at gaining information on the respondents' familiarity with and activities undertaken in Persina. In the second part we asked respondents which ESs they thought were provided by Persina and asked them to indicate which wetland type could be related to these ESs. We used pictograms that visualized the ESs to avoid scientific terminology (see supplementary material). Each respondent had to choose from the 14 ESs a maximum of 5 ESs that they thought were supplied by (a) Persina in general, (b) riparian forests, (c), marshes and (d) meadows. The different wetland types were each represented by 2 photos (see supplementary material). For each component, i.e. Persina and the three wetland types, we also asked respondents whether they could think of any other benefit that was not depicted on the cards. In addition we asked the respondents if they had any negative associations and if so, to specify them.

In the third part of the questionnaire respondents had to choose a maximum of 5 ESs that they thought were most important for their personal well-being (i.e., self-oriented values) and for the well-being of all people living in Persina (i.e., other-oriented values), assuming that all could be equally supplied by Persina.

In the fourth part of the questionnaire, we tested the specific attitude towards wetland restoration in Persina. We first showed respondents a map of Persina on which wetland restoration projects were indicated. Respondents were firstly asked whether they were aware of these wetland restoration projects. Consequently respondents had to evaluate 7 statements about wetland restoration in Persina. The statements were evaluated using a 3-point Likert scale. In the final part of the questionnaire, respondents were asked to provide information regarding their age, place of residence, occupation, income and education.

2.4 Data analysis

We first defined multi-response sets for the awareness and valuation variables. The multi-response sets were consequently used for further analysis by calculating frequency tables and chi-square statistics to analyze differences between user groups and awareness of the ESs supplied by the different wetland types.

To link ES values to attitudes towards wetland restoration we calculated a value variable that reflected how many ESs in each ES category, i.e. cultural, provisioning and regulating, the respondent had mentioned. Since the amount of services was not equal in all three ES categories, we corrected the value of each variable by dividing the amount of mentioned ESs by the total amount of ESs in each category.

For instance if a respondent had mentioned one regulating service, the value was 1 divided by 5, since 5 regulating services were considered. To each respondent we assigned a value for cultural ESs, regulating ESs, and provisioning ESs. Consequently, we standardized both the value variables and attitude variables and performed an exploratory principal component analysis (PCA) using a Promax rotation. We followed the Kaiser criterion (eigenvalue >1) to identify significant components. Finally, to identify how much each respondent within the user groups could be associated with each component, we created scatter plots based on the component scores.

3. RESULTS

3.1 Use of ESs

3.1.1 Farming and Fishing

More than half of the farmers worked on a mixed farm, 27.3 % worked on an animal farm and 21.2 % worked on a crop farm (Table 1). The size of the farms differed substantially, but animal farms were much smaller (0.5–40 ha) than crop (2–400 ha) and mixed farms (1–1900 ha).

The majority of the farmers (71 %) worked on farms smaller than 50 ha, which they fully owned, partially leased or fully leased. Farmers working on lands ranging from 100 to 1000 ha (20 %) owned only a very small proportion of this land. None of the fishermen said they could live entirely of fishing, and so many had taken on other jobs to sustain their livelihoods.

Table 1 Use of ESs by different user groups. Values indicate percentages per user group

Use of ESs	Farmers	Fishermen	Residents	Total
	($n = 33$)	($n = 24$)	($n = 45$)	($n = 102$)
Work				
Type of farm				
Crop	21.2	–	–	
Animal	27.3	–	–	
Mixed	51.5	–	–	
Percentage of income dependent on farming/fishing				
Less than half	17.1	56.0	–	
About half	28.6	20.0	–	
More than half	17.1	24.0	–	
All	37.2		–	
Recreation & collection of food/materials				
Visit Persina Nature Park outside work	81.9	100	84.5	87.3
Once	15.2	16.7	17.8	16.7
2–5 times	36.4	20.8	20.0	25.5
More than 5 times	30.3	62.5	46.7	45.1
Visit Danube river and/or islands	60.6	91.7	82.2	77.4
Walking	69.7	50.0	60.0	60.8
Hunting	9.1	20.8	–	7.8
Collecting food/materials for own livelihoods	12.1	29.2	31.3	24.6

3.2 Visiting Persina

Of all respondents taken together, 12.7 % never visited Persina Nature Park, 16.7 % had visited Persina once, 25.5 % had visited Persina a few times and 45.1 % visited Persina often. Fishermen visited Persina most often outside work, while farmers visited Persina less often. The most popular recreational activity was walking and 7.8% of the respondents, only farmers and fishermen, also said they went hunting every now and then. Of all respondents, 24.6 % mentioned they went to Persina to collect food and/or materials. Four respondents mentioned they collected materials, mostly wood. Other respondents collected herbs and fruits (11.8 %) or fish (11.8 %).

3.3 Awareness of ESs

3.3.1 Do People Recognize the ESs Provided by Wetlands in Persina?

When asked whether they thought that Persina provided any of the depicted ESs, the respondents mentioned provisioning and cultural ESs the most, specifically provision of food and recreation (Fig. 3). Flood regulation was only mentioned once. Seven percent of the respondents had negative associations with Persina, which mostly had to do with mosquitoes, accessibility, and perceived degradation of the area.

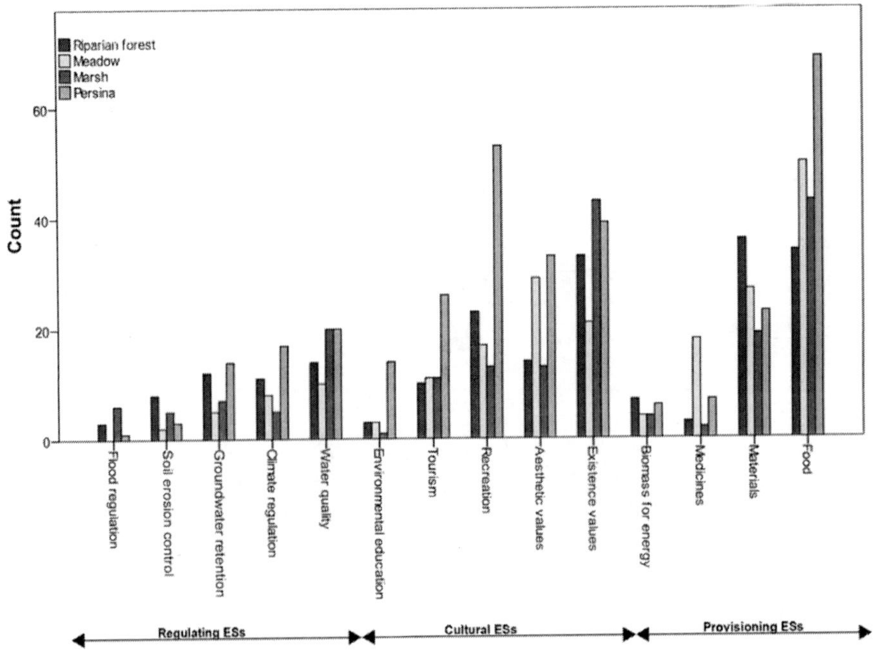

Figure. 3 The awareness of ESs expressed as the amount of times the ES was mentioned by all respondents for Persina as a whole and for each wetland type

When linking ESs to specific wetland types in Persina, cultural and provisioning ESs were mentioned the most (Fig. 3). Regulating services were not mentioned that often, but a few respondents did associate flood regulation and soil erosion control with marshes and riparian forests, whereas they did not mention these services when asked for ESs provided by Persina in general. Alternatively, several ESs were mentioned often when linked to Persina in general, but were mentioned less often when linked to specific wetland types: this was particularly the case for environmental education, tourism, recreation, and the provision of food.

There were significant differences between the awareness of ESs provided by meadows, mashes and riparian forests ($\chi^2 = 83.219$, df = 29, $p < 0.05$). Meadows were associated with the provision of medicine and food, more than the other wetland types. Biodiversity and water quality were mostly associated with marshes, and the provision of materials and recreation were mostly associated with riparian forests. Meadows were associated more with aesthetic quality than marshes and riparian forests. Twenty-two percent of the respondents had negative associations with marshes, mostly regarding the presence of mosquitoes. Two respondents associated marshes with a threat to floods. Ten percent of the respondents had negative associations with riparian forests, which surprisingly related most to a fear of flooding. In addition a few respondents mentioned that riparian forests were not easily accessible and that they are not properly managed. Two respondents had negative associations with meadows and they both mentioned improper management of the area.

3.3.2 How are Knowledge of and Visits to Persina Related to the Awareness of ESs?

As shown by previous results, most respondents were not aware of the regulating ESs provided by Persina. Since many regulating services are not visible (in contrast to the provisioning ESs) it may require additional knowledge to be aware of them. To see whether those who were aware of the regulating ESs may have had better (general) knowledge about Persina Park, we asked respondents several questions regarding their knowledge of Persina. Of all respondents, 80 % had heard about Persina before and 75 % knew that Persina had a protected area status. At the same time, however, only 37 % had heard about the wetland restoration project at Persin island. We performed a chi-square test to see whether people who knew about the wetland restoration project, were also more aware of regulating ESs provided by wetlands in Persina, but found no significant results.

Respondents who visited Persina more often may also have been more aware of particular ESs provided by Persina. We therefore assessed several linkages between awareness of ESs and the visits people made to

Persina: type of visits, frequency of visits and location of visits. We grouped all responses to each separate wetland type (i.e. marshes, meadows and riparian forests) into one larger variable, reflecting whether people had recognized any ESs for any of the wetland ecosystems in Persina. Since we studied a large number of ESs, we made a selection for further analysis based on the results in Section Do people recognize the ESs provided by wetlands in Persina?. From each category, i.e. cultural, provisioning, and regulating ESs, we selected the ES that was mentioned most for the wetland types: water quality, existence values for biodiversity and the provision of food.

Significant differences were found for all three ESs (Table 2). Respondents who visited Persina frequently, more often recognized water quality as an ES than respondents who visited Persina a few times or less, did. In addition, respondents who went walking in Persina or went to Persina to collect food were more aware of water quality than respondents who did not. Respondents who went to the Danube River during visits also recognized water quality more often than respondents who did not. Respondents who visited the Danube River also recognized the ability of wetlands to support biodiversity more often than respondents who did not visit the Danube river. Not surprisingly, respondents who collected food in Persina mentioned food as an ES more often than respondents who did not collect food.

Table 2 Chi-square statistics (χ^2) for frequency, location and type of visits to Persina and awareness of wetland ESs

Visits to Persina	Awareness of wetland ESs		
	Water quality	Biodiversity	Food
Frequency of visits	**4.867***	1.687	2.444
Visits Danube river and/or islands	**3.217***	**3.265***	0.094
Walking	**5.217****	0.253	1.234
Hunting	–	–	–
Collecting food	**5.074****	0.0008	**6.575****

For hunting the χ^2 could not be calculated due to limited responses. These fields were left blank

*Significance level at 10 %, **Significance level at 5 %

3.4 Values for ESs Provided by Persina

To account for both self-oriented values and other-oriented values, the respondents were asked to value ESs first for their own personal well-being and consequently for the well-being of all residents living in Persina. Overall, differences between the two types of value were not significant ($\chi^2 = 9.958$, df = 14, $p = 0.765$). For both self-oriented and other-oriented values, the provisioning of food and materials, recreation, existence of biodiversity, aesthetic values, and tourism were found to be the most important ESs (Fig. 4).

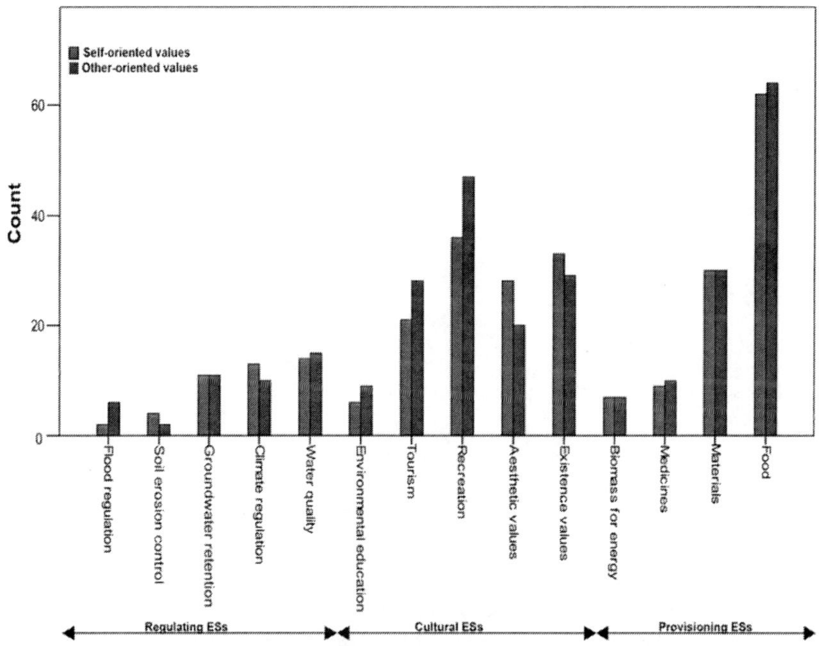

Figure. 4 The socio-cultural values for ESs expressed as the amount of times the ES was mentioned for personal well-being (self-oriented) and well-being of everybody living in Persina (other-oriented)

3.5 Differences in Awareness and Values Between Different Users

There were significant differences between user groups in their awareness of ESs (Table 3). Fishermen recognized food as an ES significantly more often than did farmers and residents. Fishermen also recognized materials as an ES significantly more than residents did. Farmers were significantly more aware of regulating ESs than residents, specifically water quality (Table 3).

Table 3 Percentages of people who recognized and valued each

Awareness of ESs					Values of ESs				
	Farmers ($n = 33$)	Fishermen ($n = 24$)	Residents ($n = 45$)	χ^2		Farmers ($n = 33$)	Fishermen ($n = 24$)	Residents ($n = 45$)	χ^2
Regulating ESs					*Regulating ESs*				
Flood regulation	3	0	0	–	Flood regulation	0	0	4.4	–
Soil erosion control	0	4.2	4.5	–	Soil erosion control	0	0	8.9	–
Ground water retention	18.2	8.3	13.6	–	Ground water retention	12.1	13.0	8.9	–
Climate regulation	24.2	20.8	9.1	3.624	Climate regulation	18.2	8.7	11.1	–
Water quality	30.3[a]	25	9.1[a]	**6.117***	Water quality	9.1	21.7	13.3	–
Cultural ESs					*Cultural ESs*				
Environmental education	15.2	4.2	18.2	–	Environmental education	6.1	0	8.9	–
Tourism	21.2	29.2	27.3	0.522	Tourism	9.1	34.8	22.2	5.126
Recreation	51.5	62.5	47.7	1.576	Recreation	24.2	52.2	35.6	4.039
Aesthetic values	30.3	45.8	27.3	2.721	Aesthetic values	24.2	21.7	33.3	1.480
Existence values of biodiversity	48.5	25	38.6	3.252	**Existence values for biodiversity**	46[a]	39.1	20[a]	**6.016***
Provisioning ESs					*Provisioning ESs*				
Provision of biomass for energy	0	8.3	9.1	–	Provision of biomass for energy	0	17.4	6.7	–
Provision of medicines	9.1	0	9.1	–	Provision of medicines	9.1	8.7	8.9	–
Provision of materials	18.2	41.7[a]	15.9[a]	**6.643***	**Provision of materials**	30.3	60.9[a]	13.3[a]	**15.285****
Provision of food	63.6[a]	91.7[a,b]	59.1[b]	**8.572***	**Provision of food**	60.6[b]	91.3[a,b]	46.7[a]	**10.949****

ES Significant values indicate that there was a significant difference between, at least two of, the user groups

*Significance level at 5 %, **Significance level at 1 %. For some services the χ^2 could not be calculated due to limited responses. These fields were left blank. Values marked with the same letter are significantly different (calculated through post-hoc pairwise comparisons, $p < 0.05$)

Despite the higher awareness of water quality, farmers did not mention this ES significantly more often than residents when asked to indicate which ESs were most important for their personal well-being. Across all user groups, the cultural and provisioning ESs were found most important. Significant differences were found for biodiversity, which was more important to farmers than to residents.

Provision of materials and food were both found most important by the fishermen. Although provisioning services were important for fishermen, they also assigned a lot of value to cultural ESs, specifically recreation. The most important cultural ESs for residents were recreation and aesthetic values.

3.6 Links Between Values for ESs and Attitudes Towards Wetland Restoration

Results from the PCA demonstrated four different perspectives on wetland restoration (Table 4). Four components with Eigen values larger than 1 were found, explaining 67.5 % of the total variance (Kaiser-Meyer-Olkin measure of sampling adequacy, KMO = 0.700; Bartlett's test of sphericity, $\chi^2 = 285.605$, df =45, $p < 0.001$). The Promax component correlation matrix indicated a slight correlation ($r^2 = 0{,}389$) between the first and the second component.

Table 4 Four perspectives were identified linking attitude toward wetland restoration and values for ES categories. Loadings between −0.4 and 0.4 are excluded from the table

Variables	Cultural perspective (32,5 %)	Eco-centric perspective (14,5 %)	Economic perspective (10,5 %)	Negative perspective (10 %)
Wetlands are important for the local population	.769			
Wetlands are important for downstream communities	.748			
Value for cultural ESs	.740			
Wetlands should not be drained for agricultural purposes		.824		
Wetlands are an important part of the landscape in this region		.636		
Wetlands are important for plants and animals		.628		−.425

Value for regulating ESs			–.803	
Value for provisioning ESs			.661	
The restoration of wetlands limits the economic growth of this region	–.511		.548	
The restoration of wetlands is a waste of valuable land				.878

The first component explained 32.5 % of the variance and can be understood as supporting wetland restoration from a *cultural perspective*, emphasizing the importance of wetlands for human well-being, especially for the provision of cultural ecosystem services. Both the statement 'Wetlands are important for the local population' and 'Wetlands are important for downstream communities' correlated positively with this component. The statement 'the restoration of wetlands limits the economic growth of this region' correlated negatively with this component, suggesting that wetland restoration and economic development do not necessarily conflict.

The second component explained 14.5 % of the variance and can be explained as supporting wetland restoration from an *eco-centric perspective*, with a focus on the importance of wetlands for plants and animals. No correlation was found with any of the ecosystem service variables, which may indicate that this perspective can be understood as seeing a wetland as 'nature for itself', without making strong links with benefits for people. This is confirmed by the positive correlation of the statement "wetlands should not be drained for agricultural purposes".

The third component explained 10,5 % of the variance and can be understood as viewing wetland restoration from an *economic perspective*. Appreciation of provisioning services correlated with this component, while appreciation of regulating services negatively correlated with this component. As the statements concerning the importance of wetlands did not strongly correlate (either positively or negatively) with this component, it is difficult to judge whether this component indicates a negative or positive attitude toward wetland restoration. The statement 'the restoration of wetlands limits the economic growth of this region' does not necessarily imply a negative attitude towards wetland restoration, but acknowledges that there is a trade-off between economic prosperity and wetland restoration.

The final component demonstrates a negative attitude towards wetland restoration. There was no correlation with appreciation for ESs, suggesting that a negative perspective may go together with low values for all ESs, i.e. seeing nature as providing no benefits.

Figure 5 shows that respondents are heterogeneously distributed across the four perspectives. Using a one-way ANOVA we tested for differences between the user groups for all components. No significant differences were found for component 2 and 4.

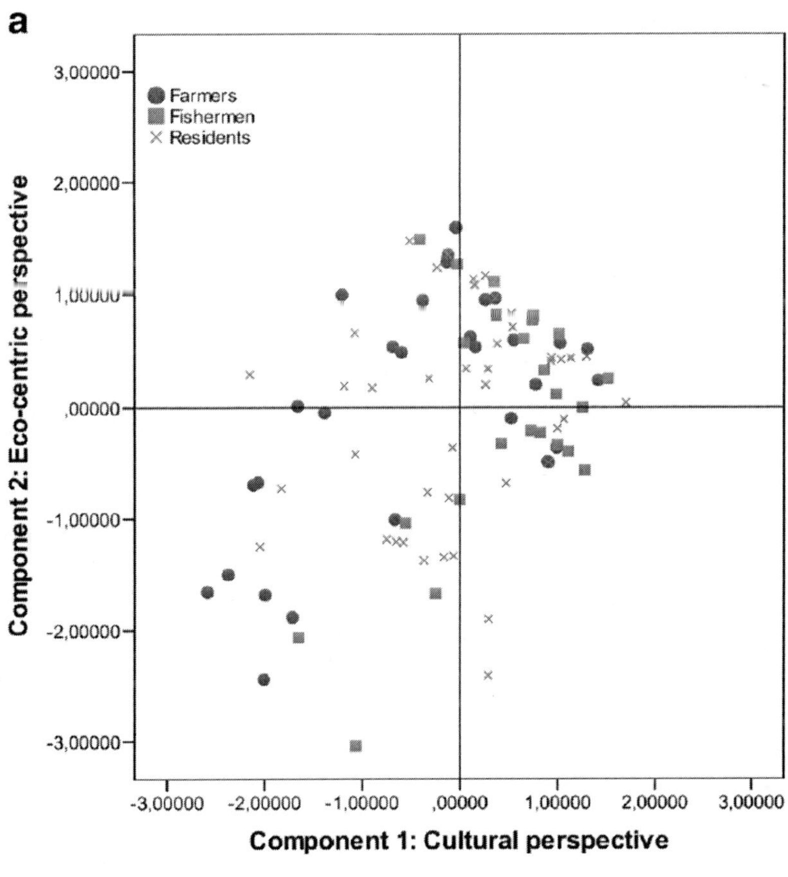

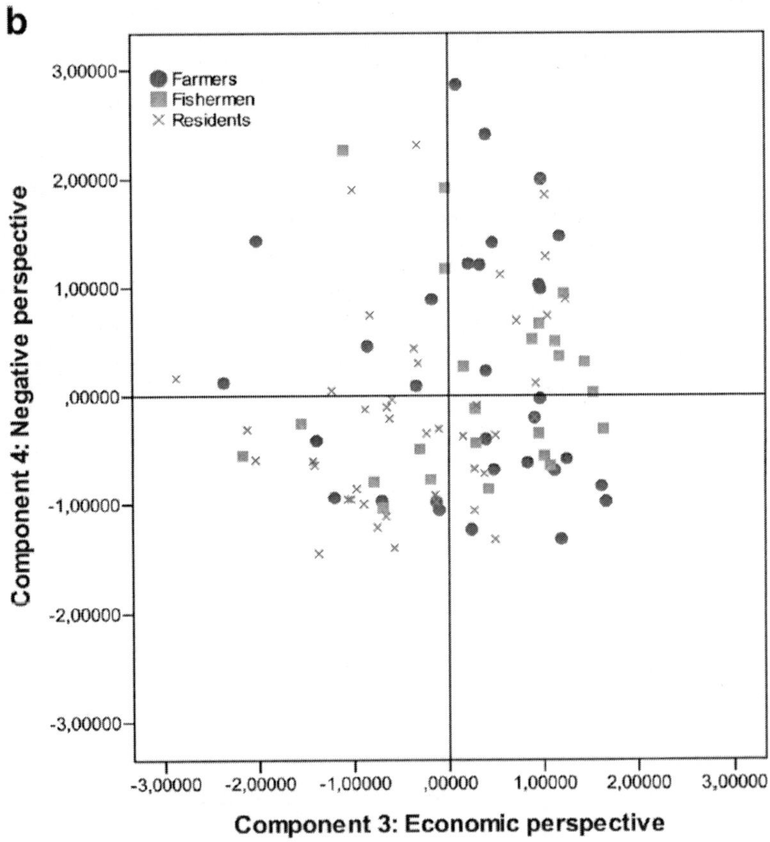

Figure. 5 Scatterplots of PCA scores for each respondent, categorized by the different user groups

We did find significant differences between the user groups for component 1 ($F(2,99) = 5.985$, $p < 0.01$) and component 3 ($F(2,99) = 4.126$, $p < 0.05$). For component 1, the Tukey post-hoc test revealed a significant difference between farmers and fishermen ($p < 0.01$): whereas farmers had mostly negative scores for the cultural perspective, fishermen had mostly positive scores for the cultural perspective. For component 3, we found a significant difference between fishermen and residents ($p < 0.05$) and farmers and residents ($p < 0.1$): residents had lower scores for the economic perspective than did farmers and fishermen.

4. DISCUSSION

4.1 Linking use, Awareness and Valuation of ESs

Across the entire sample, we found a low appreciation for regulating ESs and a higher appreciation for cultural and provisioning ESs. Agbenyega et al. (2009) suggested that regulating ESs may be overlooked because of the invisible character of these services. People may need better understanding of such services in order to appreciate them. Our results show that farmers mentioned water quality more often than the residents did. As farmers need some level of ecological knowledge to maintain their lands, thereby interacting with ecosystems in a different manner than residents do, they may indeed have a better understanding of the ecological functioning of an ecosystem. At the same time, however, our results demonstrate that despite the higher awareness of water quality as an ES, farmers did not place higher values on water quality than the residents. Regulating services are often not considered important because people do not see the end-benefit of such services or how they are relevant in their lives. An improved general awareness of ESs does not necessarily have to lead to a higher appreciation of ESs. Instead, use of and experience with the ecosystem may be more important for the appreciation of certain ESs (Setten et al. 2012; Pröpper and Haupts 2014).

The importance of provisioning ESs, especially the provision of food, is not surprising given the economic dependency of farmers and fisherman on the resources provided by Persina. An emphasis on the importance of the material benefits of an ecosystem is common in areas where social and ecological systems are tightly linked (Rönnbäck et al. 2007; Hartter 2010). This does not mean that immaterial benefits are disregarded or not considered important: our results show that cultural ESs were valued highly, by all user groups. This is in line with findings by Calvet-Mir et al. (2012), who studied home-gardens and found that, despite the importance of the provision of food, the cultural services, such as cultural heritage and aesthetic values, 'played a central role in explaining the societal importance attributed to home gardens' (p. 159). What is particularly apparent in our study is that farmers and fishermen, who had high values for provisioning services, also mentioned cultural ESs more often than residents. Residents placed the highest values on aesthetic quality of the landscape, but fishermen placed high values on recreation and tourism. In addition, farmers found the support of biodiversity significantly more important than did the residents. This suggests, that even though there may be conflicts between agriculture and biodiversity, being a farmer may promote a sense of stewardship, by which a strong connection with nature can be established.

The importance fishermen assigned to both recreation and food provision may indicate the limitations of treating ESs as separate entities. This problem has been discussed by several other scholars who indicate that ESs are often interlinked, and so it may be difficult for people to value them separately (Asah et al. 2014; Pröpper and Haupts 2014). The collection of food may be important for subsistence, but it may also contribute towards social cohesion. On a similar level, fishing as an economic activity may have multiple dimensions. Fishing provides income, but being out on the river may simultaneously have recreational aspects. Future ES research should focus on identifying the linkages between services and should acknowledge the myriad of ways with which people interact and relate with nature (Russell et al. 2013; Chan et al. 2016). Insight into how people connect with nature may provide us with the information necessary to strengthen this relationship, thereby promoting a nature conservation ethic.

4.2 Linking ESs to Wetland Types

Previous ES studies have often related socio-cultural values to ESs provided by a conservation area in general, not differentiating between specific ecosystems or ecosystem components (Castro et al. 2011; Tengberg et al. 2012; Allendorf and Yang 2013; Petrosillo et al. 2013; Sagie et al. 2013). Our results show that there is substantial difference between asking people for their awareness of ESs provided by Persina in general and the specific wetland types. Fifty-three percent of the respondents associated Persina with recreation, but less than 25 % associated recreation with any of the wetland types. Such discrepancies were also found for tourism, environmental education, and the provision of food. This could mean that people associated different types of ecosystems and/or landscapes with these services. People may have associated the provision of food most with the agricultural fields present in Persina, or the Danube River itself, rather than with the wetlands within the park. It could however, also mean that people had a general perception of the landscapes in Persina, which did not completely correspond with the landscapes they saw on the photos.

Scholars studying landscape perceptions and preferences argue that people do not only form opinions about the landscape on the basis of the biophysical characteristics of that landscape, but also by 'mental images' (Lindemann-Matthies et al. 2010) or conceptions of what a landscape should look like (Terkenli 2001). Such mental images may be based on information, experiences, knowledge, institutions, and social networks. Comparing visual versus verbal techniques to capture forest landscape preferences, Tahvanainen et al. (2001) showed that preconceptions (i.e. mental images, about landscape management measures) did not match visual perceptions of these measures. The verbal assessment revealed a negative attitude towards small clear cutting in forests, while the visual assessment revealed that respondents thought small clear cutting enhanced the scenic value of forests. Visual illustration may thus be necessary for people to properly evaluate issues related to landscape and/or ecosystem management.

While landscape visualizations are commonly used in landscape preference studies (Kohsaka and Flitner 2004; Dramstad et al. 2006; Lange et al. 2008; Arnberger and Eder 2011; Barroso et al. 2012), they are less often used in ES studies. When the goal of the research is to elucidate general environmental values this does not have to be problematic, but in the case of ESs, researchers are often interested in assigned values, related to specific ecosystem and/or landscape types. Pietrzyk-Kaszyńska et al. (2012) suggested that people are more positive towards 'nature' on a general level than they are towards specific landscapes. Our results confirm this notion because only 7 % of the respondents had negative associations when asked about Persina in general, while more respondents had negative associations with wetland ecosystems. Twenty-two percent of the respondents had negative associations with marshes. The importance of targeted information about specific areas is invaluable when it comes to promoting public support and sustainable use of conservation areas.

4.3 Understanding Attitudes Towards Wetland Restoration from Different Social Perspectives

The four perspectives we identified demonstrate the different motivations people may have to support or oppose wetland restoration. These motivations can be explained by the way people perceive and interact with wetland ecosystems (Connely et al. 2002; Aggestam 2013). Farmers, who do not participate in recreation activities much, do not find the recreational value of wetland ecosystems very important and put most emphasis on the interplay between wetland restoration and economic benefits. Similar results have been found by Aggestam (2013), who reported that landowners highlighted the economic values of wetlands. Farmers are therefore likely to support wetland restoration when they can reap some of the benefits of the restored ecosystem services. This also means that when wetland restoration brings additional costs to local farmers, they may oppose to wetland restoration efforts. Fishermen on the other hand, who spend a lot of time near and on the river, recognize the multiple purposes of wetland restoration. As indicated by the importance they assigned to both provision of food and recreation, they may support wetland restoration because wetlands support their fisheries, e.g. by providing spawning areas for fish, but also because wetlands provide places where people can come at ease and connect with nature. In addition, residents in Persina may support wetland restoration out of concern for nature itself, not thinking about the benefits that humans obtain from wetland ecosystems.

Our findings are consistent with findings from other scholars who have identified underlying motivations for biodiversity and ES conservation. Studying local motivation for biodiversity conservation, Johansson (2005) identified three personal motives for biodiversity conservation: (a) consideration of human well-being and recreation, (b) human survival, and (c) respect for nature. Similarly, Opdam et al. (2015) put forward a socio-cultural frame, a sustainability frame, and an economic frame. In ES research much effort has been spent on clarifying links between values for ESs and socio-demographic characteristics, while environmental value orientations may be more closely linked to the assigned importance to ESs.

Environmental value orientations underlie attitudes and normative beliefs (Vaske and Donnelly 1999) and can be found on a continuum between eco-centric and anthropocentric value orientations. Whereas anthropocentric value orientations stress the instrumental use of nature, corresponding to the cultural and economic perspective, eco-centric value orientations put more emphasis on the intrinsic value of nature, corresponding to the eco-centric perspective. The importance of environmental value orientations as underlying constructs for assigned values and attitudes has been put forward by several scholars in environmental psychology (Stern and Dietz 1994; Fulton et al. 1996; Vaske and Donnelly 1999; Kaltenborn and Bjerke 2002), but the link between environmental value orientations and values for ESs has been poorly studied.

The ES framework provides an anthropocentric frame for communicating the benefits of nature conservation by emphasizing the benefits provided by nature for human well-being. Opdam et al. (2015) argue, however, that even within the larger frame of the ES approach, the three ES categories, cultural, regulating and provisioning ESs, correspond with three different storylines. In their framework cultural ESs correspond with a socio-cultural frame that emphasizes cultural benefits, regulating ESs correspond with a sustainability frame that emphasizes benefits of ecological processes and provisioning ESs correspond with an economic frame that puts emphasis on the profits made from ecosystems. Our results partially confirm this conceptualization, although we have found two perspectives that did not correspond with values for ES categories.

The eco-centric perspective did not correlate with values for ESs. This may indicate that the anthropocentric character of the ES framework, with a focus on instrumental values, does not necessarily fit with eco-centric motivations for nature conservation (McCauley 2006; Norgaard 2010;

Luck et al. 2012). ESs are commonly framed as the ecological functions that contribute to human well-being. A person with a strong eco-centric value orientation may not be willing to evaluate nature in this manner. Although several scholars have argued that socio-cultural ES assessments may allow for the integration of eco-centric values (Jax et al. 2013; Schröter et al. 2014), most ES valuation studies have not attempted to explicitly address eco-centric values.

The negative perspective did also not correlate with values for ESs. An explanation for this may be that people with a dominant negative attitude towards nature may put more emphasis on the costs of nature conservation instead of the benefits. One of the critiques on the ES framework is its positive framing, implying that all outcomes of ecological processes are desirable (McCauley 2006; Schröter et al. 2014). Our results show that 22 % of the respondents had negative associations with marshes: respondents referred mostly to poor management, mosquitoes, and inaccessibility. This indicates that for a group of the respondents, wetland restoration may come with a certain cost that outweighs the associated benefits. For the success of wetland restoration efforts it is important to give attention to both the social costs and benefits of such projects. Awareness raising programs should therefore not only focus on the benefits provided by ecosystems, but also include strategies that deal with potential costs of wetland restoration.

4.4 Fostering Public Support for Wetland Restoration

In this study, most fishermen had positive values for both the cultural and economic perspective on ESs. Environmental perspectives therefore do not have to be mutually exclusive: every person is likely to have multiple perspectives, but in differing strengths (Kollmuss and Agyeman 2002). Even though people may adhere to multiple motivations for wetland restoration, when goals related to these motivations conflict, a choice has to be made. De Groot and Steg (2007) argue that 'although altruistic and environmental values may be correlated, they seem to be clearly differently related to environmental beliefs and intentions when altruistic and biospheric goals conflict' (p. 349). As an example: many respondents in this study highly valued biodiversity, but at the same time highly valued provisioning ESs. On a general level, people may be sympathetic towards the conservation of biodiversity, but when it comes to daily practices and actual environmental behavior, activities that conflict with biodiversity conservation still gain priority. This emphasizes the necessity of wetland restoration projects to identify synergies and trade-offs, making them better capable of dealing with both complementary and competing goals. As expressed by McShane et al. (2011), win-win scenarios are scarce and nature conservation involves hard choices, which need to be made explicit.

To foster support for wetland restoration and promote sustainable use wetland restoration areas, a targeted approach is necessary. Raising awareness can be a useful strategy to foster public support for wetland restoration, but different groups of people may require different kinds of information. Reflecting upon the theory of Festinger (1957), Kollmuss and Agyeman (2002) put forward the idea that people may show a resistance against non-conforming information, meaning that 'information that supports our existing values and mental frameworks is readily accepted, whereas information that contradicts or undermines our beliefs is avoided or not perceived at all' (p. 254). This means that the benefits and losses of wetland restoration should be "framed" in ways that resonate with the public (Groffman et al. 2010). The results from this paper suggest that gaining insight into how people interact with and perceive ecosystems is important to understand what aspects of ecosystem conservation and/or restoration people put most emphasis on. The perspectives we identified provide guidance for targeted information that can be used by local environmental organizations to reach a diverse set of stakeholders.

In addition, careful thought should be put into crafting opportunities to engage with local users about wetland restoration (Varner 2014). It is important to get more interested in our audience beyond treating them as 'recipients for expert knowledge' (Salmon et al. 2015), as most people do not learn about science through formal education but through informal sources (Groffman et al. 2010). When forming ideas and attitudes about ecological restoration, people often do not rely on logical arguments and reasoning, but on knowledge they draw from experiences, e.g. by being in nature or by talking with others, and/or emotions (Brody 2005; Fazey et al. 2013). Having local stakeholders actively participate in ecological restoration efforts, for example by assisting in environmental monitoring, may positively affect the way local users think about ecological restoration (Conrad and Hilchey 2011). In addition, scientists should make use of informal and non-scientific events to exchange ideas about the benefits and costs of ecological restoration with the public (Sayer et al. 2014). By actively engaging with local users, scientists and environmental managers can foster support for wetland restoration projects, which is likely to increase the success of wetland restoration and conservation (Cooke et al. 2013).

ACKNOWLEDGMENTS

We would like to thank Emilia Petkova, Vasilka Petrova, Zlatko Kamenarov and Channa Lozeva for their assistance with the interviews. This work was supported by the European Union Seventh Framework Programme (FP7/2007-2013) under grant agreement no. 303 393 "OPERAs".

APPENDIX

Table 5 Characteristics of the sample

Socio-economic variables		Respondents ($n = 102$)*
Grew up in Persina (%)		90
Gender (%)	Male	66.7
	Female	33.3
Age (M, *SD*)	45.9 (*14.4*)	
Education (%)	Primary	13.7
	Secondary	66.7
	Medium professional formation	9.8
	University	9.8
Occupation (%)	Farmer	33.3
	Fisherman	24.5
	Business owner	4.9
	Other	16.7
	Retired	10.8
	Unemployed	9.8
Monthly net household income (%)	no income	1.0
	0–50 euro	3.9
	51–255 euro	46.1
	256–510 euro	40.1
	511–1022 euro	6.9
	>1022 euro	2.0
Household size (M, *SD*)	4 (*2.8*)	

REFERENCES

1. Adams VM, Pressey RL, Naidoo R (2010) Opportunity costs: Who really pays for conservation? Biol Conserv 143:439–448. doi:10.1016/j.biocon.2009.11.011.
2. Agbenyega O, Burgess PJ, Cook M, Morris J (2009) Application of an ecosystem function framework to perceptions of community woodlands. Land Use Policy 26:551–557. doi:10.1016/j.landusepol.2008.08.011.
3. Aggestam F (2013) Wetland restoration and the involvement of stakeholders: an analysis based on value-perspectives. Landsc Res 39:680–697. doi:10.1080/ 01426397. 2013.819076.
4. Allendorf TD, Yang J (2013) The role of ecosystem services in park–people relationships: the case of gaoligongshan nature reserve in southwest China. Biol Conserv 167:187–193. doi:10.1016/j.biocon.2013.08.013.
5. Arnberger A, Eder R (2011) Exploring the heterogeneity of rural landscape preferences: an image-based latent class approach. Landsc Res 36:19–40. doi:10.1080/ 01426397.2010.536204.
6. Asah ST, Guerry AD, Blahna DJ, Lawler JJ (2014) Perception, acquisition and use of ecosystem services: human behavior, and ecosystem management and policy implications. Eco Ser. doi:10.1016/j.ecoser.2014.08.003.
7. Barroso FL, Pinto-Correia T, Ramos IL et al (2012) Dealing with landscape fuzziness in user preference studies: photo-based questionnaires in the Mediterranean context. Landsc Urban Plan 104:329–342.
8. Brockington D, Wilkie D (2015) Protected areas and poverty. Philos Trans R Soc Lond B Biol Sci 370:20140271. doi:10.1098/rstb.2014.0271.
9. Brody M (2005) Learning in nature. Environ Educ Res 11:603–621. doi:10. 1080/ 13504620500169809.
10. Calvet-Mir L, Gómez-Baggethun E, Reyes-García V (2012) Beyond food production: ecosystem services provided by home gardens. A case study in vall fosca, Catalan Pyrenees, northeastern Spain. Ecol Econ 74:153–160. doi:10.1016/j. ecolecon. 2011.12.011.
11. Castro AJ, Martín-López B, García-LLorente M et al (2011) Social preferences regarding the delivery of ecosystem services in a semiarid Mediterranean region. J Arid Environ 75:1201–1208. doi:10.1016/j.jaridenv.2011.05.013.
12. Chan KMA, Balvanera P, Benessaiah K et al (2016) Opinion: Why protect nature? rethinking values and the environment. Proc Natl Acad Sci 113:1462–1465. doi:10.1073/pnas.1525002113.
13. Čížková H, Květ J, Comín FA et al (2011) Actual state of European wetlands and their possible future in the context of global climate change. Aquat Sci 75:3–26. doi:10.1007/s00027-011-0233-4
14. Comberti C, Thornton TF, Wylliede Echeverria V, Patterson T (2015) Ecosystem services or services to ecosystems? valuing cultivation and reciprocal relationships between humans and ecosystems. Glob Environ Chang 34:247–262. doi:10.1016/ j.gloenvcha. 2015.07.007
15. Connely NA, Knuth BA, Kay DL (2002) Public support for ecosystem restoration in the Hudson River Valley, USA. Environ Manag 29:467–476. doi:10.1007/s00267-001-0033-Z.
16. Conrad CC, Hilchey KG (2011) A review of citizen science and community-based environmental monitoring: issues and opportunities. Environ Monit Assess 176:273–291. doi:10.1007/s10661-010-1582-5.
17. Cooke SJ, Lapointe NWR, Martins EG et al (2013) Failure to engage the public in issues related to inland fishes and fisheries: strategies for building public and political will to promote meaningful conservationa. J Fish Biol 83:997–1018. doi:10.1111/jfb.12222
18. Costanza R, d'Arge R, de Groot R et al (1997) The value of the world's ecosystem services and natural capital. Nature 387:253–260. doi:10.1038/387253a0.

19. Daily GC (1997) Nature's services: societal dependence on natural ecosystems. Island Press Washington DC, Washington.
20. Daw T, Brown K, Rosendo S, Pomeroy R (2011) Applying the ecosystem services concept to poverty alleviation: the need to disaggregate human well-being. Environ Conserv 38:370–379. doi:10.1017/S0376892911000506.
21. de Groot JIM, Steg L (2007) Value orientations to explain beliefs related to environmental significant behavior: How to measure egoistic, altruistic, and biospheric value orientations. Environ Behav 40:330–354. doi:10.1177/0013916506297831.
22. de Groot RS, Wilson MA, Boumans RM (2002) A typology for the classification, description and valuation of ecosystem functions, goods and services. Ecol Econ 41:393–408. doi:10.1016/S0921-8009(02)00089-7.
23. de Groot R, Fisher B, Christie M, et al. (2010) Integrating the ecological and economic dimensions in biodiversity and ecosystem service valuation. Econ. Ecosyst. Biodivers. Ecol. Econ. Found
24. de Groot R, Brander L, van der Ploeg S et al (2012) Global estimates of the value of ecosystems and their services in monetary units. Ecos Ser 1:50–61. doi:10.1016/j.ecoser.2012.07.005
25. Dobbie M, Green R (2013) Public perceptions of freshwater wetlands in Victoria, Australia. Landsc Urban Plan 110:143–154. doi:10.1016/j.landurbplan.2012.11.003
26. Dramstad WE, Tveit MS, Fjellstad WJ, Fry GLA (2006) Relationships between visual landscape preferences and map-based indicators of landscape structure. Landsc Urban Plan 78:465–474. doi:10.1016/j.landurbplan.2005.12.006
27. Ebert S, Hulea O, Strobel D (2009) Floodplain restoration along the lower Danube: a climate change adaptation case study. Climate Dev 1:212–219. doi:10.3763/cdev.2009.0022
28. Fazey I, Evely AC, Reed MS et al (2013) Knowledge exchange: a review and research agenda for environmental management. Environ Conserv 40:19–36. doi:10.1017/S037689291200029X.
29. Festinger L (1957) A theory of cognitive dissonance. Stanford University Press, Stanford.
30. Fulton DC, Manfredo MJ, Lipscomb J (1996) Wildlife value orientations: a conceptual and measurement approach. Hum Dimens Wildl 1:24–47. doi:10.1080/10871209609359060.
31. Grêt-Regamey A, Celio E, Klein TM, Wissen Hayek U (2013) Understanding ecosystem services trade-offs with interactive procedural modeling for sustainable urban planning. Landsc Urban Plan 109:107–116. doi:10.1016/j.landurbplan.2012.10.011.
32. Groffman PM, Stylinski C, Nisbet MC et al (2010) Restarting the conversation: challenges at the interface between ecology and society. Front Ecol Environ 8:284–291. doi:10.1890/090160.
33. Gumiero B, Mant J, Hein T et al (2013) Linking the restoration of rivers and riparian zones/wetlands in Europe: sharing knowledge through case studies. Ecol Eng 56:36–50
34. Haines-Young RH, Potschin MB (2010) The links between biodiversity, ecosystem services and human well-being. Ecosyst. Ecol. a new Synth
35. Hartter J (2010) Resource use and ecosystem services in a forest park landscape. Soc Nat Resour 23:207–223. doi:10.1080/08941920903360372.
36. Hulea O, Ebert S, Strobel D (2009) Floodplain restoration along the lower Danube: a climate change adaptation case study IOP conference series. Earth Environ Sci 6:402002. doi:10.1088/1755-1307/6/40/402002.
37. Jax K, Barton DN, Chan KMA et al (2013) Ecosystem services and ethics. Ecol Econ 93:260–268. doi:10.1016/j.ecolecon.2013.06.008.
38. Johansson M (2005) Local people's motives for biodiversity conservation. Hogrefe & Huber Publishers, Cambridge.
39. Junk WJ, An S, Finlayson CM et al (2012) Current state of knowledge regarding the world's wetlands and their future under global climate change: a synthesis. Aquat Sci 75:151–167. doi:10.1007/s00027-012-0278-z.
40. Kaltenborn BP, Bjerke T (2002) Associations between environmental value orientations and landscape preferences. Landsc Urban Plan 59:1–11. doi:10.1016/S0169-2046(01)00243-2.

41. Kari S, Korhonen-Kurki K (2013) Framing local outcomes of biodiversity conservation through ecosystem services: a case study from ranomafana, Madagascar. Ecos Ser 3:e32–e39. doi:10.1016/j.ecoser.2012.12.003.
42. Kohsaka R, Flitner M (2004) Exploring forest aesthetics using forestry photo contests: case studies examining Japanese and German public preferences. Forest Policy Econ 6:289–299. doi:10.1016/j.forpol.2004.03.016.
43. Kollmuss A, Agyeman J (2002) Mind the Gap : Why do people act environmentally and what are the barriers to pro- environmental behavior ? 37–41.
44. Lange E, Hehl-Lange S, Brewer MJ (2008) Scenario-visualization for the assessment of perceived green space qualities at the urban–rural fringe. J Environ Manag 89:245–56. doi:10.1016/j.jenvman.2007.01.061.
45. Lindemann-Matthies P, Junge X, Matthies D (2010) The influence of plant diversity on people's perception and aesthetic appreciation of grassland vegetation. Biol Conserv 143:195–202. doi:10.1016/j.biocon.2009.10.003.
46. Luck GW, Chan KMA, Eser U et al (2012) Ethical considerations in on-ground applications of the ecosystem services concept. Bioscience 62:1020–1029. doi:10.1525/bio.2012.62.12.4.
47. MA (Millenium Ecosystem Assessment) (2005) Ecosystems and human well-being - synthesis. Island Press, Washington.
48. McCauley DJ (2006) Selling out on nature. Nature 443:27–8. doi:10.1038/443027a.
49. McShane TO, Hirsch PD, Trung TC et al (2011) Hard choices: making trade-offs between biodiversity conservation and human well-being. Biol Conserv 144:966–972. doi:10.1016/j.biocon.2010.04.038.
50. Meindl CF (2000) Past perception of the great American wetland: Florida's everglades during the early twentieth century. Environ Hist 5:378–395
51. Minteer BA, Miller TR (2011) The New conservation debate: ethical foundations, strategic trade-offs, and policy opportunities. Biol Conserv 144:945–947. doi:10.1016/j.biocon.2010.07.027.
52. Nassauer JI (2004) Monitoring the success of metropolitan wetland restorations: cultural sustainability and ecological function. Wetlands 24:756–765. doi:10.1672/0277-5212(2004)024[0756:MTSOMW]2.0.CO;2.
53. Naughton-Treves L, Holland MB, Brandon K (2005) The role of protected areas in conserving biodiversity and sustaining local livelihoods. Annu Rev Environ Resour 30:219–252. doi:10.1146/annurev.energy.30.050504.164507.
54. Norgaard RB (2010) Ecosystem services: from eye-opening metaphor to complexity blinder. Ecol Econ 69:1219–1227. doi:10.1016/j.ecolecon.2009.11.009.
55. Opdam P, Coninx I, Dewulf A et al (2015) Framing ecosystem services: affecting behaviour of actors in collaborative landscape planning? Land Use Policy 46:223–231. doi:10.1016/j.landusepol.2015.02.008.
56. Petrosillo I, Costanza R, Aretano R et al (2013) The use of subjective indicators to assess how natural and social capital support residents' quality of life in a small volcanic island. Ecol Indic 24:609–620. doi:10.1016/j.ecolind.2012.08.021.
57. Pietrzyk-Kaszyńska A, Cent J, Grodzińska-Jurczak M, Szymańska M (2012) Factors influencing perception of protected areas—the case of natura 2000 in polish Carpathian communities. J Nat Conserv 20:284–292. doi:10.1016/j.jnc.2012.05.005.
58. Pröpper M, Haupts F (2014) The culturality of ecosystem services. Emphasizing process and transformation. Ecol Econ 108:28–35. doi:10.1016/j.ecolecon.2014.09.023.
59. Raymond CM, Singh G, Benessaiah K et al (2013) Ecosystem services and beyond. Bioscience 63:536–546. doi:10.1525/bio.2013.63.7.7
60. Rönnbäck P, Crona B, Ingwall L (2007) The return of ecosystem goods and services in replanted mangrove forests: perspectives from local communities in Kenya. Environ Conserv 34:313–324. doi:10.1017/S0376892907004225.
61. Russell R, Guerry AD, Balvanera P, et al. (2013) Humans and Nature: How Knowing and Experiencing Nature Affect Well-Being. Annual Review of Environment and Resources. 10.1146/annurev-environ-012312-110838

62. Sagie H, Morris A, Rofè Y et al (2013) Cross-cultural perceptions of ecosystem services: a social inquiry on both sides of the Israeli–Jordanian border of the Southern Arava Valley Desert. J Arid Environ 97:38–48. doi:10.1016/j.jaridenv.2013.05.007.
63. Salmon RA, Priestley RK, Goven J (2015) The reflexive scientist: an approach to transforming public engagement. J Environ Stud Sci. doi:10.1007/s13412-015-0274-4
64. Sayer EJ, Featherstone HC, Gosling WD (2014) Sex & bugs & rock'n roll--getting creative about public engagement. Trends Ecol Evol 29:65–7. doi:10.1016/j.tree.2013.12.008
65. Schneider E (2010) Floodplain Restoration of Large European Rivers with Examples from the Rhine and the Danube. In: Eiseltová M (ed) Restoration of Lakes, Streams, Floodplains, and Bogs in Europe. Springer Netherlands, pp 185–223
66. Scholte SSK, van Teeffelen AJA, Verburg PH (2015) Integrating socio-cultural perspectives into ecosystem service valuation: a review of concepts and methods. Ecol Econ 114:67–78. doi:10.1016/j.ecolecon.2015.03.007.
67. Schröter M, van der Zanden EH, van Oudenhoven APE, et al. (2014) Ecosystem services as a contested concept: a synthesis of critique and counter-arguments. Conservation Letters 7:n/a–n/a. 10.1111/conl.12091
68. Secretariat RC (2010) Designating ramsar sites: strategic framework and guidelines for the future development of the list of wetland for international importance. Gland, Switzerland.
69. Setten G, Stenseke M, Moen J (2012) Ecosystem services and landscape management: three challenges and one plea. Int J Biodivers Scie Ecos Serv Manag 8:305–312. doi:10.1080/21513732.2012.722127.
70. Southworth J, Nagendra H, Munroe DK (2006) Introduction to the special issue: Are parks working? exploring human–environment tradeoffs in protected area conservation. Appl Geogr 26:87–95. doi:10.1016/j.apgeog.2005.11.001.
71. Stern PC, Dietz T (1994) The value basis of environmental concern. J Soc Issues 50:65–84. doi:10.1111/j.1540-4560.1994.tb02420.x.
72. Tahvanainen L, Tyrväinen L, Ihalainen M et al (2001) Forest management and public perceptions — visual versus verbal information. Landsc Urban Plan 53:53–70. doi:10.1016/S0169-2046(00)00137-7.
73. Tengberg A, Fredholm S, Eliasson I et al (2012) Cultural ecosystem services provided by landscapes: assessment of heritage values and identity. Ecos Serv 2:14–26. doi:10.1016/j.ecoser.2012.07.006.
74. Terkenli TS (2001) Towards a theory of the landscape: the Aegean landscape as a cultural image. Landsc Urban Plan 57:197–208. doi:10.1016/S0169-2046(01)00204-3.
75. van Asselen S, Verburg PH, Vermaat JE, Janse JH (2013) Drivers of wetland conversion: a global meta-analysis. PLoS One 8, e81292. doi:10.1371/journal.pone.0081292.
76. Varner J (2014) Scientific outreach: toward effective public engagement with biological science. Bioscience 64:333–340. doi:10.1093/biosci/biu021.
77. Vaske JJ, Donnelly MP (1999) A value-attitude-behavior model predicting wildland preservation voting intentions. Soc Nat Resour 12:523–537. doi:10.1080/ 089419 299279425.
78. Verhoeven JTA (2014) Wetlands in Europe: perspectives for restoration of a lost paradise. Ecol Eng 6:6–9.
79. WWF (2010). Lower Green Danube Corridor. Accessed online at http://assets.panda.org/downloads/wwf_ldgc.pdf
80. Zedler JB, Kercher S (2005) Wetland resources: status, trends, ecosystem services, and restorability. Annu Rev Environ Resour 30:39–74. doi:10.1146/annurev.energy.30.050504.144248.

CHAPTER 14

Restoration of Ailing Wetlands

Oswald J. Schmitz*

School of Forestry and Environmental Studies, Yale University, New Haven, Connecticut, United States of America

ABSTRACT

It is widely held that humankind's destructive tendencies when exploiting natural resources leads to irreparable harm to the environment. Yet, this thinking runs counter to evidence that many ecological systems damaged by severe natural environmental disturbances (e.g., hurricanes) can restore themselves via processes of natural recovery. The emerging field of restoration ecology is capitalizing on the natural restorative tendencies of ecological systems to build a science of repairing the harm inflicted by humans on natural environment. Evidence for this, for example, comes from a new meta-analysis of 124 studies that synthesizes recovery of impacted wetlands worldwide. While it may take up to two human generations to see full recovery, there is promise, given human will, to restore many damaged wetlands worldwide.

In a now classic essay entitled *Round River* [1], Aldo Leopold—the father of modern environmental ethics—lamented,

"One of the penalties of an ecological education is that one lives alone in a world of wounds ... ... An ecologist must either harden his shell and make believe that the consequences of science are none of his business, or he must be the doctor who sees the marks of death in a community that believes itself well and does not want to be told otherwise."

Leopold used the metaphor of integrated medical science and practice to encourage a parallel integrated environmental science and practice in which one studied ecological processes in part to provide the means and capacity to diagnose the environment's ailments and then restore it back to health [2]. His prescience however tended to be lost on generations of ecological scientists and conservation practitioners who instead viewed the world rather dichotomously. There was the built environment where humans went about living; and then there was wild Nature where ecological science could undertake detailed analysis of the processes that shape the diversity of life and associated ecosystem functioning [3]. The business of doing ecological science became tantamount to finding cures for sickness by studying only healthy subjects [3]. The business of conservation practice became one of diagnosing and chronicling human-caused environmental destruction with the intent to spur the protection of Nature within preserves and protected areas that eschewed human presence. This reinforced an approach in ecology and conservation of forecasting perpetual gloom and doom by giving the impression that all human–environmental interactions necessarily lead to irreversible damages [4],[5].

This is not to suggest that human-caused damages to the environment are not prevalent or problematic. Indeed, an expanding human population has translated into increased demand for natural resources and environmental services [4]–[6]. The global human footprint is now so large and far-reaching that some have even begun to question the whole notion that the planet could somehow be rationally divided into places that were available to be domesticated by humans and places that safely remained wild [5]. The reality is that as the biotic and biophysical conditions of the environment become degraded in places where humans exist, they often tend to abandon those places and search for new ones to exploit. But, living on a finite planet with finite space and finite resources means that there is limited if any recourse to continue to abandon degraded areas and shift exploitation to nondegraded ones [5]–[7]. The time has come to operationalize Leopold's vision of an integrated environmental science and practice that provides the scientific understanding and means to restore degraded environments back to health.

That vision, embodied in the idea of "restoration ecology," connects basic ecological research with the mission to develop techniques for rehabilitating the environment by encouraging natural processes or by translating scientific insights into management to speed up the processes [2],[4],[8]. In some respects, this is a logical outgrowth of classic scientific understanding of the way ecosystems have assembled themselves over time. Ecosystems throughout the globe originated from natural development processes of primary succession or natural restorative processes of secondary succession [4],[9]. Primary succession follows when biotic components of ecosystems become established on barren substrates like lava flows or glacial remains and then build up to form a complex ecosystem. Secondary succession arises on substrates previously occupied by biotic species after major disturbances like fires and floods denuded the areas of the biota. Widespread evidence of ecological succession shows the power of natural processes to re-create ecosystems without help [4]. These principles now form the basis of a new framework for systematic study and reconstruction of ecosystems. The goal of restoration ecology is to raise and answer questions through synthetic analysis of the restorative process [3]. The application of the science involves harnessing this natural capacity by introducing interventions that reverse the effects of long-term problems and steer ecological systems back to their original, natural state. The promise of restoration ecology is that it can create a tool kit of management options to balance environmental protection and providing environmental services for a burgeoning human population.

Realizing the promise, however, requires addressing two looming issues that pose important scientific challenges. First, the idea of designing restoration to emulate or enhance successional processes implies that ecosystems can recover gradually from disturbances [7],[10]. But, there is the potential that disturbances could cause ecosystems to reach critical thresholds causing catastrophic shifts in their state (Figure 1). Restoration efforts may then become a proverbial difficult up-hill climb, if they can be restored at all [7],[10]. Second, ecological science has not been conducted for a sufficient length of time to be able to catalogue what the natural states of the myriad ecological systems of the globe in fact are. Thus, to transfer ecological science into practice, ecologists must first wrestle with defining what it means for an ecosystem to be fully recovered and then, through synthesis of ecological studies, identify conditions likely to lead to full recovery [2],[4],[7],[9],[11],[12].

A case in point concerns the global need to restore wetlands such as marshes, peatlands, floodplains, mangroves, and brackish estuaries [9]. Relative to their low representation globally (1.5% of the Earth's surface), wetlands provide huge services to humans, valued in multiple trillions of dollars [9]. These highly important ecosystems have, however, suffered some of the greatest levels of destruction of all ecosystem types [13]. These facts necessarily make wetlands important candidates for restoration efforts. But, their successful recovery may be highly contingent upon the landscape context, including surrounding habitat type and land development, hydrological regime and topography, nutrient inputs, and natural disturbance regimes [9]. In this issue of *PLoS Biology*, Moreno-Mateos et al. [14] report on a synthesis of how landscape context influences wetland restoration success in order to establish a scientific prognosis for their recovery.

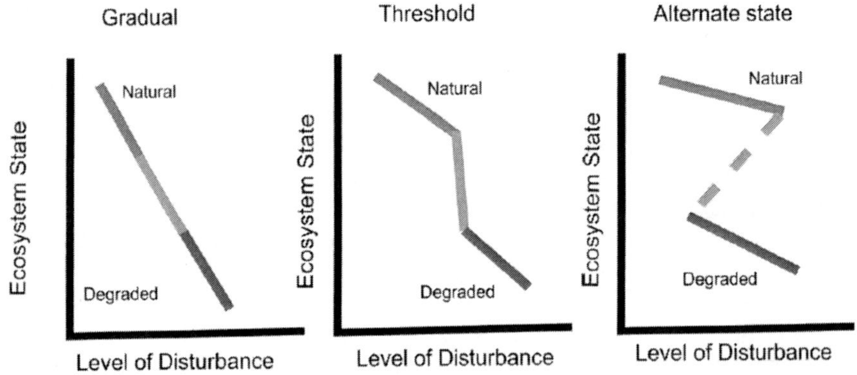

Figure 1. Different potential ways that ecosystem state may change in relation to the level of environmental disturbance. Solid lines denote pathways of state changes from natural to degraded conditions, and the dashed line indicates a transition where the system jumps from a natural to a degraded state. The figure illustrates three general scenarios. Ecosystems may undergo gradual degradation with a rise in disturbance level and may recovery gradually as the disturbance is abated. Ecosystems may exhibit threshold-like behavior in which a certain level of disturbance causes an abrupt change in state and disturbance abatement causes an abrupt "up-hill" return in ecosystem state. Finally, an ecosystem may exhibit a threshold shift in ecosystem state that may only be recoverable with a large turnaround of the critical environmental parameter or disturbance that caused the system to shift from the original state. The scientific challenges in restoration ecology are: characterizing what a natural "green" state is; identifying how long a perturbation must be in place to determine whether the system changes gradually or abruptly; and how long it will take to reverse the effects of a disturbance. Figure is adapted from [10].

Moreno-Mateos et al. [14] conducted an exhaustive search of the scientific literature and identified about 3,000 studies that report on wetland restoration efforts. They then filtered this list of studies using stringent criteria needed to judge restoration success. Foremost, the study had to have an undisturbed reference to serve as a natural state against which to compare the degree of wetland damage and recovery. The study also had to focus on natural wetlands, as opposed to highly engineered artificial systems. Finally, the study had to be conducted over long time periods in order to determine if the wetland is undergoing either gradual recovery, threshold-like recovery, or is locked in an alternative state. These criteria were met in only 4% of the approximately 3,000 studies. Granted, that 4% (124 studies) is a sufficient number to undertake a scientifically defensible synthetic analysis; but, this limitation in the number of rigorous studies faced here, with similar constraints faced by other syntheses of ecological restoration [7],[12], highlights that the science of restoration ecology is still very much in its infancy in its ability to gauge restoration successes.

The synthesis of the 124 wetland restoration studies revealed that recovery of the physical and biotic properties and the functioning of wetland ecosystems proceeded on different time scales. Active restoration of wetland physical features like topography, soil permeability, surface and ground water flows lead to immediate recovery. The abundance and composition of wetland vertebrate species recovered to reference levels usually within 5 years. Large aquatic invertebrates took 5 to 10 years to approach reference levels, but in many cases did not reach absolute reference levels. Plant assemblages took on average 30 years to converge on reference states. Finally, it took 50 to 100 years for wetlands to recover normal nutrient cycling. Interestingly, these time scales are on par with the time course of secondary succession following natural disturbances [7]. Consistent with expectations for wetland systems [9], the rate of recovery varied with the environmental context. Larger wetlands recovered more quickly than smaller wetlands. Wetlands in warmer climates recovered more rapidly than in colder climates. Wetlands connected to other wetlands via intact hydrological structure tended to recover more rapidly than isolated wetlands. All told, in most cases the systems tended to recover rather than be locked in an alternate state.

In an ideal world, society would exploit ecological systems in ways that ensure long-term sustainability of their structure and function rather than degrade them. But, with even the best sustainable practices in place, unforeseen outcomes and damages can happen accidentally [4],[5]. Moreno-Mateos et al.'s study provides evidence that given human will, it is possible to restore human-damaged ecosystems on timescales of one to two human generations. On a societal level, the promise of restoration ecology demonstrated in this synthesis and other recent syntheses [7],[12] first helps to dispel the notion that human activity necessarily has irreversible negative impacts on ecosystems [3],[4],[7] and, second, shows that with ecological know-how and application, it is possible to cure some ailing environments.

REFERENCES

1. Leopold A. S (1953) Round river. New York: Oxford University Press. A. S. Leopold1953Round riverNew YorkOxford University Press
2. Palmer M. A, Filoso S (2009) Restoration of ecosystem services for environmental markets. Science 325: 575–576.M. A. PalmerS. Filoso2009Restoration of ecosystem services for environmental markets.Science325575576

3. Aber J. D, Jordan W. R (1985) Restoration ecology: an environmental middle ground. Bio Science 35: 399–399.J. D. AberW. R. Jordan1985Restoration ecology: an environmental middle ground.Bio Science35399399
4. Dobson A. P, Bradshaw A. D, Baker A. J. M (1997) Hopes for the future: restoration ecology and conservation biology. Science 277: 515–522.A. P. DobsonA. D. BradshawA. J. M. Baker1997Hopes for the future: restoration ecology and conservation biology.Science277515522
5. Kareiva P, Watts S, McDonald R, Boucher T (2007) Domesticated nature: shaping landscapes and ecosystems for human welfare. Science 316: 1866–1869.P. KareivaS. WattsR. McDonaldT. Boucher2007Domesticated nature: shaping landscapes and ecosystems for human welfare.Science31618661869
6. Foley J. A, DeFries R, Asner G. P, Barford C, Bonan G, et al. (2005) Global consequences of land use. Science 309: 570–574.J. A. FoleyR. DeFriesG. P. AsnerC. BarfordG. Bonan2005Global consequences of land use.Science309570574
7. Jones H. P, Schmitz O. J (2009) Rapid recovery of damaged ecosystems. Plos ONE 4: e5653.H. P. JonesO. J. Schmitz2009Rapid recovery of damaged ecosystems.Plos ONE4e5653
8. Hobbs R. J, Harris J. A (2001) Restoration ecology: repairing the Earth's ecosystems in the new millennium. Restor Ecol 9: 239–246.R. J. HobbsJ. A. Harris2001Restoration ecology: repairing the Earth's ecosystems in the new millennium.Restor Ecol9239246
9. Zedler J. B (2000) Progress in wetland restoration ecology. Trends Ecol Evol 15: 402–407.J. B. Zedler2000Progress in wetland restoration ecology.Trends Ecol Evol15402407
10. Suding K. N, Hobbs R. J (2009) Threshold models in restoration and conservation: a developing framework. Trends Ecol Evol 24: 271–279.K. N. SudingR. J. Hobbs 2009Threshold models in restoration and conservation: a developing framework.Trends Ecol Evol24271279
11. Hobbs R. J, Norton D. A (1996) Towards a conceptual framework for restoration ecology. Restor Ecol 4: 93–110.R. J. HobbsD. A. Norton1996Towards a conceptual framework for restoration ecology.Restor Ecol493110
12. Rey Benayas J. M, Newton A. C, Diaz A, Bullock J. M (2009) Enhancement of biodiversity and ecosystem services by ecological restoration: a meta-analysis. Science 325: 1121–1124.J. M. Rey BenayasA. C. NewtonA. DiazJ. M. Bullock2009Enhancement of biodiversity and ecosystem services by ecological restoration: a meta-analysis.Science32511211124
13. Mitsch W. J, Gosselink J. G (2007) Wetlands. Hoboken: John Wiley & Sons, Inc. pp. 177–183.W. J. MitschJ. G. Gosselink2007WetlandsHobokenJohn Wiley & Sons, Inc177183
14. Mateos-Moreno D, Power M. E, Comin F. A, Yockteng R (2012) Structural and functional loss in restored wetland ecosystems. PLoS Biol 10: e1001247.D. Mateos-MorenoM. E. PowerF. A. CominR. Yockteng2012Structural and functional loss in restored wetland ecosystems.PLoS Biol10e1001247

CHAPTER 15

Structural and Functional Loss in Restored Wetland Ecosystems

David Moreno-Mateos[1,2*], Mary E. Power[1], Francisco A. Comı´n[3], Roxana Yockteng[4]

[1]*Integrative Biology Department, University of California at Berkeley, Berkeley, California, United States of America*

[2]*Jasper Ridge Biological Preserve, Stanford University, Woodside, California, United States of America,*

[3]*Department of Conservation of Biodiversity and Ecosystem Restoration, Pyrenean Institute of Ecology – CSIC, Zaragoza, Spain,*

[4]*UMR CNRS 7205, Muse´um National d'Histoire Naturelle, Paris, France*

ABSTRACT

Wetlands are among the most productive and economically valuable ecosystems in the world. However, because of human activities, over half of the wetland ecosystems existing in North America, Europe, Australia, and China in the early 20th century have been lost. Ecological restoration to recover critical ecosystem services has been widely attempted, but the degree of actual recovery of ecosystem functioning and structure from these efforts remains uncertain. Our results from a meta-analysis of 621 wetland sites from throughout the world show that even a century after restoration efforts, biological structure (driven mostly by plant assemblages), and biogeochemical functioning (driven primarily by the storage of carbon in wetland soils), remained on average 26% and 23%

lower, respectively, than in reference sites. Either recovery has been very slow, or postdisturbance systems have moved towards alternative states that differ from reference conditions. We also found significant effects of environmental settings on the rate and degree of recovery. Large wetland areas (>100 ha) and wetlands restored in warm (temperate and tropical) climates recovered more rapidly than smaller wetlands and wetlands restored in cold climates.

Also, wetlands experiencing more (riverine and tidal) hydrologic exchange recovered more rapidly than depressional wetlands. Restoration performance is limited: current restoration practice fails to recover original levels of wetland ecosystem functions, even after many decades. If restoration as currently practiced is used to justify further degradation, global loss of wetland ecosystem function and structure will spread.

1. INTRODUCTION

From tropical mangroves to boreal peatlands, wetlands are amongst the most productive and economically valuable ecosystems in the world [1]. They provide critical ecosystem goods and services, including carbon storage, biodiversity conservation, fish production, fuel production, water purification, flood and shoreline surge protection and erosion control, and recreation [1]–[3]. However, owing to human activities, over half of the wetland ecosystems existing in the early 20th century have been lost in North America, Europe, Australia, and China [2]. Over the last century, restoration of degraded wetlands and creation of new ones have been attempted, in efforts to recover physical, chemical, and biological processes and entities lost because of wetland destruction or degradation [4]. Frequently, however, this approach does not restore ecosystem structure and functions to preimpact levels [5]–[8].

In North America (including Canada, United States, and Mexico) alone, over US$70 billion have been spent attempting to restore more than 3,000,000 ha of wetlands in the last 20 y (see Text S1) [9], but the recovery trajectories of structure and functions in restored wetlands have not yet been globally assessed [10],[11].

After degradation or natural perturbation, ecosystem structure and functions recover towards reference levels [7],[12], but recovery rates might be affected by the physical characteristics of the ecosystem, the degrading activity, or the environmental setting [7],[12]. Abiotic factors, such as size of restored ecosystems and climate, might affect recovery rates. It could be expected that intensely engineered small (few hectares) wetlands might recover faster than less manipulated, large wetlands (hundreds of hectares) to their original characteristics, but this prediction remains unconfirmed. Higher recovery rates could also be expected in warmer climates than in cold ones, because of accelerated ecosystem processes [7],[13].

Restoration efforts during the recovery process may lead ecosystems to reference states or redirect them towards alternative states [14]–[16] that could also be initiated by prerestoration disturbance itself. If recovery is slow, it could be difficult to distinguish between these alternatives. We surveyed long-term (up to 100 y, available for some but not all of the studied variables) chronosequences of restored wetland ecosystems from 621 restored and created wetlands relative to 556 reference wetlands (Figure S1). Following Article 1.1 of the Ramsar Convention of Wetlands [17], we considered wetlands to be marshes, peatlands, floodplains, mangroves, depressional wetlands, and lacustrine wetlands—submerged permanently or periodically under flowing or still fresh, salty, or brackish water. We compared structure and function of 401 wetlands restored on sites where they had been previously degraded and 220 newly created wetlands (wetland creation de novo is currently accepted for environmental mitigation [4]).

We also examined how size of ecosystem and its environmental setting (climate regime and hydrologic connectivity) affected recovery. Using a standardized method (see Materials and Methods), we selected 124 studies (see Text S2) in which ecological responses were measured at known time intervals since restoration. From the selected studies, we extracted 1,501 data points (Tables 1, S1, and S1S2) comparing hydrologic, biological, and biogeochemical variables in restored or created and reference wetlands. Response ratios (see Materials and Methods) were calculated for each data point. Variables selected from the same studies were not necessarily independent (see Materials and Methods), so statistical inferences must be interpreted cautiously.

We compared recovery trajectories of hydrologic, biological, and biogeochemical variables of restored and created wetlands to address three questions: (a) How fast are biological and biogeochemical components of restored ecosystems changing relative to less perturbed reference ecosystems?; (b) Do these changes trend towards or away from the predisturbed ecosystem or parallel control ecosystems?; and (c) Does wetland size or environmental setting (regional climate, hydrologic connectivity) affect recovery?

Table 1. Variables measured simultaneously in restored or created and reference wetlands to estimate wetland restoration performance over time.

Wetland Structure and Functions	n[a]	Variables Measured
Hydrology	32	Water level, flooding regime, water storage
Biological components	809	
Vertebrates	166	Abundance, density, species richness, occupancy
Macroinvertebrates	161	Density, abundance, species richness
Plants	439	Plant cover, species richness, biomass, abundance
Biogeochemistry	692	
Carbon storage and cycling	103	Soil total and organic carbon, respiration rate, mineralization rate
Nitrogen storage and cycling	102	Soil total and organic nitrogen, denitrification, and nitrification
Phosphorus storage	103	Soil total and organic phosphorus, Ca-Fe-Al bounded phosphorus
Other elements storage	106	Salinity, soil Fe, Al, Ca, K, Mn, Mg, water dissolved oxygen
Organic matter accumulation	177	Soil organic matter, bulk density, soil texture, soil moisture

Only the most frequently measured variables were included (see Tables S1 and S2, for full description of the variables measuring restoration performance).
[a]*n* = number of variables used to plot each chronosequence.
doi:10.1371/journal.pbio.1001247.t001

2. RESULTS/DISCUSSION

2.1 Hydrologic and Biological Recovery

Some hydrologic features can often be restored by manipulating local topography, soil permeability, surface and ground water flows—physical features that are usually engineered in wetland restoration projects. Hydrological features defined for these analyses (Table 1) appeared to be recovered immediately after restoration (Figure 1A), but see Cole [18], Hunt et al. [19], Ahn and Dee [20], and Kumar and Zhao [21] for deeper considerations of challenges to hydrologic restoration in wetlands (from factors like climate variation [20] or complex flow paths of water through heterogeneous vegetation and soils [21]). In addition, all hydrologic variables reported in studies we reviewed were followed only for 10 y to 15 y, so longer-term changes remain unknown.

In contrast to reported hydrologic performance, biological structure (as defined in Table 1) in restored or created wetlands, recovered to only 77% (on average) of reference values (Figure 1A and 1B; Table S3), even 100 y after restoration, when data on 14 taxa from two studies of three wetland sites are available [22],[23]. Abundance, species richness, and diversity of native animals and plants in wetlands were severely reduced following degradation. After restoration, recovery proceeded at different rates, and trajectories plateaued at different levels. Vertebrate assemblages reached similar structural values to those in reference wetlands within 5 y (Figure 1B). Vertebrate richness recovered more slowly than abundance (p=0.021; Figure 2A), possibly reflecting responses by a few highly mobile vertebrate species [24],[25] once hydrological connectivity was restored. Macroinvertebrates (64% noninsects) took 5 y to 10 y to statistically converge with reference assemblages in restored and created wetlands (Figure 1B), and average values never reached absolute reference levels. Many macroinvertebrates cannot recolonize new or restored wetlands by themselves, but are carried in by flowing water or other organisms [26],[27]; however, their short life cycles (often annual or semi-annual) could accelerate population recovery after they arrive [28],[29].

Figure 1. Recovery trajectories of created and restored wetlands. Chronosequences of the means (±standard error [SE]) of the response ratios (see Materials and Methods) of restored and created wetlands at successive age classes of 5 y or 10 consecutive y for hydrology, biological structure, and biogeochemical functions (A) and for the main biological structural components (B). Chronosequences of the means (±SE) of the element loss in soils of restored or created wetlands at successive age classes of 5 y or 10 consecutive y (C). The zero value dashed line represents reference wetlands. Only trend lines for those variables for which we had enough data points (see Materials and Methods) were plotted (N, number of data points used to calculate the mean [±SE] per age class; Y, years after restoration. Subscripts are as follows: bp, biogeochemical processes; bs, biological structure; C, carbon; hf, hydrological features; m, macroinvertebrates; N, nitrogen; p, plants; P, phosphorus; v, vertebrates).

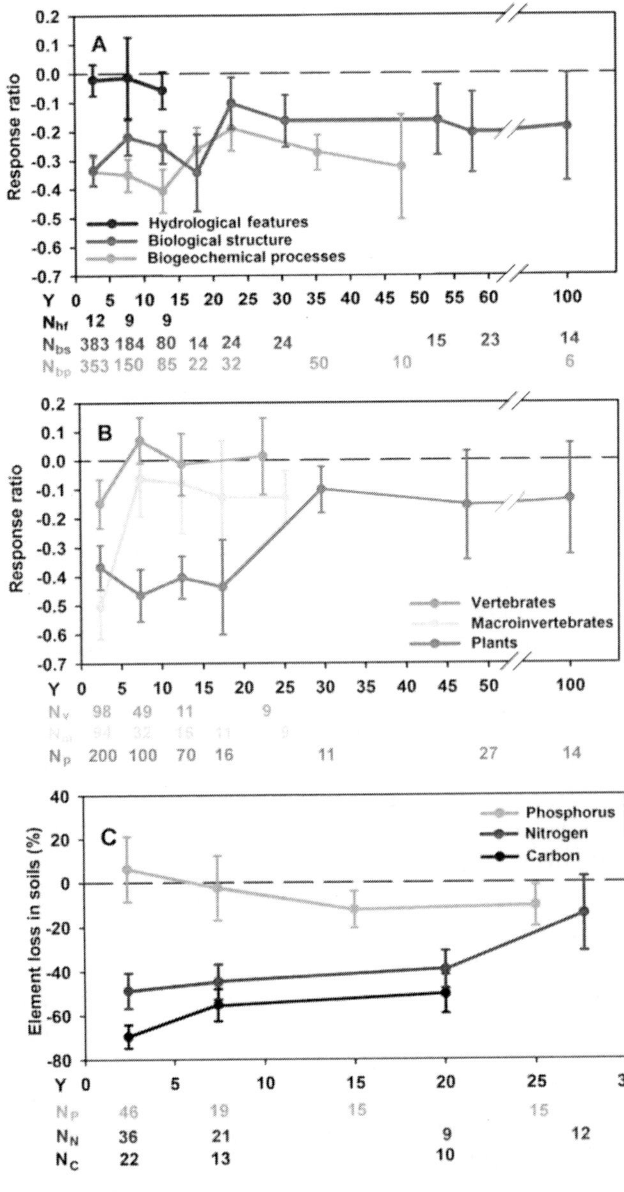

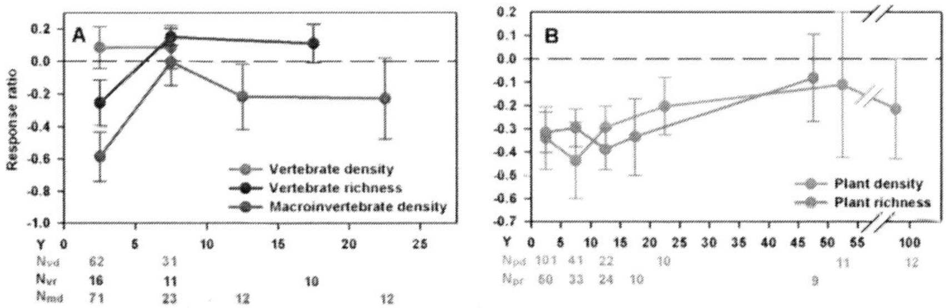

Figure 2. Recovery trajectories of animal and plant richness and density. Chronosequences of the means (±standard error [SE]) of the response ratios (see Materials and Methods) of restored or created wetlands at successive age classes of 5 y or 10 consecutive y for vertebrates and macroinvertebrates density and richness (A) and for plant density and richness (B). Insufficient data points meeting our plotting criteria (see Materials and Methods) were available to plot for macroinvertebrate richness. The zero value dashed line represents reference wetlands (N, number of data points used to calculate the mean [±SE] per age class; Y, years after restoration. Subscripts are as follows: md, macroinvertebrates density; pd, plant density; pr, plant richness; vd, vertebrate density; vr, vertebrates richness).

Plant assemblages in restored and created wetlands were slowest to recover. Plants took on average 30 y to converge statistically with reference states; although again, absolute average values of structural features of plant assemblages remained lower than reference levels even after 100 y following restoration (Figures 1B and 2B). The slow and incomplete recovery of plant assemblage might be due to dispersal limitation, vulnerable early life history stages, or sensitivity of any life stage to altered conditions (e.g., reduced organic content of soils, discussed below) during early succession following disturbance [30],[31]. Other factors, such as exotic colonists, subsequent disturbance or altered disturbance regimes, priority effects (historical legacies), and nonlinear interactions may also lead to delayed recovery or persistent differences between restored biota and those in reference wetlands [6],[31],[32].

2.2 Biogeochemical Recovery

Four biogeochemical responses were sufficiently well documented in some studies we reviewed to examine trends over time: these were the storage of carbon, nitrogen, and phosphorus (Figure 1C) (see also storage and cycling combined for carbon and nitrogen in Figure S2A), and the accumulation of organic matter in soil (Figure S2B). The storage and cycling of carbon and nitrogen were drastically reduced from preimpact levels after degradation. In contrast, phosphorus storage seemed unaffected. After restoration, responses were variable. Initially, carbon storage increased slightly but then plateaued below reference levels; nitrogen storage and cycling increased slowly but continuously; and phosphorus storage remained unaffected. Wetland degradation notoriously oxidizes stores of accumulated organic carbon and releases CO_2 to the atmosphere, as aerobic conditions accelerate microbial respiration [2],. After wetland hydrologic regimes are recovered, more anaerobic conditions allow stores of organic carbon to slowly reaccumulate in the soil. After 20 y, however, carbon storage in restored and created wetland soils was still significantly lower (by 50%; p=0.008) than in reference wetlands (Figure 1C; Table S3; Text S1; data from six studies of 21 wetlands) Organic matter accumulated slowly [34],[35], so that average values remained only 62% of the value at the reference wetlands 20–30 y following restoration (Figure S2B; data from seven studies of 21 wetlands).

Aerobic conditions in degraded wetlands also perturb nitrogen storage and cycling, allowing mineralization of organic N and transformation of ammonium to nitrate [2]. Nitrate is quickly processed by microorganisms and plants, leaving the original pool of nitrogen in the soil depleted or unavailable. Nitrogen storage remained significantly lower in restored wetlands for 30 y after the wetlands were restored or created (Figure 1C; Table S3). Depleted or unavailable soil nitrogen can limit wetland productivity, retarding carbon storage [33],[36]. In contrast, total phosphorus decreased only slightly in restored or created wetlands and did not show significant differences with reference wetlands (Figure 1C). Although, phosphorus chemical fractions could change in representation, the amount of total phosphorus did not change significantly [37]. This lack of variation in phosphorus might be explained because of the more conservative cycling by phosphorus (lack of exchange with the atmosphere) [38]. In addition, without extrinsic inputs, phosphorus levels would be geologically determined.

After 50 y to 100 y, restored wetlands recovered only to an average of 74% of their biogeochemical functioning relative to reference wetlands (Figure 1A; data from two studies of seven wetlands; data of wetlands recovering for more than 50 y after restoration were not plotted in Figure 1A because the sample size did not meet our criteria for average points, see Materials and Methods section, on this graph).

Since phosphorus storage appeared only slightly changed, the overall lack of recovery of biogeochemical functioning may have been driven largely by the low recovery of the carbon storage and the low accumulation of soil organic matter (see Text S1).

2.3 Effects of Size and Environmental Setting

Comparing wetland recovery trajectories under different conditions may shed light on factors that impede or facilitate recovery. Although biogeochemical responses in both restored and created wetlands were similar, biological structure in created wetlands approached reference conditions more quickly (Figure S3A and S3B; Table S5). Created wetlands may have been engineered to force the initial system towards defined reference conditions [39].

Ecosystem size and local and regional context affect wetland recovery. Large wetlands (>100 ha) appeared to recover their biological structure and biogeochemical functions sooner after restoration or creation than smaller wetlands (Figures 3 and S4; Table S4; data from 13 studies of 25 wetlands). This differential recovery suggests that small wetlands may not provide adequate local resources or connectivity for local biota to restore preimpact functioning.

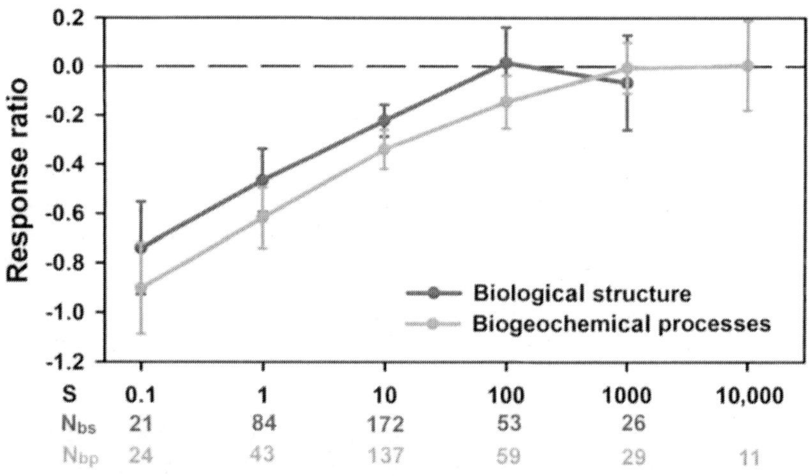

Figure 3. Effect of size on wetland recovery. Evolution of the mean (±standard error [SE]) of the response ratios (see Materials and Methods) of restored or created wetlands at successive size categories for wetlands between 0 y to 5 y after restoration or creation. The zero value dashed line represents reference wetlands. Mean (±SE) at 0.1 ha was estimated for wetlands with sizes ≤0.1 ha. Means (±SE) at 1 ha were estimated for wetlands in which sizes ranged between 0.1 ha and 1 ha. The same approach was used to estimate the means (±SE) at 10, 100, 1,000, and 10,000 ha (N, number of data points used to calculate the mean [±SE] per age class; size, size in hectares of the restored wetlands. Subscripts are as follows: bp, biogeochemical processes; bs, biological structure).

Restored and created wetlands, particularly if small, may have become more isolated and surrounded by more fragmented landscapes than they had been before impact [40]. Also, small wetlands would only be able to support a limited number of individuals, and thus, will not be able to support all the species, particularly taxa with large body sizes, formerly capable of occupying the area [41].

Regional climate had a strong effect on the sequence and rate of wetland recovery following restoration. As expected, warm temperatures accelerate ecosystem processes [7],[13],[42], including those mediating biological and biogeochemical recovery after wetland restoration or creation. In tropical and summer-warm temperate climates, wetlands approached reference conditions relatively rapidly, while wetlands restored in cold climates had not recovered to reference conditions after 50 y (Figure 4A and 4B; Tables S3 and S5). In tropical climates only, biogeochemical variables recovered to reference levels before biological structure did (data from eight studies of eight wetlands). Whether this difference in recovery sequence is a real aspect of tropical wetlands, or an artifact of small sample size, remains to be seen. In a much larger sample of studies from temperate climates this sequence was reversed, and biogeochemical recovery was slower. Biological structural variables appeared recovered 5 y after restoration, while even 30 y after restoration, biogeochemical functions had only recovered to 79% of reference levels (data from 83 studies of 302 wetlands). In cold climates, corresponding biogeochemical recovery was only 53% 50 y after restoration; both biogeochemical functions and biological structure variables remained statistically distinct from reference conditions for the entire (50-y) chronosequence (Figure 4A and 4B; Tables S3 and S5; data from 33 studies of 311 wetlands).

Hydrologic setting [43] also affected recovery (Figure 4C and 4D; Tables S3 and S5). Riverine and tidal wetlands, linked to larger hydrologic regimes by natural flow variation, recovered biogeochemical functions and biological structure after 20 y and 30 y, respectively (data from 73 studies of 210 wetlands). These results are similar to those (15 y to 25 y to recover the original biotic composition and diversity) found by Borja et al. [8] in 51 globally distributed estuarine and coastal ecosystems. In contrast, wetlands in inland depressions that were watered by precipitation or groundwater flow had not recovered to reference conditions even after 50 y following restoration (data from 36 studies of 358 wetlands). Peatlands (usually only the upper layer [<1 m] of peat was removed) recovered biological structure immediately, but 30 y after restoration, biogeochemical functioning in peatlands remained statistically lower than in reference wetlands (data from 11 studies of 18 wetlands).

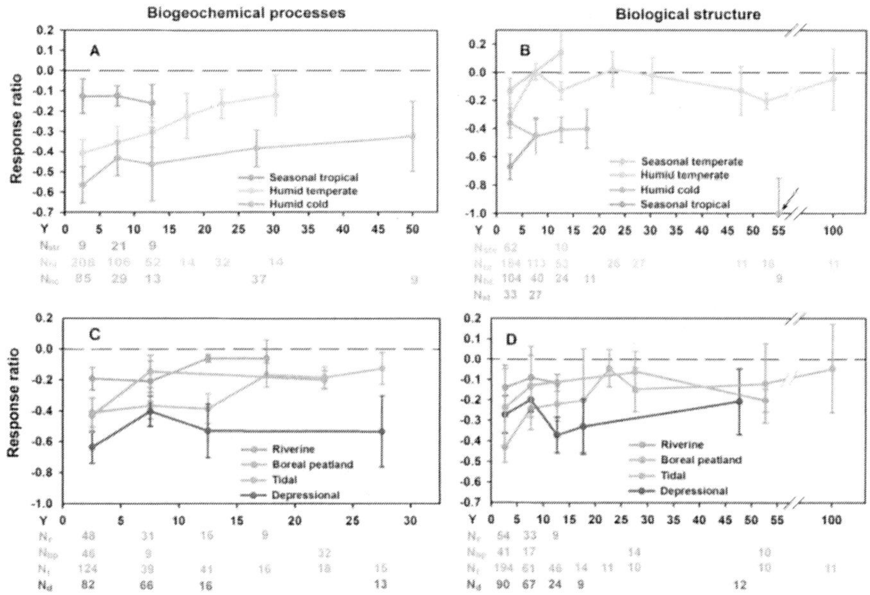

Figure 4. Effects of climate and hydrology on wetland recovery trajectories. Chronosequences of the means (±standard error [SE]) of the response ratios (see Materials and Methods) of restored and created wetlands at successive age classes of 5 y or 10 consecutive y for biogeochemical functions and for biological structures under contrasting climates (A and B), and under different hydrologic connectivity (C and D) [31]. The zero value dashed line represents reference wetlands. The arrow (B) indicates the outlier mean value of two restoration studies with extremely low recovery rates (N, number of data points used to calculate the mean [±SE] per age class; Y, years after restoration. Subscripts are as follows: bp, boreal peatland; d, depressional; hc, humid cold; ht, humid temperate; r, riverine; str, seasonal tropical; ste, seasonal temperate; t, tidal).

2.4 Slow Recovery or Alternative States?

Two hypotheses could explain the lag in biological and biogeochemical recovery of the biological structure and biogeochemical functioning. First, the chronosequences we examined may be too short (<30 y) for full recovery, especially of carbon and nitrogen storage [44]. Second, restored wetlands may have shifted to alternative states, different from their condition before degradation [14],[15]. The subreference plateaus of soil organic accumulation, carbon storage, and general biogeochemical functioning could support the second hypothesis of alternative states in restored systems. Slow recovery of plant density and richness might be linked to lags in carbon storage. Mutualist symbionts critical for plant productivity (e.g., N-fixing bacteria [2] or mycorrhizal fungi [45]) may be absent in recently (<50 y) restored wetland soils. Alternatively, fast-growing, early successional terrestrial plants, and potentially also wetland plants, usually allocate most of their carbon to photosynthetically active structures of low density and high nutrient content, which are easily grazed or rapidly decomposed, retarding local storage of carbon [46],[47].

2.5 Comparison with Other Findings

Two other studies have assessed recovery rates of large scale natural ecosystems following disturbance or perturbations [7],[12]. Both of these studies examined a broad range of ecosystem types (terrestrial, freshwater, and marine), including wetlands. Jones and Schmitz [12] found that across ecosystems and perturbation types (natural and human-caused), about half of the tracked response variables were considered by original authors to have recovered to preimpact states. Jones and Schmitz computed averaged recovery times for the subsamples of variables and cases that primary authors considered to have recovered over the course of their studies. These recovery times ranged from about 10 y to 40 y, and were longer for forests, and following human-caused, rather than natural perturbations. To assess whether systems had recovered or not, Jones and Schmitz used authors' expert opinion, return to historic initial conditions, or approach to parallel reference states (our study evaluated recovery only for studies using the last of these criteria). Given the narrower scope of our study (assessing wetlands only), and our different analysis approach, estimated recovery times from these two reviews are surprisingly similar. Rey Benayas et al. [7] studied recovery across a wide range of human-perturbed ecosystems, including wetlands. Using (as we did) the response ratio of restored to reference ecosystems, Benayas et al. found biodiversity and selected ecosystems services to be 86% and 80% recovered in a sample of 89 cases pooled over all age categories since perturbation. Interestingly, they reported slightly (6%) higher recovery in biological variables compared to ecosystems services (nutrient cycling; primary production; provisioning of timber, fish, and food crops; and regulation of climate, water supply, and soil). These ecosystem services overlap in part with categories of biogeochemical variables in our study (e.g., carbon and nutrient storage and cycling). The similarity between their results and our finding (that biological variables were 9% more recovered than these biogeochemical responses) suggests that structural recovery might often be necessary to achieve functional recovery.

3. CONCLUSIONS

Our meta-analysis suggests that recovery of wetlands following restoration as currently practiced is often slow and incomplete. In warm climates, and in settings linked to riverine or tidal flows, recovery may proceed more rapidly. Recovery may also be more likely and more rapid if >100 contiguous ha are restored. In many wetlands, however, ecosystem services may not be fully recovered even when wetlands appear to be biologically restored. If markets for ecosystem services and mitigation offsets from restored or created wetlands are used to justify further wetland degradation, net loss of global wetland services will continue and likely accelerate (see also Race and Fonseca [48]). We join other wetland ecologists and restoration scientists in calling for better scientific understanding of biotic and abiotic factors that constrain ecosystem restoration. For our common future, we need more realistic, long-term evaluations to find better ways to alleviate constraints limiting the recovery of wetland ecosystems.

4. MATERIALS AND METHODS

4.1 Literature Search

On the 22nd of December 2010 a reference search was done in the scientific database ISI Web of Science – SCI-Expanded. The terms used were "(wetland* or floodplain*or peatland* or marsh* or mangrove*) same (restor* or creat* or re-creat* or rehabilit*)." We used these terms to cover a wide variety of wetlands as defined in the Article 1.1 of the Ramsar Convention text [7]. For this analysis, we considered restored wetlands to be wetlands recreated on sites where wetlands had formerly existed but been drained or otherwise severely degraded. Created wetlands were described by authors as wetlands built on sites that lacked previous wetland history. We selected studies of wetlands under natural hydrological regimes, planted with native species, and in which no allochthonous substrates were imported during the restoration or creation activities. For this reason, the term "construct*" was not included in the search terms, because we found in an independent search that >99% of the studies of constructed wetlands were of highly artificial systems not maintained under natural conditions. The search produced 2,959 selected articles. We applied the general selection criterion: "Articles must compare measurements of structural components and biogeochemical processes in restored or created and reference wetlands at a known age." Under this criterion we selected 172 articles. These articles were read, and those in which data were averaged over time intervals larger than 5 y, those in which sizes differing by more than one order of magnitude were averaged, and those lacking reliable measurements or comparable restored and reference conditions were discarded, leaving 124 articles (see Text S2). Reference wetlands were usually adjacent to restored or created wetlands, although in some cases they were separated by several kilometers (maximum distance found was ~100 km). In all cases, restored or created wetlands were of the same wetland hydrogeomorphic type [17] as reference wetlands with which they were compared. From the selected articles, six were carried out on experimental wetlands, the rest were carried out on wetland restoration or creation projects. Studies either described measurements at a known age after wetlands were restored or created, or a chronosequence of the progression during the wetland restoration process. Restored and created wetlands were located in 12 countries and totaled >21,294 ha in area and reference wetlands >19,694 ha. The exact total area is not known because it was not reported in 23 out of the 124 selected studies.

4.2 Data Extraction

Measurements of structural components and biogeochemical processes were extracted from the main text, tables, and figures of the articles. When abundance of one species was measured at different life stages, only the adult abundance of each species was selected. Variables describing hydrological structure, biological structure, element storage and cycling, and organic matter accumulation were classified as structural components or biogeochemical processes according to wetland functions described by Smith et al. [42], and as ecosystem services described in the Millennium Ecosystem Assessment (MEA) (organic matter accumulation was sometimes designated as "soil formation" in the MEA but not in other soil science references) [49].

Element storage and cycling variables measured processes (mineralization or denitrification) and concentration of elements in different pools (total content in soil, organic content in soil, or content in roots), which suggest how nutrients are moving between pools through biotic and abiotic processes (Tables S1 and S2). The studies presented enough data points to plot recovery of storage of carbon, nitrogen, and phosphorus.

4.3 Response Ratio Calculation

To standardize and compare data, we used standard response ratios used in meta-analysis, $ln(Xrest+1/Xref+1)$ [3], where *Xrest* is the value of the measured variable in the restored or created wetland and *Xref* is the value of the measured variable in the reference wetland. To avoid the value "0" in the natural logarithm of the equation, "1" was added to both values in restored or created and reference wetlands. The effect of adding "1" to the values in the response ratio equation has been demonstrated to have little effect on conclusions [50]. The effect size was not weighted because variance was reported for only 64% of the variables. Differences between weighted and unweighted meta-analysis statistics are generally small [7].

As variables depicting structural components and biogeochemical processes in restored or created wetlands converged to values in the reference wetlands, recovery of function was generally enhanced. But for some variables, such as soil bulk density [51],[52], or proportions of exotic species [53],[54], higher values are associated with lower levels of wetland recovery. In some cases, the specific context of a study made variables negative for recovery of a particular restored wetland, e.g., the presence of woody species where none had occurred in the reference wetlands [55],[56]. In these cases (11% of the collected variables), we changed the sign to reverse the value of the response ratio.

4.4 Data Classification

For each variable we recorded the age of the restored or created wetland, the wetland hydrogeomorphic type, the number of restored or created and reference wetlands considered in a given study, the size (ha) of the restored or created and reference wetlands, the initial condition (restored or created), the geographic location, and the climate. Most data (49%) were from wetlands that had been restored or created for less than 5 y (Figure S1). If data from several wetlands of different sizes were averaged in the study, then we also averaged the sizes for our analysis. The geographic location was registered as the latitude and longitude in degrees of the center of the wetland or group of wetlands. The climate was classified according to the last revision of the Köppen-Geiger climate classification [57]. We used the name humid temperate climate for Cf climate, humid cold climate for Df climate, seasonal temperate climate for Cs climate (with dry summer), and seasonal tropical for A climates. Two of our sampled studies were done in seasonal temperate climate with dry winter (Köppen-Geiger climate classification Cw), and were not considered in our climate study. Wetland hydrogeomorphic type was classified according to Brinson [58] and Smith et al. [42] as depressional, riverine, tidal, peatland, lacustrine, and seeping slope. Only three studies were on lacustrine wetlands and one on seeping slope wetlands, so these types were not considered in our study of differences among wetland types.

In studies where more than one wetland was studied and data were available for each individual wetland, data were collected for each wetland.

In 27 studies, more than one wetland was compared with the same reference wetland, and in 11 studies, restored or created wetlands were compared with more than one reference wetland. All studies where more reference rather than restored or created wetlands were studied provided only averaged data for both groups of wetlands. We calculated contingency tables between the wetland size, the initial conditions (created versus restored), and the covariates included in the environmental setting section (climate and wetland hydrogeomorphic type), using contingency coefficients (*C*), to test for independence between them. Wetland type showed relevant degrees of association with the climate (C=0.63) and wetland size (C=0.58), the rest of variables had coefficients below 0.5, indicating low degree of association. These associations may be explained by the influence of the climate on wetland types, e.g., peatlands are usually associated to cold climates, and mangroves to tropical climates. Also, peatlands usually extend over vast surfaces (hundreds or thousands of hectares) and depressional wetlands are usually small basins (less than 10 ha or few tens of hectares).

4.5 Statistical Analysis

Because data were non-normally distributed (according to the Kolmogorov-Smirnoff test for normality), we used Wilcoxon signed rank tests to test for significant deviations from zero (no difference from reference conditions) for each estimated mean of the response ratios for variables at each age interval of a restored or created wetland. To test for differences between the same variable measured under two different environmental settings at a given recovery time, we used Kruskal-Wallis tests.

4.6 Chronosequences Plotting

To plot the temporal trends, the mean values and the standard error of each variable with every age class of 5 y (0–4.9, 5–9.9, etc) were used. The criterion for a mean for a certain age class to be used in the plot was that it must have been derived from at least nine different data points obtained from at least two different studies. When this criterion was not fulfilled, the mean values and standard error of age classes of 10 y (e.g., 10–19.9, or 20–30) were used. Temporal trend lines were fitted when enough data to calculate means for two or more age classes were available.

ACKNOWLEDGMENTS

We acknowledge the comments from Joy B. Zedler, Oswald Schmidt, J.M. Rey Benayas, William E. Dietrich, Andrea Swei, and one anonymous reviewer on earlier versions of the manuscript, and the collaboration of Maribel Vara Rodriguez.

REFERENCES

1. Costanza R, dArge R, deGroot R, Farber S, Grasso M, et al. (1997) The value of the world's ecosystem services and natural capital. Nature 387: 253–260.R. CostanzaR. dArgeR. deGrootS. FarberM. Grasso1997The value of the world's ecosystem services and natural capital.Nature387253260
2. Mitsch W. J, Gosselink J. G (2007) Wetlands. Hoboken (New Jersey): John Wiley & Sons, Inc. pp. 177–183.W. J. MitschJ. G. Gosselink2007WetlandsHoboken (New Jersey)John Wiley & Sons, Inc177183
3. Millenium Ecosystem Assessment (2005) Ecosystems and human well-being: wetlands and water. Washington (D.C.): World Resources Institute. Millenium Ecosystem Assessment2005Ecosystems and human well-being: wetlands and waterWashington (D.C.)World Resources Institute
4. Palmer M. A (2009) Reforming watershed restoration: science in need of application and applications in need of science. Estuaries Coasts 32: 1–17.M. A. Palmer2009Reforming watershed restoration: science in need of application and applications in need of science.Estuaries Coasts32117
5. Matthews J. W, Endress A. G (2008) Performance criteria, compliance success, and vegetation development in compensatory mitigation wetlands. Environ Manage 41: 130–141.J. W. MatthewsA. G. Endress2008Performance criteria, compliance success, and vegetation development in compensatory mitigation wetlands.Environ Manage41130141
6. Zedler J. B, West J. M (2008) Declining diversity in natural and restored salt marshes: a 30-year study of Tijuana Estuary. Restor Ecol 16: 249–262.J. B. ZedlerJ. M. West2008Declining diversity in natural and restored salt marshes: a 30-year study of Tijuana Estuary.Restor Ecol16249262
7. Rey Benayas J. M, Newton A. C, Diaz A, Bullock J. M (2009) Enhancement of biodiversity and ecosystem services by ecological restoration: a meta-analysis. Science 325: 1121–1124.J. M. Rey BenayasA. C. NewtonA. DiazJ. M. Bullock2009Enhancement of biodiversity and ecosystem services by ecological restoration: a meta-analysis.Science32511211124
8. Borja A, Dauer D. M, Elliott M, Simenstad C. A (2010) Medium- and long-term recovery of estuarine and coastal ecosystems: patterns, rates and restoration effectiveness. Estuaries Coasts 33: 1249–1260.A. BorjaD. M. DauerM. ElliottC. A. Simenstad2010Medium- and long-term recovery of estuarine and coastal ecosystems: patterns, rates and restoration effectiveness.Estuaries Coasts3312491260
9. Copeland C (2010) Wetlands: an overview of issues. Washington (D.C.): Congressional Research Service. RL33483. C. Copeland2010Wetlands: an overview of issuesWashington (D.C.)Congressional Research Service. RL33483
10. Palmer M. A, Filoso S (2009) Restoration of ecosystem services for environmental markets. Science 325: 575–576.M. A. PalmerS. Filoso2009Restoration of ecosystem services for environmental markets.Science325575576
11. Zedler J. B (2000) Progress in wetland restoration ecology. Trends Ecol Evol 15: 402–407.J. B. Zedler2000Progress in wetland restoration ecology.Trends Ecol Evol15402407

12. Jones H. P, Schmitz O. J (2009) Rapid recovery of damaged ecosystems. Plos One 4: e5653.H. P. JonesO. J. Schmitz2009Rapid recovery of damaged ecosystems.Plos One4e5653
13. Rustad L. E, Campbell J. L, Marion G. M, Norby R. J, Mitchell M. J, et al. (2001) A meta-analysis of the response of soil respiration, net nitrogen mineralization, and aboveground plant growth to experimental ecosystem warming. Oecologia 126: 543–562.L. E. RustadJ. L. CampbellG. M. MarionR. J. NorbyM. J. Mitchell2001A meta-analysis of the response of soil respiration, net nitrogen mineralization, and aboveground plant growth to experimental ecosystem warming.Oecologia126543562
14. Hobbs R. J, Higgs E, Harris J. A (2009) Novel ecosystems: implications for conservation and restoration. Trends Ecol Evol 24: 599–605.R. J. HobbsE. HiggsJ. A. Harris2009Novel ecosystems: implications for conservation and restoration.Trends Ecol Evol24599605
15. Suding K. N, Gross K. L, Houseman G. R (2004) Alternative states and positive feedbacks in restoration ecology. Trends Ecol Evol 19: 46–53.K. N. SudingK. L. GrossG. R. Houseman2004Alternative states and positive feedbacks in restoration ecology.Trends Ecol Evol194653
16. Suding K. N, Hobbs R. J (2009) Threshold models in restoration and conservation: a developing framework. Trends Ecol Evol 24: 271–279.K. N. SudingR. J. Hobbs 2009Threshold models in restoration and conservation: a developing framework.Trends Ecol Evol24271279
17. Ramsar Convention Secretariat (2006) The Ramsar Convention Manual: a guide to the Convention on Wetlands (Ramsar, Iran, 1971), 4th edition. Gland, Switzerland: Ramsar Convention Secretariat. Ramsar Convention Secretariat2006The Ramsar Convention Manual: a guide to the Convention on Wetlands (Ramsar, Iran, 1971), 4th editionGland, SwitzerlandRamsar Convention Secretariat
18. Cole C. A (1999) Ecological theory and its role in the rehabilitation of wetlands. In: Streever W. R, editor. An international perspective on wetland rehabilitation. Dordrecht, The Netherlands: Kluwer Academic Publishers. pp. 232–243.C. A. Cole1999Ecological theory and its role in the rehabilitation of wetlands.W. R. StreeverAn international perspective on wetland rehabilitationDordrecht, The NetherlandsKluwer Academic Publishers232243
19. Hunt R. J, Walker J, Krabbenhoft D (1999) Characterizing hydrology and the importance of ground-water discharge in natural and constructed wetlands. Wetlands 19: 458–472.R. J. HuntJ. WalkerD. Krabbenhoft1999Characterizing hydrology and the importance of ground-water discharge in natural and constructed wetlands.Wetlands19458472
20. Ahn C, Dee S (2011) Early development of plant community in a created mitigation wetland as affected by introduced hydrologic design elements. Ecol Eng 37: 1324–1333.C. AhnS. Dee2011Early development of plant community in a created mitigation wetland as affected by introduced hydrologic design elements.Ecol Eng3713241333
21. Kumar J. L. G, Zhao Y. Q (2011) A review on numerous modeling approaches for effective, economical and ecological treatment wetlands. J Environ Manage 92: 400–406.J. L. G. KumarY. Q. Zhao2011A review on numerous modeling approaches for effective, economical and ecological treatment wetlands.J Environ Manage92400406
22. Collins B. D, Montgomery D. R (2002) Forest development, wood jams, and restoration of floodplain rivers in the Puget Lowland, Washington. Restor Ecol 10: 237–247.B. D. CollinsD. R. Montgomery2002Forest development, wood jams, and restoration of floodplain rivers in the Puget Lowland, Washington.Restor Ecol10237247
23. Crooks S, Schutten J, Sheern G. D, Pye K, Davy A. J (2002) Drainage and Elevation as Factors in the Restoration of Salt Marsh in Britain. Restor Ecol 10: 591–602.S. CrooksJ. SchuttenG. D. SheernK. PyeA. J. Davy2002Drainage and Elevation as Factors in the Restoration of Salt Marsh in Britain.Restor Ecol10591602
24. Lesbarreres D, Fowler M. S, Pagano A, Lode T (2010) Recovery of anuran community diversity following habitat replacement. J Appl Ecol 47: 148–156.D. LesbarreresM. S. FowlerA. PaganoT. Lode2010Recovery of anuran community diversity following habitat replacement.J Appl Ecol47148156

25. Warren R. S, Fell P. E, Rozsa R, Brawley A. H, Orsted A. C, et al. (2002) Salt marsh restoration in Connecticut: 20 years of science and management. Restor Ecol 10: 497–513.R. S. WarrenP. E. FellR. RozsaA. H. BrawleyA. C. Orsted2002Salt marsh restoration in Connecticut: 20 years of science and management.Restor Ecol10497513
26. Levin L. A, Talley T. S (2002) Natural and manipulated sources of heterogeneity controlling early faunal development of a salt marsh. Ecol Appl 12: 1785–1802.L. A. LevinT. S. Talley2002Natural and manipulated sources of heterogeneity controlling early faunal development of a salt marsh.Ecol Appl1217851802
27. Figuerola J, Green A. J, Michot T. C (2005) Invertebrate eggs can fly: evidence of waterfowl-mediated gene flow in aquatic invertebrates. Am Nat 165: 274–280.J. FiguerolaA. J. GreenT. C. Michot2005Invertebrate eggs can fly: evidence of waterfowl-mediated gene flow in aquatic invertebrates.Am Nat165274280
28. Badosa A, Frisch D, Arechederra A, Serrano L, Green A. J (2010) Recovery of zooplankton diversity in a restored Mediterranean temporary marsh in Doñana National Park (SW Spain). Hydrobiologia 654: 67–82.A. BadosaD. FrischA. ArechederraL. SerranoA. J. Green2010Recovery of zooplankton diversity in a restored Mediterranean temporary marsh in Doñana National Park (SW Spain).Hydrobiologia6546782
29. Meyer C. K, Whiles M. R (2008) Macroinvertebrate communities in restored and natural Platte River slough wetlands. J N Am Benthol Soc 27: 626–639.C. K. MeyerM. R. Whiles2008Macroinvertebrate communities in restored and natural Platte River slough wetlands.J N Am Benthol Soc27626639
30. Donath T. W, Holzel N, Otte A (2003) The impact of site conditions and seed dispersal on restoration success in alluvial meadows. Appl Veg Sci 6: 13–22.T. W. DonathN. HolzelA. Otte2003The impact of site conditions and seed dispersal on restoration success in alluvial meadows.Appl Veg Sci61322
31. Matthews J. W, Endress A. G (2010) Rate of succession in restored wetlands and the role of site context. Appl Veg Sci 13: 346–355.J. W. MatthewsA. G. Endress2010Rate of succession in restored wetlands and the role of site context.Appl Veg Sci13346355
32. Zedler J. B, Callaway J. C (1999) Tracking wetland restoration: do mitigation sites follow desired trajectories? Restor Ecol 7: 69–73.J. B. ZedlerJ. C. Callaway1999Tracking wetland restoration: do mitigation sites follow desired trajectories?Restor Ecol76973
33. Knops J. M. H, Tilman D (2000) Dynamics of soil nitrogen and carbon accumulation for 61 years after agricultural abandonment. Ecology 81: 88–98.J. M. H. KnopsD. Tilman2000Dynamics of soil nitrogen and carbon accumulation for 61 years after agricultural abandonment.Ecology818898
34. Ballantine K, Schneider R (2009) Fifty-five years of soil development in restored freshwater depressional wetlands. Ecol Appl 19: 1467–1480.K. BallantineR. Schneider2009Fifty-five years of soil development in restored freshwater depressional wetlands.Ecol Appl1914671480
35. Bechtold J. S, Naiman R. J (2009) A quantitative model of soil organic matter accumulation during floodplain primary succession. Ecosystems 12: 1352–1368.J. S. BechtoldR. J. Naiman2009A quantitative model of soil organic matter accumulation during floodplain primary succession.Ecosystems1213521368
36. van Groenigen K. J, Six J, Hungate B. A, de Graaff M. A, van Breemen N, et al. (2006) Element interactions limit soil carbon storage. Proc Natl Acad Sci U S A 103: 6571–6574.K. J. van GroenigenJ. SixB. A. HungateM. A. de GraaffN. van Breemen2006Element interactions limit soil carbon storage.Proc Natl Acad Sci U S A10365716574
37. Lawrence D, Schlesinger W. H (2001) Changes in soil phosphorus during 200 years of shifting cultivation in Indonesia. Ecology 82: 2769–2780.D. LawrenceW. H. Schlesinger2001Changes in soil phosphorus during 200 years of shifting cultivation in Indonesia.Ecology8227692780
38. Smil V (2000) Phosphorus in the environment: natural flows and human interferences. Annu Rev Energ Env 25: 53–88.V. Smil2000Phosphorus in the environment: natural flows and human interferences.Annu Rev Energ Env255388
39. Korfel C. A, Mitsch W. J, Hetherington T. E, Mack J. J (2010) Hydrology, physiochemistry, and amphibians in natural and created vernal pool wetlands. Restor Ecol 18: 843–854.C.

A. KorfelW. J. MitschT. E. HetheringtonJ. J. Mack2010Hydrology, physiochemistry, and amphibians in natural and created vernal pool wetlands.Restor Ecol18843854

40. Maurer B. A (2006) Ecological restoration from a macroscopic perspective. In: Falk D. A, Palmer M. A, Zedler J. D, editors. Foundations of restoration ecology. Washington (D.C.): Island Press. B. A. Maurer2006Ecological restoration from a macroscopic perspective.D. A. FalkM. A. PalmerJ. D. ZedlerFoundations of restoration ecologyWashington (D.C.)Island Press
41. Whittaker R. J (2006) Island biogeography: ecology, evolution, and conservation. Oxford, UK: Oxford University Press. R. J. Whittaker2006Island biogeography: ecology, evolution, and conservationOxford, UKOxford University Press
42. Anderson-Teixeira K. J, Vitousek P. M, Brown J. H (2008) Amplified temperature dependence in ecosystems developing on the lava flows of Mauna Loa, Hawai'i. Proc Natl Acad Sci U S A 105: 228–233.K. J. Anderson-TeixeiraP. M. VitousekJ. H. Brown 2008Amplified temperature dependence in ecosystems developing on the lava flows of Mauna Loa, Hawai'i.Proc Natl Acad Sci U S A105228233
43. Brinson M. M (1993) A hydrogeomorphic classification for wetlands. Washington (D.C.): US Army Corps of Engineers. WRP-DE-4. 103 p.M. M. Brinson1993A hydrogeomorphic classification for wetlandsWashington (D.C.)US Army Corps of Engineers. WRP-DE-4103
44. Craft C, Megonigal P, Broome S, Stevenson J, Freese R, et al. (2003) The pace of ecosystem development of constructed Spartina alterniflora marshes. Ecol Appl 13: 1417–1432.C. CraftP. MegonigalS. BroomeJ. StevensonR. Freese2003The pace of ecosystem development of constructed Spartina alterniflora marshes.Ecol Appl1314171432
45. Dolinar N, Gaberscik A (2010) Mycorrhizal colonization and growth of Phragmites australis in an intermittent wetland. Aquat Bot 93: 93–98.N. DolinarA. Gaberscik2010Mycorrhizal colonization and growth of Phragmites australis in an intermittent wetland.Aquat Bot939398
46. Kardol P, Wardle D. A (2010) How understanding aboveground belowground linkages can assist restoration ecology. Trends Ecol Evol 25: 670–679.P. KardolD. A. Wardle2010How understanding aboveground belowground linkages can assist restoration ecology.Trends Ecol Evol25670679
47. De Deyn G. B, Cornelissen J. H. C, Bardgett R. D (2008) Plant functional traits and soil carbon sequestration in contrasting biomes. Ecol Lett 11: 516–531.G. B. De DeynJ. H. C. CornelissenR. D. Bardgett2008Plant functional traits and soil carbon sequestration in contrasting biomes.Ecol Lett11516531
48. Race M. S, Fonseca M. S (1996) Fixing compensatory mitigation: what will it take? Ecol Appl 6: 94–101.M. S. RaceM. S. Fonseca1996Fixing compensatory mitigation: what will it take?Ecol Appl694101
49. Smith R. D, Ammann A, Bartoldus C, Brinson M. M (1995) An approach for assessing wetland functions using hydrogeomorphic classification, reference wetlands, and functional indices. Washington (D.C.): US Army Corps of Engineers. R. D. SmithA. AmmannC. BartoldusM. M. Brinson1995An approach for assessing wetland functions using hydrogeomorphic classification, reference wetlands, and functional indicesWashington (D.C.)US Army Corps of Engineers
50. Scheiner S. M, Gurevitch J (2001) Design and analysis of ecological experiments. Cary, North Carolina: Oxford University Press, Incorporated. S. M. ScheinerJ. Gurevitch2001Design and analysis of ecological experimentsCary, North CarolinaOxford University Press, Incorporated
51. Cardinale B. J, Srivastava D. S, Duffy J. E, Wright J. P, Downing A. L, et al. (2006) Effects of biodiversity on the functioning of trophic groups and ecosystems. Nature 443: 989–992.B. J. CardinaleD. S. SrivastavaJ. E. DuffyJ. P. WrightA. L. Downing2006Effects of biodiversity on the functioning of trophic groups and ecosystems.Nature443989992
52. Marvier M, McCreedy C, Regetz J, Kareiva P (2007) A meta-analysis of effects of Bt cotton and maize on nontarget invertebrates. Science 316: 1475–1477.M. MarvierC. McCreedyJ. RegetzP. Kareiva2007A meta-analysis of effects of Bt cotton and maize on nontarget invertebrates.Science31614751477

53. Craft C, Broome S, Campbell C (2002) Fifteen years of vegetation and soil development after brackish-water marsh creation. Restor Ecol 10: 248–258.C. CraftS. BroomeC. Campbell2002Fifteen years of vegetation and soil development after brackish-water marsh creation.Restor Ecol10248258
54. Ballantine K, Schneider R (2009) Fifty-five years of soil development in restored freshwater depressional wetlands. Ecol Appl 19: 1467–1480.K. BallantineR. Schneider2009Fifty-five years of soil development in restored freshwater depressional wetlands.Ecol Appl1914671480
55. Gutrich J. J, Taylor K. J, Fennessy M. S (2009) Restoration of vegetation communities of created depressional marshes in Ohio and Colorado (USA): the importance of initial effort for mitigation success. Ecol Eng 35: 351–368.J. J. GutrichK. J. TaylorM. S. Fennessy2009Restoration of vegetation communities of created depressional marshes in Ohio and Colorado (USA): the importance of initial effort for mitigation success.Ecol Eng35351368
56. Parikh A, Gale N (1998) Vegetation monitoring of created dune swale wetlands, Vandenberg Air Force Base, California. Restor Ecol 6: 83–93.A. ParikhN. Gale1998Vegetation monitoring of created dune swale wetlands, Vandenberg Air Force Base, California.Restor Ecol68393
57. Zampella R. A, Laidig K. J (2003) Functional equivalency of natural and excavated coastal plain ponds. Wetlands 23: 860–876.R. A. ZampellaK. J. Laidig2003Functional equivalency of natural and excavated coastal plain ponds.Wetlands23860876
58. Kottek M, Grieser J, Beck C, Rudolf B, Rubel F (2006) World Map of the Köppen-Geiger climate classification updated. Meteorologische Zeitschrift 15: 259–263.M. KottekJ. GrieserC. BeckB. RudolfF. Rubel2006World Map of the Köppen-Geiger climate classification updated.Meteorologische Zeitschrift15259263

CHAPTER 16

Relationships Between Salinity and Short-Term Soil Carbon Accumulation Rates from Marsh Types Across a Landscape in the Mississippi River Delta

Melissa M. Baustian[1], Camille L. Stagg[2], Carey L. Perry[3,4], Leland C. Moss[1], Tim J. B. Carruthers[1], Mead Allison[1,5]

[1]*The Water Institute of the Gulf, One American Place, 301 Main St, Ste 2000, Baton Rouge, LA 70825, USA*

[2]*U.S. Geological Survey, Wetland and Aquatic Research Center, 700 Cajundome Blvd, Lafayette, LA 70506, USA*

[3]*Gulf South Research Corporation, 8081 Innovation Park Drive, Baton Rouge, LA 70820, USA*

[4]*Coalition to Restore Coastal Louisiana, 6160 Perkins Road, Ste. 225, Baton Rouge, LA 70808, USA*

[5]*Department of Earth & Environmental Sciences, Tulane University, New Orleans, LA 70118, USA*

ABSTRACT

Salinity alterations will likely change the plant and environmental characteristics in coastal marshes thereby influencing soil carbon accumulation rates. Coastal Louisiana marshes have been historically classified as fresh, intermediate, brackish, or saline based on resident plant community and position along a salinity gradient. Short-term total carbon accumulation rates were assessed by collecting 10-cm deep soil cores at 24 sites located in marshes spanning the salinity gradient. Bulk density, total carbon content, and the short-term accretion rates obtained with feldspar horizon markers were measured to determine total carbon accumulation rates. Despite some significant differences in soil properties among marsh types, the mean total carbon accumulation rates among marsh types were not significantly different (mean ± std. err. of 190 ± 27 g TC m^{-2} $year^{-1}$). However, regression analysis indicated that mean annual surface salinity had a significant negative relationship with total carbon accumulation rates. Based on both analyses, the coastal Louisiana total marsh area (1,433,700 ha) accumulates about 2.7 to 3.3 Tg C $year^{-1}$. Changing salinities due to increasing relative sea level or resulting from restoration activities may alter carbon accumulation rates in the short term and significantly influence the global carbon cycle.

Keywords: Soil, Organic matter, Carbon accumulation, Salinity, Wetland, Louisiana

1. INTRODUCTION

Wetlands significantly impact the global carbon cycle because they contain 20–25% of the world's soil organic carbon pool (Roulet 2000; Mitra et al. 2005). Carbon accumulation in marsh soils is an important ecosystem process that contributes to the burial of carbon, also known as "blue carbon" (Chmura et al. 2003; Mitra et al. 2005; Hopkinson et al. 2012). These carbon accumulation dynamics in marsh soils will likely be altered in response to future environmental conditions driven by climate change and coastal restoration activities. For example, it is predicted that as air and sea temperatures increase, some herbaceous wetland vegetation will be replaced with tropical woody mangrove species (Comeaux et al. 2012; Osland et al. 2012; Doughty et al. 2015) and that these community shifts may influence carbon production and accumulation rates in wetland soils. Sea-level rise is also predicted to produce more saline environmental conditions that could drive transitions of fresh marshes to more saline marsh types (Herbert et al. 2015) or drive salt marsh transgression inland (Kirwan et al. 2016). In addition, large-scale restoration projects could introduce more fresh water to some estuaries, resulting in fresher marsh types (Morris et al. 2013).

Coastal marshes depend on the burial of organic matter for their vertical accretion to help keep pace with sea-level rise (Reed 1995; Neubauer 2008; Nellemann and Corcoran 2009; McLeod et al. 2011). Marsh vegetation influences accretionary processes by contributing autochthonous carbon to the soil and enhancing sediment trapping of allochthonous particulate carbon mineral sediment and deposition with their stems (Morris et al. 2002; Baustian et al. 2012). Accretion is commonly measured via marker horizons, such as feldspar, and particle-reactive radiotracers, such as ^{7}Be, ^{137}Cs, and ^{210}Pb (Callaway et al. 1996; Taylor et al. 2013). These methodologies are complementary: feldspar marker horizons measure accretion on a short time scale, months to years, (Cahoon and Turner 1989; Bianchette et al. 2015) whereas ^{137}Cs and ^{210}Pb measure accretionary processes over decadal timescales to a century or more (Delaune et al. 1978). Primary production processes dominate over decomposition processes in the surface layers (top ~ 10 cm) of the wetland soil column, which significantly influences the accumulation of organic matter and carbon content of marsh soils (Hatton et al. 1983; Nyman et al. 2006; Turner et al. 2006). Most of the root system for these species is concentrated in the upper 10 cm of the soil column (Gallagher et al. 1984); therefore, these short-term biological (e.g., production, decomposition) and physical (e.g., erosion, bioturbation) processes, which overall influence short-term carbon accumulation rates (Hansen and Nestlerode 2014), are captured by feldspar horizon methodologies.

Short-term carbon accumulation rates in marsh habitats depend mainly on the net balance of primary production and microbial decomposition (Odum et al. 1984; De Deyn et al. 2008). Both production and decomposition of organic matter are influenced by biological drivers and environmental conditions. Environmental conditions in surface waters and soil pore water, including salinity, temperature, inorganic nutrients, mineral sediment, and inundation frequency, can strongly influence organic matter accumulation in wetland soils (Valiela et al. 1976; Morris and Bradley 1999; Craft et al. 2002; Turner 2004; Marín-Spiotta et al. 2014). Salinity is an important factor in structuring coastal wetland communities and influences biogeochemical pathways (Craft 2007; Craft et al. 2009; Neubauer and Craft 2009). For example, increasing salinity can increase the concentration of terminal electron acceptors (such as iron, manganese and sulfate) that can increase microbial mineralization of organic matter and thus could result in a reduction in carbon accumulation (Herbert et al. 2015). Therefore, better understanding of how these environmental conditions influence carbon accumulation rates in various coastal wetland types is needed to predict how climate change through sea-level rise and future coastal restoration actions will influence the carbon accumulation rates (Markewich et al. 2007; Couvillion et al. 2013). In addition, these results may help support wetland restoration and management activities for carbon storage/burial capabilities and carbon credits (McLeod et al. 2011; Callaway et al. 2012; Hansen and Nestlerode 2014).

We hypothesized that short-term carbon accumulation rates varied along a gradient of increasing salinity. The objective of the present study was to examine the short-term total carbon accumulation rates in four marsh types across a salinity gradient in coastal Louisiana. We addressed the following questions: 1) how do short-term total organic carbon accumulation rates differ among marsh types in coastal wetlands, and 2) how do the environmental conditions influence the total organic carbon accumulation rates in these marsh types?

2. METHODS

2.1 Site Description

A total of 24 marsh sites were selected for this study in the Mississippi and Atchafalaya delta regions of Louisiana within four marsh types. We defined marsh types based on plant community composition and salinity values (sensu Visser et al. 2002). Each marsh type (fresh, intermediate, brackish, and saline, Chabreck 1972) was associated with a particular salinity range (Cowardin et al. 1979; Odum 1984). However, because multiple factors influence the distribution of marsh plants, and many species occur over a wide range of salinity conditions, some sites may exist outside of the defined salinity range for a particular wetland type. Six sites each were sampled within four marsh types: fresh, intermediate, brackish, and saline (Odum et al. 1984; Visser et al. 1998) spanning two hydrologic basins, Terrebonne and Barataria (Fig. 1).

Dominant vegetation included: *Panicum hemitomon* and *Typha latifolia* at fresh marshes, *Sagittaria lancifolia* and *Scheonoplectus americanus* at intermediate marsh types, *Spartina patens* and *Scheonoplectus americanus* at brackish marsh types, *Spartina alterniflora* and *Juncus romerianus* at saline marsh types. The 24 sites, which have been characterized in previous ecological studies (Stagg et al. 2016) encompassed a relatively broad geographical landscape (~65 km) (Fig. 1) and were co-located with Coastwide Reference Monitoring Systems (CRMS) stations (http://lacoast.gov/crms2/home.aspx). Since 2006, continuous hydrological (salinity, temperature, water level) and discrete vegetation, soil, and elevation data have been collected at the CRMS sites (Steyer et al. 2003).

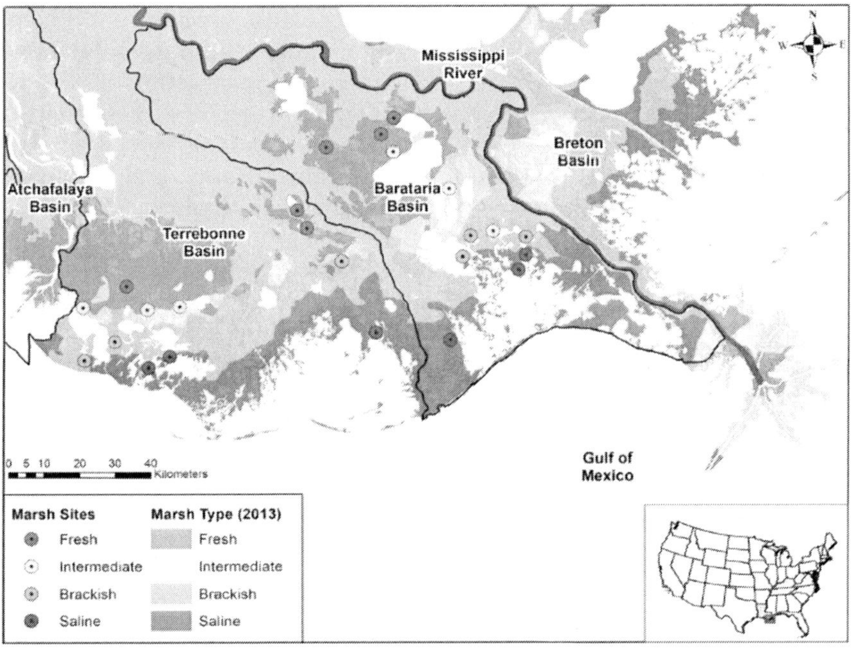

Figure. 1 Study site locations within four marsh types (fresh, intermediate, brackish and saline) of Barataria ($n = 12$) and Terrebonne basins ($n = 12$) in coastal Louisiana. Marsh type boundaries were defined using vegetation classification data from the Coastwide Reference Monitoring System

2.2 Collection of Soil Cores

Soil cores were collected at the 24 sites in February 2015. The coring location was approximately 25 m inland from the shoreline at each site and cores were collected to a depth of 100 cm with a McCaulley corer (inside diameter = 5.1 cm). Only the top 10 cm of the core was utilized for the present study. Each core was sectioned into 2-cm depth intervals in the field, placed in whirlpaks®, and stored on ice until transported to a laboratory for post-processing and further analysis.

2.3 Organic Matter Conversion

Soil samples from each 2-cm increment were freeze-dried to a constant weight to determine % moisture and dry bulk density and then ground using a mortar and pestle. Loss on ignition methodology was used to determine soil organic matter content and inorganic matter via combustion at 550 °C for 14 h (Wilson and Allison 2008). Supplemental soil cores from a previous study (Baustian et al. unpublished data) helped to develop a conversion rate from soil organic matter to total carbon content for soil samples with these region-specific characteristics (Craft et al. 1991; Morris et al. 2016). Supplementary soil cores were collected with a PVC corer (10.16 cm inner diameter, 30 cm long, sliced at 4 cm increments) at various CRMS sites in Barataria, Breton, and Atchafalaya basins. These soil samples were simultaneously analyzed for organic matter (via Loss on Ignition, Heiri et al. 2001), total organic carbon, and total carbon according to USEPA methodology 440. The total organic carbon (%) and total carbon (mineral + organic, %) of the soil layers collected previously in fresh to saline marsh types were significantly correlated to the soil organic matter (%) (Fig. 2). The soil total organic carbon fraction on average contributed about 97% of the soil total carbon, therefore these parameters are near identical and both can be used to estimate carbon content. A conversion factor of 0.47 obtained from the regression coefficient of the total carbon vs. soil organic matter was applied to the mean soil organic matter values obtained at the 24 sites in this study (Fig. 1, colored circles) to provide informed estimates of carbon content. This conversion rate is similar to other studies that reported a slope of 0.45 between organic matter and organic carbon (50 cm length cores including samples from forested wetlands) in coastal Louisiana (based on $n = 1{,}142$, $r^2 = 0.93$, $p < 0.001$) (Piazza et al. 2011; Steyer et al. 2012) and 0.50 between organic matter and total carbon based on $n = 287$, $r^2 = 0.98$, $p < 0.01$) for tidal marsh soils (50 cm length cores) in Connecticut and New York (Hill and Anisfeld 2015).

2.4 Accretion Estimates

To estimate short-term total carbon accumulation rates within each marsh type, accretion estimates were calculated using feldspar marker horizon data previously collected at 16 CRMS sites ($n = 4$ for intermediate, $n = 6$ for brackish and $n = 6$ for saline marsh types). The use of feldspar horizon markers was not recommended in the fresh marsh sites ($n = 6$) and some of the intermediate sites ($n = 2$) because they were considered floating marshes (Folse et al. 2014). The accretion measurements began in 2007 and continued through 2015 (Table 1). At each of the sites, an accretion monitoring plot was established and managed by CRMS program using a white feldspar clay as a marker horizon (Folse et al. 2014). Each plot was then sampled bi-annually for 2 years and less frequently in subsequent years (Folse et al. 2014). See other literature (Cahoon and Turner 1989; Folse et al. 2014; Bianchette et al. 2015) for more details on the feldspar methodology including limitations and benefits as well as details about downloading accretion data from the Coastal Information Management System (CIMS).

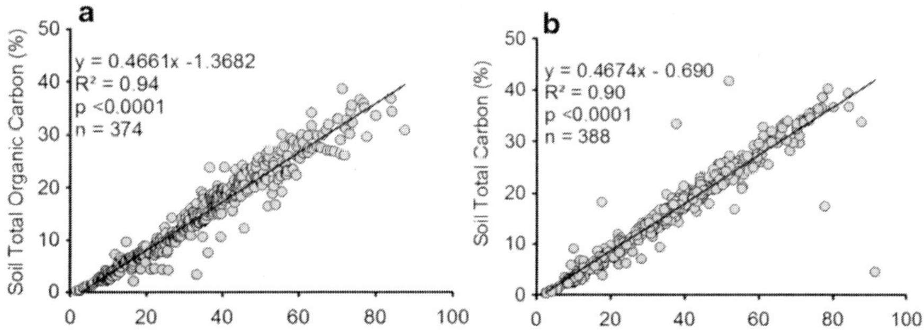

Figure. 2 Comparison of soil organic matter with (**a**) soil total organic carbon and (**b**) total carbon in soil layers (0 to 30 cm) sampled in fresh to saline marsh types in coastal Louisiana. Total carbon content was estimated from this relationship (*panel b*) for calculations of short-term total carbon accumulation rates. Outliers ($n = 14$) were removed from *panel a*

2.5 Environmental Conditions

Separate soil cores (inner diameter = 5 cm, length = 15 cm) were collected to characterize some of the environmental conditions of the marshes in December 2014. For each core, soils were homogenized, weighed, dried at 60 °C, and re-weighed to determine bulk density and % moisture. A portion of the dried soil was used to measure electrical conductivity, by mixing with distilled water (1:1 slurry ratio) and analyzed using a YSI model 30 conductivity meter (Rhoades et al. 1996), organic matter, total carbon, and total nitrogen (Nelson and Sommers 1982), and total phosphorus (Aspila et al. 1976). Elevation was another environmental condition that was measured at various locations per site with a Trimble real-time kinematic (RTK) system via GEOID99. Lastly, available annual surface water salinity data between calendar years of 2000 and 2015 were used to calculate the mean annual salinity at each site (http://www.lacoast.gov/chart2/charting.aspx?laf=crms&tab=2).

2.6 Total Carbon Accumulation Estimates

The calculation of total carbon accumulation (g TC m^{-2} $year^{-1}$) in the top 10 cm of the soil core was determined by multiplying the carbon density (g C cm^{-3}) by the accretion rate (cm $year^{-1}$) following Bernal and Mitsch (2008; 2012) and Bianchi et al. (2013).

Table 1 Field sites in various marsh types where feldspar accretion (years 2007 to 2015) and mean annual salinity (years 2010 to 2015) data were collected. Feldspar accretion estimates were not measured at fresh or intermediate floating marshes

Basin	Marsh Type	Floating Type	Site ID	No. Obs.	Begin Sample Date	End Sample Date	R^2	Feldspar Accretion rate (cm $year^{-1}$)	No. Obs.	Mean Annual Salinity (ppt)
BA	Fresh	Floating	CRMS0211	ND	ND	ND	ND	ND	7	0.20
BA	Fresh	Floating	CRMS0273	ND	ND	ND	ND	ND	10	0.62
BA	Fresh	Floating	CRMS0327	ND	ND	ND	ND	ND	10	0.61
TE	Fresh	Floating	CRMS0331	ND	ND	ND	ND	ND	10	0.29
TE	Fresh	Floating	CRMS0367	ND	ND	ND	ND	ND	10	0.27
BA	Fresh	Floating	CRMS3166	ND	ND	ND	ND	ND	9	0.25
BA	Intermediate	Floating	BA-01-04	ND	ND	ND	ND	ND	10	0.46
BA	Intermediate	Non-Floating	CRMS0225	21	11/6/2008	9/23/2015	0.51	1.78	10	5.25
TE	Intermediate	Non-Floating	CRMS0305	22	4/26/2007	10/29/2015	0.87	1.75	10	1.98
TE	Intermediate	Non-Floating	CRMS0398	24	3/22/2007	10/23/2015	0.82	1.60	7	1.95
TE	Intermediate	Non-Floating	CRMS4045	18	9/25/2009	9/25/2015	0.76	0.80	9	1.26
BA	Intermediate	Floating	CRMS4245	ND	ND	ND	ND	ND	7	1.67
BA	Brackish	Non-Floating	CRMS0253	23	4/5/2007	10/12/2015	0.63	2.25	10	4.95
TE	Brackish	Non-Floating	CRMS0309	22	4/26/2007	9/23/2015	0.52	0.64	10	9.37
TE	Intermediate	Non-Floating	CRMS0398	24	3/22/2007	10/23/2015	0.82	1.60	7	1.95
TE	Intermediate	Non-Floating	CRMS4045	18	9/25/2009	9/25/2015	0.76	0.80	9	1.26
BA	Intermediate	Floating	CRMS4245	ND	ND	ND	ND	ND	7	1.67
BA	Brackish	Non-Floating	CRMS0253	23	4/5/2007	10/12/2015	0.63	2.25	10	4.95
TE	Brackish	Non-Floating	CRMS0309	22	4/26/2007	9/23/2015	0.52	0.64	10	9.37
TE	Brackish	Non-Floating	CRMS0399	20	4/26/2007	10/8/2015	0.53	1.10	10	5.44
TE	Brackish	Non-Floating	CRMS2825	18	10/26/2009	10/29/2015	0.43	1.02	9	5.42
BA	Brackish	Non-Floating	CRMS3565	24	4/16/2007	9/22/2015	0.73	1.64	5	4.82
BA	Brackish	Non-Floating	CRMS3617	19	10/15/2009	9/14/2015	0.41	0.90	10	8.08
BA	Saline	Non-Floating	CRMS0175	18	2/4/2009	10/2/2015	0.63	0.65	10	15.54
BA	Saline	Non-Floating	CRMS0224	20	5/5/2008	10/22/2015	0.65	0.34	10	9.45
BA	Saline	Non-Floating	CRMS0237	17	5/28/2008	10/22/2015	0.40	1.86	10	8.90
TE	Saline	Non-Floating	CRMS0337	19	10/21/2009	10/9/2015	0.48	0.54	10	18.23
TE	Saline	Non-Floating	CRMS0377	18	2/3/2009	10/5/2015	0.21	0.23	10	14.33
TE	Saline	Non-Floating	CRMS4455	15	2/3/2009	10/6/2015	0.84	1.01	10	9.64

BA Barataria, *TE* Terrebonne, *ND* no data

Coastwide salinity values were obtained from the Ecohydrology and Wetland Morphology simulation models (Meselhe et al. 2013) through publicly available output underlying the 2012 Louisiana Coastal Master Plan (Coastal Protection and Restoration Authority 2012). The model output consists of a spatial distribution of surface water salinity representing year 2010 with centroids of a 500 × 500 m grid size. The salinity model outputs were used in a regression equation (with $n = 16$ sites, see Fig. 6e) from this study to estimate the total carbon (g TC $m^{-2}year^{-1}$) per grid cell and then multiplied by the grid size to get the total mass of carbon (g TC $year^{-1}$). The grid cells were summed across coastal Louisiana marsh area (2013 coastal marsh layer, including floating marshes, delineated by Sasser et al. (2014)) to obtain a grand total estimate in units of Tg TC $year^{-1}$. The total marsh area was estimated at ~1,433,700 ha.

2.7 Statistical Analysis

Statistical analyses and data manipulations were performed in SAS 9.4. Our null hypothesis for all tests was that there was no difference in soil properties (carbon content, bulk density, total carbon density), feldspar accretion rates, and short-term carbon accumulation rates among marsh types and slopes did not differ from zero with $\alpha = 0.05$. Slopes of the time-series accretion data as well as the soil properties (%TC and bulk density were square root transformed), feldspar accretion rates, and short-term accumulation rates versus mean annual salinity were estimated with PROC REG. Extreme outliers in the time-series accretion and mean annual salinity linear regressions were defined and identified if criteria were met by both influence techniques of DFFITS and DBETAS (Belsley et al. 1980; Aguinis et al. 2013) and were removed from further analyses. Approximately 5% of the feldspar measurements (mean value per plot set per date) were identified as outliers when regressed against time and removed prior to calculating the slopes or accretion rate estimates. Differences among marsh types were evaluated with an analysis of variance (ANOVA) using PROC GLMMIX. Exploratory analysis related to how the environmental conditions (represented as vectors) may vary across (represented as symbols) marsh types was conducted with a correlation biplot based on principal component analysis (PCA). A correlation matrix on standardized data and α decomposition of 0.5 for a symmetrical plot with the PROC PRINCOMP statement and biplot macro was used in SAS. Using R software (R Core Team 2013), we conducted an analysis of similarities (ANOSIM) to test statistically whether there was a significant difference in principal components factor scores among the four marsh types.

3. RESULTS

3.1 Soil Properties per Marsh Type

The mean percent total carbon (TC) in the layers of the top 10 cm varied among marsh types (Fig. 3). Overall, by pooling the layers per marsh type the mean percent TC was about 32 ± 1, 17 ± 2, 16 ± 1 and 9 ± 1% in the fresh, intermediate, brackish and saline marsh types, respectively. A mean of the top four layers per soil core was calculated to estimate the soil properties of the sites including the total carbon accumulation rate.

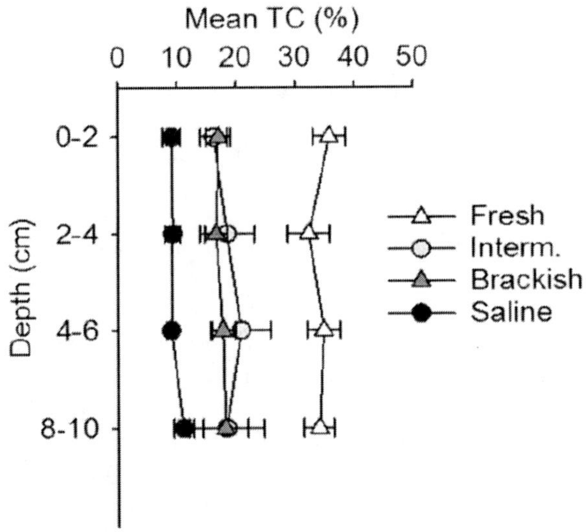

Figure. 3 Mean (± std. error) total carbon content (%) per soil layer in the top 10 cm ($n = 6$ per layer per marsh type) representing four marsh types, fresh, intermediate, brackish and saline in coastal Louisiana

The mean percent TC (±std. error) and bulk density among the four marsh types significantly differed ($p = 0.0001$, Fig. 4a, b). The mean percent TC of the fresh marshes was highest, with intermediate and brackish types similar between each other but significantly higher than the saline marshes (Fig. 4a). Dry soil bulk density in the upper 10 cm ranged between 0.07 to 0.19 g cm^{-3} among the marsh types and compared to the percent TC the bulk density had a similar significant but opposite trend with increasing salinity of marsh types (Fig. 4b). The mean total carbon density (g cm^{-3}) was higher in the fresh marsh type (0.022 ± 0.001, $p = 0.002$) compared to the other three types: intermediate (0.017 ± 0.002), brackish (0.015 ± 0.001), and saline (0.016 ± 0.001) (Fig. 4c).

3.2 Accretion Estimates per Marsh Type

The mean feldspar estimated accretion rates were nearly significantly different among the four marsh types ($p = 0.07$, Fig. 4d). The mean feldspar accretion rate was not available for the fresh marsh type while the intermediate marsh type was estimated at 1.71 ± 0.05 cm $year^{-1}$, brackish at 1.26 ± 0.24 cm $year^{-1}$, and saline marsh type at 0.77 ± 0.24 cm $year^{-1}$. Accretion rates at brackish and saline sites were highly variable (Fig. 4d).

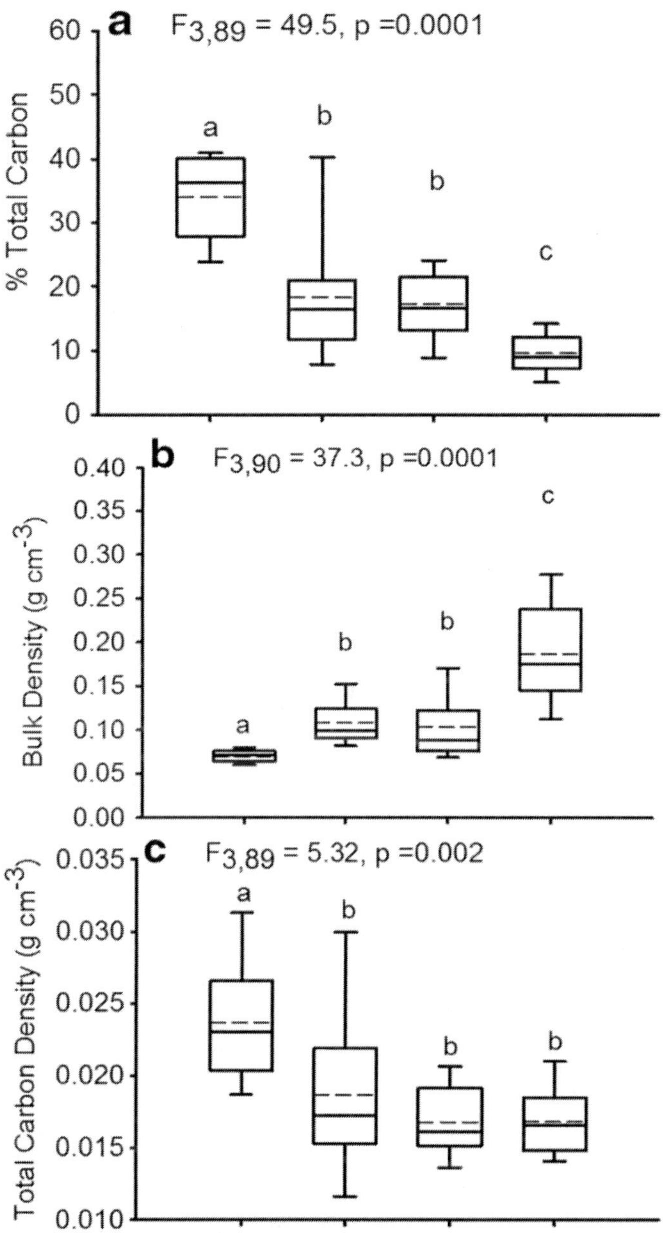
a
F3,89 = 49.5, p =0.0001
% Total Carbon
0
10
20
30
40
50
60
a
b
b
c
b
F3,90 = 37.3, p =0.0001
Bulk Density (g cm-3)
0.00
0.05
0.10
0.15
0.20
0.25
0.30
0.35
0.40
a
b
b
c
c
F3,89 = 5.32, p =0.002
Total Carbon Density (g cm-3)
0.010
0.015
0.020
0.025
0.030
0.035
a
b
b
b

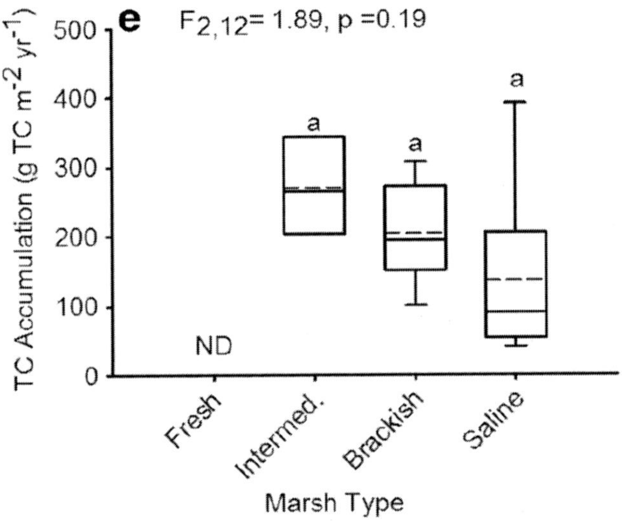

Figure. 4 Box plots of top 10 cm soil characteristics of (**a**) total carbon ($n = 22$ fresh, $n = 23$ intermediate, $n = 24$ brackish, $n = 24$ saline), (**b**) bulk density ($n = 23$ fresh, $n = 23$ intermediate, $n = 24$ brackish, $n = 24$ saline), (**c**) total carbon density ($n = 22$ fresh, $n = 23$ intermediate, $n = 24$ brackish, $n = 24$ saline), (**d**) feldspar accretion rates ($n = 0$ fresh, $n = 4$ intermediate, $n = 6$ brackish and $n = 6$ saline) and **e** short-term total carbon accumulation rates ($n = 0$ fresh, $n = 4$ intermediate, $n = 6$ brackish, and $n = 6$ saline) representing four marsh types, fresh, intermediate, brackish and saline in coastal Louisiana. Different letters indicate significant difference. The box plot dashed line = mean, solid line = median and whiskers above and below the box indicate the 90th and 10th percentiles. ND = no data

3.3 TC Accumulation Rates per Marsh Type

The short-term TC accumulation rates ranged between 41 to 391 g TC m^{-2} $year^{-1}$ ($n = 16$, Fig. 4e). One intermediate site (CRMS 4045, 137 g TC m^{-2} $year^{-1}$) was identified and removed as an outlier when regressed on mean annual salinities. The short-term TC accumulation rates for the remaining sites ($n = 15$) were then used an ANOVA and these rates were not significantly different across the three marsh types ($F_{2,12} = 1.89$, $p = 0.19$). Overall, the grand mean (±std. err.) of the short-term TC accumulation rates used in the ANOVA ($n = 15$) was 190 ± 27 g TC m^{-2} $year^{-1}$.

3.4 Environmental Variables of Marsh Types

The PCA (Fig. 5) indicated that the first two dimensions accounted for 88% of the variation of the field samples. From the PCA biplot (Fig. 5), it was observed that the field sites tended to cluster into their respective four marsh types irrespective of their basin location with fresh sites on the far right and saline sites on the far left leaving the intermediate and brackish sites intermixed in the middle. The soil properties of organic matter, total carbon, total phosphorus, and total nitrogen had high positive correlation (i.e., small angles between vectors) among each other. For PC1, the compositional dissimilarities among the four wetland types were greater than those within the wetland types, indicating that there was a significant separation of PC1 parameters among the four wetland types ($R = 0.22$, $p = 0.002$).

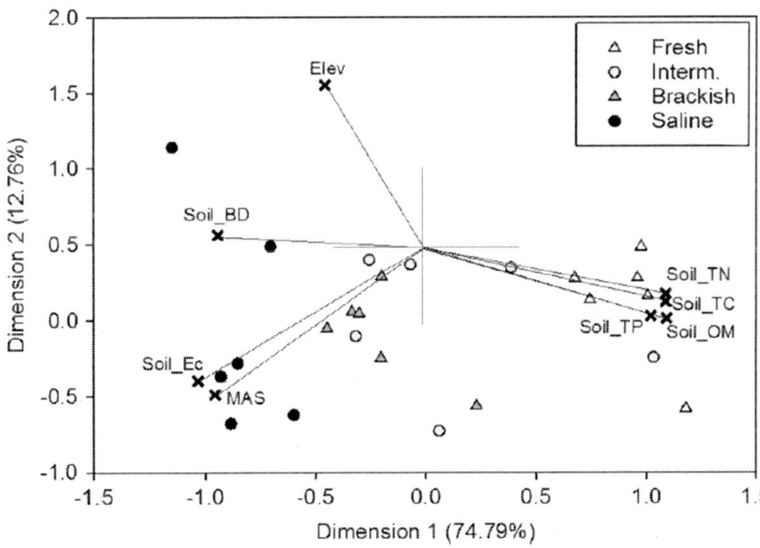

Figure. 5 Principal component analysis (PCA) biplot of soil and surface water variables as vectors ($n = 8$) and marsh sites as points color coded per marsh type ($n = 24$). The soil variables included total carbon (soil_TC), organic matter (soil_OM), bulk density (soil_BD), total phosphorus (soil_TP), total nitrogen (soil_TN), electroconductivity (soil_Ec), and elevation (Elev). The surface water variable included the mean annual salinity (years 2000–2015, MAS). Perpendicular vectors are uncorrelated, vectors with small angles between them are positively correlated and opposite vectors are negatively correlated. Longer lengths of vectors indicate higher variability

High bulk density and electrical conductivity of the soils as well as the mean annual salinity of the surface waters were positively correlated with the saline marsh sites. Elevation seemed to be uncorrelated with the other soil properties. The saline marsh sites had the lowest concentrations of the soil properties of % TN, % TC and % TP (i.e., furthest location from the variable).

The soil properties and short-term TC accumulation rates were all significantly related to the mean annual salinities (years 2000 to 2015) of the marsh sites (Fig. 6). The % total carbon, total carbon density, and feldspar accretion rates all had significant negative relationships with mean annual salinity (Fig. 6a, c, d). Bulk density was the only soil property that had a positive relationship (Fig. 6b). Mean annual salinity was negatively correlated with the TC accumulation rates (Fig. 6e) and accounted for about 40% of the variation of the short-term TC accumulation rates ($R^2 = 0.40$, $p = 0.01$, $n = 15$).

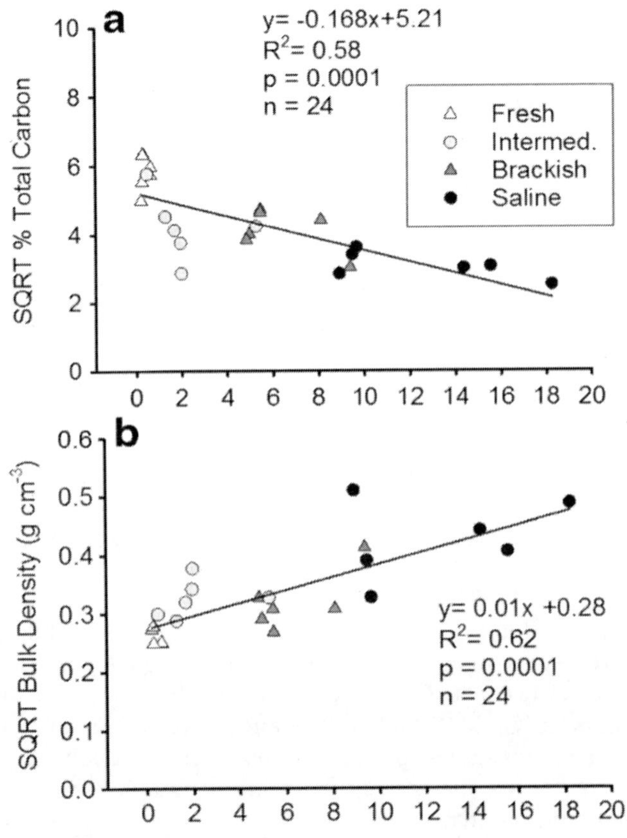

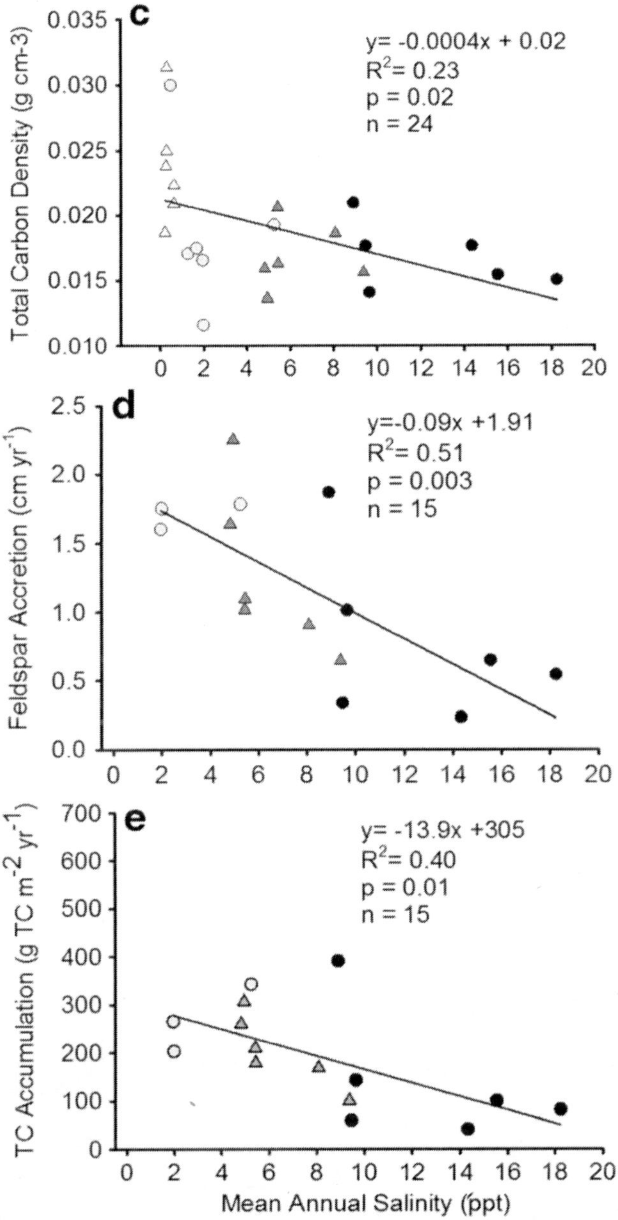

Figure. 6 Soil properties (**a** to **c**), feldspar accretion rates (**d**) and short-term total carbon (TC) accumulation rates (**e**) in relation to mean annual salinity within four marsh types, fresh (*open triangle*), intermediate (*light grey filled circle*), brackish (*dark grey filled triangle*) and saline (*black filled circle*)

3.5 Coastwide Estimate of Short-Term Carbon Pool

The total short-term carbon pool of fresh, intermediate, brackish and saline marshes in coastal Louisiana was estimated based on two approaches, the mean annual salinity and short-term TC accumulation regression equation (see Fig. 6e) and the mean value when comparing the marsh types provides a global context on the amount of marsh soil carbon coastal Louisiana contributes. Based on previously modeled salinity output and the regression equation, the coastal marshes on the central and western region tend to have the highest short-term TC accumulation rates with the eastern area having the lowest and overall the marshes store about 3.3 Tg TC year^{-1} (Fig. 7). By using the mean value of short-term carbon accumulation rate from comparing the marsh types and the estimated area of marsh in coastal Louisiana (~1,433,700 ha) these marshes were estimated to store in the short term about 2.7 Tg TC year^{-1}.

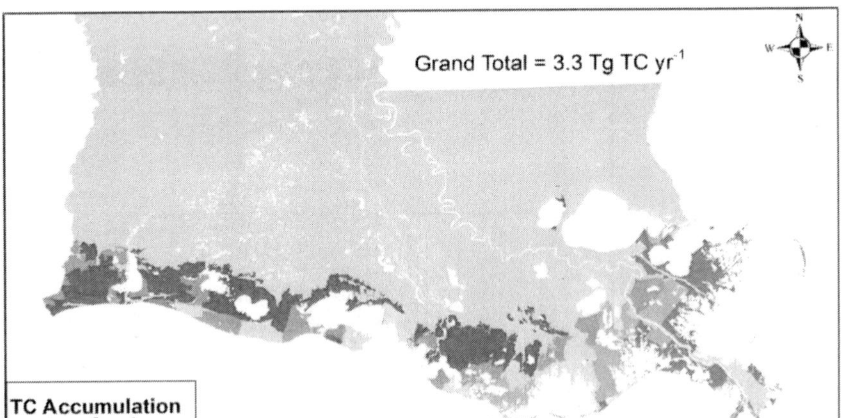

Figure. 7 Map of short-term TC accumulation rate estimates in the marshes of coastal Louisiana, USA, based on previously modeled spatial salinity output and regression analysis (see Fig. 6e). Grand total of short-term carbon pool for the entire marsh area of coastal Louisiana is listed. Marsh type comparison approach suggested a lower grand total of short-term carbon pool of 2.7 Tg TC year^{-1}

4. DISCUSSION

4.1 Short-Term TC Accumulation Rates

Significant declines in short-term carbon accumulation with increasing salinity were found when rates were regressed against mean annual salinity of surface waters; however, categorical analysis of short-term carbon accumulation rates showed no significant differences among marsh types, reflecting the high variability of short-term total carbon accumulation within marsh types (fresh, intermediate, brackish and saline). This pattern was also observed with the feldspar accretion rates. Given the regression results, it is clear that salinity, per se, is an important predictor of short-term carbon accumulation rates (Craft 2007). However, because marsh types are not exclusively defined by salinity, analysis of variance among marsh types allows us to identify potential mechanisms associated with marsh type that are not directly related to salinity or may not be explicitly defined in a regression analysis. Therefore, we considered patterns across the salinity gradient as well as among different marsh types. For example, the separation of the marsh sites along the Dimension 1 axis of the PCA helps generalize characteristics about the marsh types. The fresh marshes had higher soil carbon, nitrogen and phosphorus, while the saline marshes had higher bulk density, soil and surface water salinities. As expected, the intermediate and brackish sites had values in between those end members. The fresh marsh sites, with their associated higher soil carbon content, may be associated with higher primary production and perhaps higher short-term carbon accumulation rates. Alternatively, decomposition rates may be slower in the fresher sites compared to the more saline sites, as mentioned previously and supported by Williams and Rosenheim (2015). Estimates of production and decomposition of organic matter both above and belowground within these four marsh types would help to better explain the net effect that produces these short-term carbon accumulation rates (Neubauer 2008).

Short-term TC accumulation rates from the Gulf region integrated over the top 10–15 cm of the soil cores were reported by Hansen and Nestlerode (2014) using ^{137}Cs based estimates of accretion. They found the brackish marshes dominated by *S. patens* had short-term TC accumulation rates of 290 ± 170.3 g TC m^{-2} $year^{-1}$, salt marshes in Louisiana had 329 ± 217 g TC m^{-2} $year^{-1}$ and salt marshes outside of Louisiana as 173 ± 131 g TC m^{-2} $year^{-1}$. They also reported similar mean accretion rates 0.72, 0.94 and 0.57 cm $year^{-1}$ respectively across the three marsh types in their study (Hansen and Nestlerode 2014). Overall, these mean accretion rates and TC accumulation rates are within the range of this study, but brackish marshes in the current study had accretion rates (1.26 cm $year^{-1}$) that exceed Hansen and Nestlerode (2014) estimates (0.72 cm $year^{-1}$), and the saline marshes were similar (current study = 0.77 cm $year^{-1}$, Hansen and Nestlerode (2014) = 0.94 cm $year^{-1}$). However, in this study saline marsh types had lower mean short-term TC accumulation rates (~135 g TC m^{-2} $year^{-1}$) compared to Hansen and Nestlerode (2014) reported rates, which is likely because their mean bulk density values (an essential variable for calculating accumulation) for Louisiana salt marshes were twice as high as the marshes sampled in this study (current study = 0.19 g cm^{-3}, Hansen and Nestlerode (2014) = 0.4 g cm^{-3}). The differences in bulk density could be due to spatial variation within the marsh, for example marsh edge (0.35 g cm^{-3}) versus interior (0.29 g cm^{-3}) as observed in another study (Hatton et al. 1983). In addition, the use of ^{137}Cs versus feldspar to estimate short-term carbon accumulation rates could also be adding to the variability in results among these two studies.

Short-term total carbon accumulation rates calculated from feldspar methodology from various marsh types along a landscape gradient are not readily available. This study provides a novel contribution to the growing body of literature on carbon accumulation rates in wetlands by providing insight on how short-term processes (net effect from production, decomposition, bioturbation, etc.) captured in these accumulation rates vary across the coastal landscape, which in turn will likely influence the quality and quantity of carbon being accumulated for the long-term.

4.2 Environmental Influences on Short-Term TC Accumulation Rates

Variation in environmental drivers may help to explain the high variation in TC accumulation rates. Williams and Rosenheim (2015) found, in agreement with this study, that the total organic carbon content (in these soils total carbon percentages are similar) varied significantly among fresh, brackish and saline marshes in Barataria Basin, Louisiana, with fresh marsh soils containing the highest total organic carbon content (%). They also found from pyrolysis methodology that organic carbon in fresh marsh soils tended to be most stable, or least degradable, among the marsh types and attributed this to a decrease in remineralization due to high water content of the soils and higher recalcitrance of soil organic matter. This study also found a negative relationship between salinity and short-term TC accumulation rates, and ~ 40% of the variation of the short-term TC accumulation rates was explained by the mean annual salinity from years 2000 to 2015. Craft (2007) and Neubauer (2008) reported negative correlations between vertical accretion rates, which influence short-term TC accumulation rates, and salinity. They suggest that freshwater has positive effects on organic matter production and preservation while saline water stimulated root decomposition resulting in lower rates of organic matter accumulation. Seawater sulfate plays a critical role in regulating decomposition and productivity by enhancing decomposition via sulfate reduction, a process which also generates sulfides that are toxic to plants (Neubauer and Craft 2009).

4.3 Coastwide TC Short-Term Accumulation

The fresh, intermediate, brackish and saline marshes of coastal Louisiana overall accumulate between 2.7 to 3.3 Tg C year^{-1}. This regional value alone comprises a notable fraction (~5–60%) of global estimates of carbon accumulation of salt marshes (5–87 Tg C year^{-1}) (Chmura et al. 2003; McLeod et al. 2011) and emphasizes the important role that coastal and deltaic wetlands play in the global carbon cycle. In addition, Mcleod et al. (2011) used an average salt marsh accumulation rate of 218 ± 24 g TC m^{-2} year^{-1} to calculate their global accumulation estimates, similar to the current mean estimate across the three marsh types (190 ± 27 g TC m^{-2} year^{-1}, $n = 15$). However, it should be noted that the global estimates of areal extent of salt marshes is not well constrained and likely drives the high variability in the global burial estimates (Chmura 2013). Better estimates on areal extent of these marshes would help to refine global carbon budget estimates. These comparisons suggest that, not only do salt marshes play an important role in carbon accumulation and burial, but also, global estimates of carbon accumulation that do not consider salinity changes across coastal landscape gradients may not be accurate. Furthermore, if these marshes transition to more saline marshes due to climate change, or towards fresh marshes due to large scale restoration actions, the short-term global carbon pool may change. Balancing the greenhouse gas emissions from each of these four marsh types with the short-term carbon accumulation rates may provide a comprehensive understanding of the role of these marshes in overall carbon sequestration (Poffenbarger et al. 2011; Holm et al. 2016; Krauss et al. 2016). For example, fresh marshes may accumulate a large amount of carbon, but they produce methane gas during anaeraboic metabolism, which has a global warming potential that is 25 times greater than carbon dioxide (Whiting and Chanton 2001).

Short-term total carbon accumulation estimates from marsh types that span a salinity gradient illustrate potential implications for ecological responses to sea-level rise and coastal restoration. If surface waters become more saline due to sea-level rise or freshwater limitations, coastal wetlands may transition to a more salt-tolerant plant communities and accumulate less total carbon in their soils (Williams and Rosenheim 2015; Herbert et al. 2015). Conversely, if more freshwater reaches the coast via riverine discharge, restoration management actions such as large river diversions, or increased precipitation events, coastal wetlands may become fresher and accumulate more soil carbon in the short term (Morris et al. 2013). These results also may help guide where along the salinity gradient to place marsh restoration and creation projects to maximize short-term carbon accumulation capacities and potential carbon credits.

ACKNOWLEDGEMENTS

This project was supported by the Science and Engineering Program of The Water Institute of the Gulf with funds from the Louisiana Coastal Protection and Restoration Authority (CPRA) and the Baton Rouge Area Foundation (BRAF) as well as U.S. Geological Survey Ecosystems Program, and Restore the Mississippi River Delta Campaign via Coalition to Restore Coastal Louisiana (CRCL). Many helping hands assisted in retrieving and processing the soil samples. We thank the following researchers at the USGS Wetland and Aquatic Research Center: Ada Diaz, Alyssa Mitchell, Brett Patton, Courtney Lee, Evan Bergeron, Jacy Reynolds, Lindsey Hebert, Sarai Piazza and Rachel Villani. In addition, Caitlin Pinsonat from the Water Institute of the Gulf assisted in field and laboratory work. Jackson Rollings helped CRCL through AmeriCorps' Louisiana Delta Service Corps Program in field data collection. Special thanks to Diana Di Leonardo and Amanda Cuesta at Tulane University for assisting in soil organic matter laboratory analyses and to Ioannis Georgiou at University of New Orleans for letting us borrow the corer. Lastly, thank you to the three anonymous reviewers that provided excellent comments that helped strengthen our paper.

REFERENCES

1. Aguinis H, Gottfredson RK, Joo H (2013) Best-practice recommendations for defining, identifying, and handling outliers. Organizational Research Methods 16:270–301. doi:10.1177/1094428112470848
2. Aspila KI, Agemian H, Chau ASY (1976) A semi-automated method for the determination of inorganic, organic and total phosphate in sediments. Analyst 101:187–197. doi:10.1039/AN9760100187
3. Baustian JJ, Mendelssohn IA, Hester MW (2012) Vegetation's importance in regulating surface elevation in a coastal salt marsh facing elevated rates of sea level rise. Global Change Biology 18:3377–3382. doi:10.1111/j.1365-2486.2012.02792.x
4. Belsley DA, Kuh E, Welsch RE (1980) Regression diagnostics: Identifying influential data and sources of collinearity. John Wiley & Sons, Inc
5. Bernal B, Mitsch WJ (2008) A comparison of soil carbon pools and profiles in wetlands in Costa Rica and Ohio. Ecological Engineering 34:311–323. doi:10.1016/j. ecoleng. 2008.09.005

6. Bernal B, Mitsch WJ (2012) Comparing carbon sequestration in temperate freshwater wetland communities. Global Change Biology 18:1636–1647. doi:10.1111/j.1365-2486.2011.02619.x
7. Bianchette T, Liu K, Qiang Y, Lam N (2015) Wetland accretion rates along coastal Louisiana: spatial and temporal variability in light of Hurricane Isaac's impacts. Water 8:1. doi:10.3390/w8010001
8. Bianchi TS, Allison MA, Zhao J et al (2013) Historical reconstruction of mangrove expansion in the Gulf of Mexico: Linking climate change with carbon sequestration in coastal wetlands. Estuarine, Coastal and Shelf Science 119:7–16. doi:10.1016/j.ecss.2012.12.007
9. Cahoon DR, Turner RE (1989) Accretion and canal impacts in a rapidly subsiding wetland II. Feldspar marker horizon technique. Estuaries 12:260–268
10. Callaway JC, Borgnis EL, Turner RE, Milan CS (2012) Carbon sequestration and sediment accretion in San Francisco Bay tidal wetlands. Estuaries and Coasts 35:1163–1181. doi:10.1007/s12237-012-9508-9
11. Callaway JC, Nyman JA, DeLaune RD (1996) Sediment accretion in coastal wetlands: a review and a simulation model of processes. Current Topics in Wetland Biogeochemistry 2:2–23
12. Chabreck RH (1972) Vegetation, water, and soil characteristics of the Louisiana coastal region. Louisiana State University Agricultural Experiment Station. Bulletin No. 664. Baton Rouge, LA
13. Chmura GL (2013) What do we need to assess the sustainability of the tidal salt marsh carbon sink? Ocean and Coastal Management 83:25–31. doi:10.1016/j.ocecoaman.2011.09.006
14. Chmura GL, Anisfeld SC, Cahoon DR, Lynch JC (2003) Global carbon sequestration in tidal, saline wetland soils. Global Biogeochemical Cycles 17:1111. doi:10.1029/2002GB001917
15. Coastal Protection and Restoration Authority (2012) Louisiana's 2012 Coastal Master Plan | Committed to our Coast. p 189
16. Comeaux RS, Allison MA, Bianchi TS (2012) Mangrove expansion in the Gulf of Mexico with climate change: Implications for wetland health and resistance to rising sea levels. Estuarine, Coastal and Shelf Science 96:81–95. doi:10.1016/j.ecss.2011.10.003
17. Couvillion BR, Steyer GD, Wang H et al (2013) Forecasting the effects of coastal protection and restoration projects on wetland morphology in coastal Louisiana under multiple environmental uncertainty scenarios. Journal of Coastal Research 67:29–50. doi:10.2112/SI_67_3
18. Cowardin LM, Carter V, Golet FC, LaRoe ET (1979) Classification of wetlands and deepwater habitats of the United States. Fish and Wildlife Service, US Department of the Interior Washington, DC
19. Craft C (2007) Freshwater input structures soil properties, vertical accretion, and nutrient accumulation of Georgia and U.S tidal marshes. Limnology and Oceanography 52:1220–1230. doi:10.4319/lo.2007.52.3.1220
20. Craft C, Broome S, Campbell C (2002) Fifteen years of vegetation and soil development after brackish-water marsh creation. Restoration Ecology 10:248–258. doi:10.1046/j.1526-100X.2002.01020.x
21. Craft CB, Clough J, Ehman J et al (2009) Forecasting the effects of accelerated sea-level rise on tidal marsh ecosystem services. Frontiers in Ecology and the Environment 7:73–78. doi:10.1890/070219
22. Craft CB, Seneca ED, Broome SW (1991) Loss on ignition and Kjeldahl digestion for estimating organic carbon and total nitrogen in estuarine marsh soils: calibration with dry combustion. Estuaries 14:175. doi:10.2307/1351691
23. De Deyn GB, Cornelissen JHC, Bardgett RD (2008) Plant functional traits and soil carbon sequestration in contrasting biomes. Ecology Letters 11:516–531. doi:10.1111/j.1461-0248.2008.01164.x
24. Delaune RD, Patrick WH, Buresh RJ (1978) Sedimentation rates determined by ^{137}Cs dating in a rapidly accreting salt marsh. Nature 275:532–533. doi:10.1038/275532a0

25. Doughty CL, Langley JA, Walker WS et al (2015) Mangrove range expansion rapidly increases coastal wetland carbon storage. Estuaries and Coasts. doi:10.1007/s12237-015-9993-8
26. Folse TM, West JL, Hymel MK, et al (2014) A Standard Operating Procedures Manual for the Coast-wide Reference Monitoring System - Wetlands: Methods for Site Establishment, Data Collection, and Quality Assurance/Quality Control. Coastal Protection and Restoration Authority. Baton Rouge, LA. p 228
27. Gallagher JL, Wolf PL, Pfeiffer WJ (1984) Rhizome and root growth rates and cycles in protein and carbohydrate concentrations in Georgia *Spartina alterniflora* Loisel. plants. American Journal of Botany 71:165–169. doi:10.2307/2443742
28. Hansen V, Nestlerode J (2014) Carbon sequestration in wetland soils of the northern Gulf of Mexico coastal region. Wetlands Ecology and Management 22:289–303. doi:10.1007/s11273-013-9330-6
29. Hatton RS, DeLaune RD, Patrick WH (1983) Sedimentation, accretion, and subsidence in marshes of Barataria Basin, Louisiana. Limnology and Oceanography 28:494–502. doi:10.4319/lo.1983.28.3.0494
30. Heiri O, Lotter AF, Lemcke G (2001) Loss on ignition as a method for estimating organic and carbonate content in sediments: reproducibility and comparability of results. Journal of Paleolimnology 25:101–110. doi:10.1023/A:1008119611481
31. Herbert ER, Boon P, Burgin AJ et al (2015) A global perspective on wetland salinization: ecological consequences of a growing threat to freshwater wetlands. Ecosphere 6:1–43. doi:10.1890/ES14-00534.1
32. Hill TD, Anisfeld SC (2015) Coastal wetland response to sea level rise in Connecticut and New York. Estuarine, Coastal and Shelf Science 163:185–193. doi:10.1016/j.ecss.2015.06.004
33. Holm GO, Perez BC, McWhorter DE et al (2016) Ecosystem level methane fluxes from tidal freshwater and brackish marshes of the Mississippi river delta: implications for coastal wetland carbon projects. Wetlands. doi:10.1007/s13157-016-0746-7
34. Hopkinson CS, Cai W-J, Hu X (2012) Carbon sequestration in wetland dominated coastal systems — a global sink of rapidly diminishing magnitude. Current Opinion in Environmental Sustainability 4:186–194. doi:10.1016/j.cosust.2012.03.005
35. Kirwan ML, Temmerman S, Skeehan EE et al (2016) Overestimation of marsh vulnerability to sea level rise. Nature Climate Change 6:253–260. doi:10. 1038/nclimate2909
36. Krauss KW, Holm GO, Perez BC et al (2016) Component greenhouse gas fluxes and radiative balance from two deltaic marshes in Louisiana: pairing chamber techniques and eddy covariance. Journal of Geophysical Research Biogeosciences 121:1503–1521. doi:10.1002/2015JG003224
37. Marín-Spiotta E, Gruley KE, Crawford J et al (2014) Paradigm shifts in soil organic matter research affect interpretations of aquatic carbon cycling: transcending disciplinary and ecosystem boundaries. Biogeochemistry 117:279–297. doi:10.1007/s10533-013-9949-7
38. Markewich HW, Buell GR, Britsch LD, et al. (2007) Organic-carbon sequestration in soil/sediment of the Mississippi River deltaic plain— data; landscape distribution, storage, and inventory; accumulation rates; and recent loss, including a post-Katrina preliminary analysis. Soil - carbon storage and inventory for the Continental United States. U.S. Geological Survey, Reston, Virginia, p 257
39. McLeod E, Chmura GL, Bouillon S et al (2011) A blueprint for blue carbon: toward an improved understanding of the role of vegetated coastal habitats in sequestering CO_2. Frontiers in Ecology and the Environment 9:552–560
40. Meselhe E, McCorquodale JA, Shelden J et al (2013) Ecohydrology component of Louisiana's 2012 Coastal Master Plan: Mass-balance compartment model. Journal of Coastal Research 67:16–28. doi:10.2112/SI_67_2.1
41. Mitra S, Wassmann R, Vlek PLG (2005) An appraisal of global wetland area and its organic carbon stock. Current Science 88:25–35

42. Morris JT, Barber DC, Callaway JC et al (2016) Contributions of organic and inorganic matter to sediment volume and accretion in tidal wetlands at steady state. Earth's Future 4:110–121. doi:10.1002/2015EF000334
43. Morris JT, Bradley PM (1999) Effects of nutrient loading on the carbon balance of coastal wetland sediments. Limnology and Oceanography 44:699–702. doi:10.4319/lo.1999.44.3.0699
44. Morris JT, Shaffer GP, Nyman JA (2013) Brinson Review: Perspectives on the influence of nutrients on the sustainability of coastal wetlands. Wetlands 33:975–988. doi:10.1007/s13157-013-0480-3
45. Morris JT, Sundareshwar PV, Nietch CT et al (2002) Responses of coastal wetlands to rising sea level. Ecology 83:2869–2877
46. Nellemann C, Corcoran E (2009) Blue carbon: The role of healthy oceans in binding carbon: A rapid response assessment. UNEP/Earthprint
47. Nelson DW, Sommers LE (1982) Total carbon, organic carbon, and organic matter. In: Page AL (ed) Methods of Soil Analysis. Part 2. Chemical and Microbiological Properties. Soil Science Society of America, Madison, WI, pp 539–579
48. Neubauer SC (2008) Contributions of mineral and organic components to tidal freshwater marsh accretion. Estuarine, Coastal and Shelf Science 78:78–88. doi:10.1016/j.ecss.2007.11.011
49. Neubauer SC, Craft C (2009) Global change and tidal freshwater wetlands: scenarios and impacts. In: Barendregt A, Whigham D, Baldwin A (eds) Tidal Freshwater Wetlands. Backhuys Publishers, Leiden, pp 253–266
50. Nyman JA, Walters RJ, Delaune RD, Patrick WH Jr (2006) Marsh vertical accretion via vegetative growth. Estuarine, Coastal and Shelf Science 69:370–380. doi:10.1016/j.ecss.2006.05.041
51. Odum WE (1984) Dual-gradient concept of detritus transport and processing in estuaries. Bulletin of Marine Science 35:510–521
52. Odum WE, Smith III TJ, Hoover JK, McIcor CC (1984) The ecology of tidal freshwater marshes of the United States east coast: a community profile. p 186
53. Osland MJ, Spivak AC, Nestlerode JA et al (2012) Ecosystem development after mangrove wetland creation: Plant–soil change across a 20-years chronosequence. Ecosystems 15:848–866. doi:10.1007/s10021-012-9551-1
54. Piazza SC, Steyer GD, Cretini KF, et al. (2011) Geomorphic and ecological effects of Hurricanes Katrina and Rita on coastal Louisiana marsh communities. p 126
55. Poffenbarger HJ, Needelman BA, Megonigal JP (2011) Salinity influence on methane emissions from tidal marshes. Wetlands 31:831–842. doi:10.1007/s13157-011-0197-0
56. R Core Team (2013) R: A language and environment for statistical computing. R Foundation for Statistical Computing, Vienna, Austria
57. Reed DJ (1995) The response of coastal marshes to sea-level rise: survival or submergence? Earth Surface Processes and Landforms 20:39–48. doi:10.1002/esp.3290200105
58. Rhoades JD, Sparks DL, Page AL et al (1996) Salinity: Electrical conductivity and total dissolved solids. In: Sparks DL, Page AL, Helmke PA, Loeppert RH (eds) Methods of Soil Analysis. Part 3-Chemical Methods. Soil Science Society of America, American Society of Agronomy, Madison, WI, pp 417–435
59. Roulet NT (2000) Peatlands, carbon storage, greenhouse gases, and the Kyoto protocol: prospects and significance for Canada. Wetlands 20:605–615
60. Sasser CE, Visser JM, Mouton E et al (2014) Vegetation types in coastal Louisiana in 2013. U.S. Geological Survey, Reston, Virginia
61. Stagg CL, Schoolmaster DR, Piazza SC, Snedden G, Steyer GD, Fischenich CJ, McComas RW (2016) A landscape-scale assessment of above- and belowground primary production in coastal wetlands: implications for climate change-induced community shifts. Estuaries and Coasts. 1–24. doi:10.1007/s12237-016-0177-y
62. Steyer G, Couvillion B, Wang H et al (2012) Appendix D2 - Wetland morphology model technical report. Louisiana's Comprehensive Master Plan for a Sustainable Coast. Coastal Protection and Restoration Authority, Baton Rouge, LA, p 108

63. Steyer GD, Sasser CE, Visser JM, Swenson EM, Nyman JA, Raynie RC (2003) A proposed coast-wide reference monitoring system for evaluating wetland restoration trajectories in Louisiana. Environmental Monitoring and Assessment 81:107–117
64. Taylor A, Blake WH, Smith HG et al (2013) Assumptions and challenges in the use of fallout beryllium-7 as a soil and sediment tracer in river basins. Earth-Science Reviews 126:85–95. doi:10.1016/j.earscirev.2013.08.002
65. Turner RE (2004) Coastal wetland subsidence arising from local hydrologic manipulations. Estuaries 27:265–272. doi:10.1007/BF02803383
66. Turner RE, Milan CS, Swenson EM (2006) Recent volumetric changes in salt marsh soils. Estuarine, Coastal and Shelf Science 69:352–359. doi:10.1016/j.ecss.2006.05.005
67. Valiela I, Teal JM, Persson NY (1976) Production and dynamics of experimentally enriched salt marsh vegetation: belowground biomass. Limnology and Oceanography 21:245–252
68. Visser JM, Sasser CE, Chabreck RH, Linscombe RG (1998) Marsh vegetation types of the Mississippi River Deltaic Plain. Estuaries 21:818–828
69. Visser JM, Sasser CE, Chabreck RH, Linscombe, RG (2002) The impact of a severe drought on the vegetation of a subtropical estuary. Estuaries 25:1184–1195. doi:10.1007/BF02692215
70. Whiting GJ, Chanton JP (2001) Greenhouse carbon balance of wetlands: methane emission versus carbon sequestration. Tellus Series B: Chemical and Physical Meteorology 53:521–528. doi:10.1034/j.1600-0889.2001.530501.x
71. Williams EK, Rosenheim BE (2015) What happens to soil organic carbon as coastal marsh ecosystems change in response to increasing salinity? An exploration using ramped pyrolysis: Fate of coastal SOC with salinity. Geochemistry, Geophysics, Geosystems 16:2322–2335. doi:10.1002/2015GC005839
72. Wilson CA, Allison MA (2008) An equilibrium profile model for retreating marsh shorelines in southeast Louisiana. Estuarine, Coastal and Shelf Science 80:483–494. doi:10.1016/j.ecss.2008.09.004

CHAPTER 17

Carbon and Hydrogen Isotopes as Tracers of Methane Dynamic in Wetlands

Romina Sanci*, Héctor O. Panarello

Instituto de Geocronología y Geología Isotópica (INGEIS, UBA-CONICET), Argentina

ABSTRACT

This work presents a review of the main processes leading to the production and sinking of methane in wetlands and how they affect the stable isotope composition of carbon and hydrogen. Isotope fractionation factors associated to diffusion, ebullition, bacterial oxidation, etc., have been gathered from relevant literature in an intent of providing researchers in this thematic with practical procedures and tools for the interpretation of experimental data. Also it is presented guidelines of the most reliable field and laboratory methods used at present for the correct sampling and analyzing methane in different stages of occurrence, as well the most common tools used in their interpretation. Each statement is fully referenced to a long updated list of publications.

Keywords: C and H Stable Isotopes, CH_4 and CO_2 Emissions, Wetlands

1. INTRODUCTION

CH_4 is a chemically and radioactively important atmospheric trace gas and is the second most important greenhouse gas after CO_2. CH_4 concentration increased from ~800 ppb in the 1850s to ~1700 ppb in the late 2000s [1] , suggesting an imbalance between estimated global emissions and removals. Decadal budgets for CH_4 sources and sinks between 1980 and 2010 reflect average imbalance of 6 Tg CH_4 a^{-1} in the 2000s, 17 Tg CH_4 a^{-1} in the 1990s and 34 Tg CH_4 a^{-1} in the 1980s [2] . The total amount of global CH_4 emissions is currently estimated as 500 - 600 Tg a^{-1} [3] and this CH_4 is emitted from a range of natural and anthropogenic sources as a result of the anaerobic decomposition of organic matter, land use changes and fossil fuel related emissions. Roughly, natural CH_4 sources (wetlands, termite activity, other wild ruminants, geological, fresh water systems, permafrost, hydrates, wild fires) are considered responsible for 40% of total emissions while 60% of emissions are related to human activities (waste decomposition, fossil fuels, rice cultivation, agriculture, livestock and the burning of biomass) [4] . Recently, living vegetation has also been suggested as an important natural source of CH_4 [5] . On the other hand, CH_4 is removed from the atmosphere by a variety of chemical and biological processes, which occur in different regions of the atmosphere. These include tropospheric and stratospheric oxidation and uptake by soils.

Although most sources and sinks of CH_4 have been identified, their relative contributions to atmospheric CH_4 levels are still highly uncertain. The quantification of individual contributions to global emissions has proved to be difficult, despite they are crucial to predict climate change and managing Earth's carbon reservoirs. The net rate of CH_4 emissions is generally estimated from three approaches: 1) extrapolation from direct flux measurements and observations; 2) process-based modelling (bottom-up approach) and 3) inverse modelling that relies on spatially distributed, temporally continuous observations of concentration, and in some cases isotopic composition in the atmosphere (top-down approach) [6] -[9] . Observations of the $^{13}C/^{12}C$ ratio of atmospheric CH_4 was used in isotope mass balance models to define the magnitudes of different sources and sinks and to constrain the present-day CH_4 budget using optimal estimation. Three CH_4 emission scenarios are estimated: 64% - 76% biogenic, 19% - 30% fossil, and 4% - 6% pyrogenic sources [10] . Biogenic sources contain CH_4-generating microbes (methanogens) and comprise anaerobic environments such as natural wetlands and rice paddies, oxygen-poor freshwater systems (lakes, rivers, streams, reservoirs such as dams), digestive systems of ruminants and termites, and organic waste deposits (such as manure, sewage and landfills). Thermogenic CH_4, formed over millions of years through geological processes, is a fossil fuel. It is vented from the subsurface into the atmosphere through natural features (such as terrestrial seeps, marine seeps and mud volcanoes), and through the exploitation of fossil fuels, that is, through the exploitation of coal, oil and natural gas. Pyrogenic CH_4 is produced by the incomplete combustion of biomass and soil carbon during wildfires, and of biofuels and fossil fuels. Table 1 gives a data set of the main isotopic values of carbon that involved CH_4 sources.

Wetlands contribute as much as 70% of the total natural emissions [16] [17] . Their emissions from most types (marshes, swamps, bogs, and fens according to reference [18]) can vary by a few orders of magnitude over just a small number of meters, because they are influenced by a wide range of environmental parameters. These include soil characteristics such as the availability of organic carbon and nutrients, plant physiology, community composition and cover and water table depth and soil temperature. Seasonal and inter-annual variations of CH_4wetland emissions are reported by focusing on three determinants keys: temperature, substrate availability and water table's level [19] -[21] . Besides these factors, the actual amount of CH_4emitted to the atmosphere or rates of CH_4 emission depends on the balance between CH_4production and consumption and the mode of methane transport. Wetlands transfer CH_4 to the atmosphere by diffusion, ebullition, and by transport through arenchymous vascular plants but only part of the CH_4 produced is emitted to the atmosphere [22] . Considerable CH_4 amounts are consumed or oxidized by methane-trophic bacteria in the rhizosphere and in surficial oxic layers during diffusive transport to the soil surface [23] [24] .

The accuracy of predictions of CH_4 and ultimately the extent of climate change that can be expected in the earth system depend on a better understanding of CH_4 dynamics from wetlands for each region. Most studied, considering all processes, were done in wetlands from Northern Hemisphere [25] . However, until quite recently, ebullition was considered to be only locally important, and most attention was focused on matrix diffusion of dissolved CH_4 and plant-mediated transport. For instance, reference [26] used measurements to argue that ebullition has been temporally and spatially very variable and it could exceed diffusive fluxes by two orders of magnitude, stating that heterogeneity in ebullition (bubbling) is a major contribution to variability of emissions. So, it is necessary to study these systems at different regions as well as to know the suitable tools that would meet this purpose. In this review we present literature data that show how stable carbon isotopic signatures, in combination with other parameters, can be used to quantify the relative contribution of the major methanogenic pathways, CH_4 oxidation and transport mechanisms.

Table 1. Isotopic composition range from different sources of CH_4.

Biogenic CH_4 (‰)	Thermogenic CH_4 (‰)	Pyrogenic CH_4 (‰)
$\delta^{13}C$-CH_4: −110 to −50 $\delta^{2}H$-CH_4: −275 to −100	$\delta^{13}C$-CH_4: −50 to −20 $\delta^{2}H$-CH_4: −275 to −100	$\delta^{13}C$-CH_4: −25 to −13
[11] [12] [13]	[11] [13]	[14] [15]

2. WHY TO APPLY THE STABLE ISOTOPES?

Wetlands, where conditions are more favourable to the accumulation of organic matter than to its decay, are characterized by a subsurface, anaerobic zone of CH_4 production, and organic soils (as peat) often submerged or water-saturated. The generation of CH_4 by bacteria requires a fully saturated environments that excludes atmospheric O_2 in the absence of other free-energy electron-acceptors such as NO_3^- and SO_4^{2-} and low redox (Eh < ?200 mV), as can be observed in wetlands [27] . Methane-genesis is the final step in anaerobic decomposition and represents a complex process that proceed via two major metabolic pathways: acetate fermentation (1) and CO_2 reduction (2)

$$CH_3COOH \rightarrow CH_4 + CO_2$$

$$CO_2 + 4H_2 \rightarrow 2H_2O + CH_4 \quad \text{or} \quad HCO_3^- + 4H_2 \rightarrow 2H_2O + CH_4 + OH^-$$

The measurement of stable isotopes of carbon on CH_4 and CO_2 is an effective way to identify the different phases of biodegradation of organic matter [28] [29] . According to these authors, $CO_2$is isotopically light during the initial aerobic and anaerobic oxidation phases of biodegradation with δ13C values that varies from −35‰ to −10‰, thus covering the range of most terrestrial plants [27] . The initial input of isotopically light CO_2 associated with earlier biodegradation phases is soon overcome during the methanogen phase by the constant input of isotopically heavy CO_2 associated with acetate fermentation and microbial CO_2 reduction (the two primary metabolic pathways by which microbial CH_4 is produced) (Table 2). During methanogenesis, $CH_4$is enriched in the lighter carbon isotope (^{12}C) and the CO_2 associated with microbial CH_4production is enriched in the heavier isotope (^{13}C). Thus, in a semiclosed environment such as a wetland, the $\delta^{13}C$ of CO_2 is strongly affected by methanogen reactions with reported values between −10‰ and +20‰. Values of $\delta^{13}C$-CH_4 due to fermentation of acetate and CO_2reduction are −65‰ to −50‰ and −110‰ to −60‰, respectively [11] [30] .

CH_4 production is regulated mainly by O_2 concentration as mentioned above, but also of pH, temperature, salinity, organic substrates and nutrient availability [31] . Some authors think that once anaerobic conditions are attained, the quality and supply of substrates is the major factor in CH_4 production: substantial amounts of CH_4 are only produced when labile carbon substrates are largely available and old (recalcitrant) wetland plays only a subordinate role as a substrate for CH_4 production [32] . Reach conclusions about the relative importance of methanogen pathways for CH_4 production are difficult because $\delta^{13}C-CH_4$ values of the upper organic horizon (above 1 m) are substantiality affected by the processes of CH_4 transport and oxidation [33] . So, based on differences in $\delta^{13}C-CH_4$among organic layer and isotopic fractionation factors associated with gas transport and oxidation, these authors assessed the potential quotas of transported/oxidized CH_4 Also, reference [34] used a simple mixing model to estimate the relative proportion of the acetate fermentation and CO_2 reduction pathway.

CH_4 is produced at depth and lost to the atmosphere from wetlands via three mechanisms: 1) diffusion through pore water to the water table and hence through the zone above the water table (if one exists) to the wetland surface; 2) diffusion or active transport through vascular plants; and 3) ebullition, that refers to methane released to the atmosphere in form of bubbles moving to the wetland surface. CH_4 bubbles commonly occur in water saturated organic layers where they remain trapped and grow in size. CH_4 is only sparingly soluble in water [35] and Henry's law predicts that their solubility in a solvent is directly proportional to the partial pressure. Other gases, particularly nitrogen, could contribute with their partial pressure and exceed the total hydrostatic force. If this pressure is exceeded, gas bubbles, consisting of a mixture of CH_4, CO_2 and N_2, could travel upward toward groundwater surface, perhaps along fractures [36] . This process of ebullition by which bubbles transport CH_4 upward, occurs when the buoyancy of bubbles overcomes the forces that keep them in place (in particular, surface tension) when a certain threshold pressure is reached. Often this release is associated with changes in water level [37] -[39] , barometric pressure [40] -[42] and temperature [43] , fracture of the confining layer as well as mechanical disturbance [44] . The rapid transfer of CH_4 bubbles through the aerobic near surface layer means there will be little or no consumption by methanotrophs. Some ebullition conceptual models suggest that bubbles form more readily in shallow layers-within 1 m-[45] , while others as sure that they form at greater depth and accumulate within the wetland profile according physical properties of subsoil [26] or presence of zones of overpressure due to transient confining layers [46] [47] . Also, these last authors affirm the occurrence of CH_4 deep ebullition because they found large concentrations of dissolved CH_4 and inorganic carbon (DIC) in the deeper pore water related to CO_2 produced by fermentative and methanogen bacteria. Others suggest that bubbles are not trapped in subsoil but are lost directly to the water table, after which the gas moves slowly via diffusion through the zone above the water table where CH_4 is oxidized [48] .

Table 2. Enrichment factors (e) to different processes for carbon.

Processes	Enrichment factors (ε)	Reference cites
Acetate fermentation pathway	ε = 40‰ - 50‰	[11]
CO_2 reduction pathway	ε = 55‰ - 90‰	[11]
CH_4 oxidation	ε = 3‰ - 31‰ ε = 22‰	[54] [55]
Diffusion CH_4 across	ε = 20‰ (cover soil) ε = 18‰ (cover soil) ε = 2‰ - 3‰ (interface water-air)	[56] [49] [57]
Ebullition	No fractionation	[53]

Molecular diffusion occurs through the water or air filled soil pores and the atmosphere and depends on the vertical CH_4 concentration gradient following Fick's Law: $J = -D\frac{dC}{dz}$, where J is diffusive flux, D is the diffusion coefficient, C the solute concentration and Z the depth. Fractionation (a) during diffusion is caused by differences the diffusion coefficients of the two isotope species. The stable CH_4isotopes ($^{12}CH_4$, $^{13}CH_4^{12}$and CH_3^2H) differ in molecular weight and their diffusion coefficient, faster diffusive transport of the lighter isotope on hence in an enrichment of the heavier isotope in the remaining gas phase. Reference [49] demonstrated that diffusion is an important transport mechanism for CH_4 cover soils. They found that the fractionation factor due to transport, a transport, can be as high as 1.0178, due to the difference in molecular diffusion coefficients of CH_4 isotopes (ε = 18‰). Diffusive transport of CH_4 across the air-water interface was associated with anisotopic fractionation to 2‰ - 3‰ or ε = 2‰ - 3‰ [50].

Diffusion affects isotopic composition of CH_4 differently than ebullition and CH_4 transport from anoxic sediments and flooded soils by emergent plants. $\delta^{13}C$-CH_4 value of released bubbles is not different from the reservoir of bubble gas held within the sediment because ebullition bypasses oxidation in the aerobic zone [33] . Many wetland plants possess aerenchymous tissue that allows for transport of oxygen into the root zone as an adaptation to rooting in waterlogged soils. Whereas this oxygen allows for oxidation of CH_4 in the root zone [51] , at the same time, CH_4 is transported through the aerenchyma out into the atmosphere. It was influenced by the type species and functional plant groups that affect these processes differently [52] . In vegetated wetlands, plant-mediated transport involves several steps that may further fractionate isotopes, i.e. diffusion to the roots, partial oxidation by methanotrophic bacteria in rhizosphere, diffusion into the roots, transport by convective bulk flow and emission [53] . At the moment, only the isotope effects of CH4 oxidation and convective bulk flow have been well studied. The last one, results in the little isotopic fractionation [53] whereas CH_4 oxidation in the rhizosphere and in surficial oxic layers during diffusive transport to the soil surface (0 - 1 m) is notorious. Previous reports indicate that almost 60% of CH_4 production is oxidized to CO_2 [48] . Bacteria oxidize the ^{12}C- isotopeslightly faster than the ^{13}C-isotope, then, the result is an increase of the $^{13}C/^{12}C$ ratio of the remaining CH_4. The enrichment factors can be seen at Table 2. This increase can be used to estimate CH_4 oxidation. By the other way, variations on $\delta^{13}C$-DIC of water can reveal the importance of different biochemical pathways of methanogenesis and CH_4 oxidation [27] . It was remarked that acetate fermentation that occur chiefly in fresh water environment is responsible for the great fractionation of ^{13}C between CO_2 and CH_4. $CO_2$with much enriched signature incorporates to the DIC, but this signal is often masked by the simultaneous oxidation processes releasing isotope depleted CO_2 to surrounding water.

3. MEASUREMENTS AND ANALYTICAL METHODS

3.1. CH_4 and CO_2 Fluxes and Isotopes

Closed chamber methods are used to measure surface emissions. These methods are independent of the physical properties of soil and allow measuring the CH_4 and CO_2 released to the atmosphere. The closed chamber methods are applied during fieldwork, after calibration in laboratory [58] [59] . The accumulation chamber is placed on a collar that had been previously installed on the ground. Surface emissions are estimated as the increase of CH_4 and CO_2concentrations over time. Flux rates are calculated fitting by linear regression the variation of concentration (C) over time and adjusting for chamber volume and covered area, according to the following eq-uation: $F = VA\frac{dC}{dt}$ where F is the surface emission rate ($mg \cdot m^{-2} \cdot day^{-1}$), V is the chamber volume and A is thesoil area under the chamber and dC/dt is the variation of C with t within the chamber. In the closed static chamber method, soil gases are extracted with plastic syringes and analyzed by gas chromatography (GC). In flooded soil the fluxes are estimated using floating static chambers. Sometimes chambers are covered with reflective material to minimize the increase of inside temperature during deployments. Gas samples are then stored by needle injection in pre-evacuated glass flasks sealed with butyl rubber stoppers for later concentration determination.

3.2. Methane Concentration in Sediment Gas Bubbles and Water

Bubbles are sampled by stirring up the sediments. Arising bubbles are collected with an entirely submerged 25-cm diameter inverted plastic funnel at sites devoid of vascular vegetation. The open narrow end of the funnel was sealed with a rubber stopper of 3-cm diameter. With a needle, bubble gas samples are pumped into a plastic syringe adapted with a luer-lock valve and thereafter injected in pre-evacuated glass flasks sealed with butyl rubber stoppers for later GC analysis (CH_4, N_2, CO_2 concentration). Also, bubbles samples from natural ebullition are collected into floating traps without disturbing sediments [34] . Bubbles traps (~1 diameter) placed permanently under the surface of water bodies. Bubble N_2 content may be indicative of rates of ebullition from sediments [53] . As increasing volumes of CH_4 bubble from the sediments, they strip out N_2, a gas that is not produced within the sediments. Thus, bubbles with low N_2 concentration indicate high rates of bubble stripping or high ebullition. Continuous bubble fluxes of CH_4 (mg CH_4 $m^{-2}\cdot d^{-1}$) at shallow depths (20, 40, 60 cm below the surface and water table) are measured by gas trap funnels [45] . Other method to measure CH_4 entrapped can be seen in reference [60] .

If ebullition is episodic it is very difficult to quantify its importance. The role of large ebullition events needs further research and efforts to quantify this mode of emission. Methods that can quantify the distribution (and temporal variability) of free phase gas in the peat profile are needed. CH_4 bubbles accumulated at depth, episodically, can break through the confining layer to the surface causing very large rates of CH_4 loss to atmosphere (in $g\cdot m^{-2}$ as direct emission, no diffusive). Nondestructive geophysical methods that have been used in peatlands [41] such as ground penetrating radar and electrical resistivity could also be used to examine the sediment structure and gas contents within sediments. The determination of dissolved CH_4 in soil solution (and DIC) is an alternative non-destructive method, which simplifies the sampling procedure and allows associating the CH_4 and DIC dissolved with episodic ebullition (less DIC after ebullition).Reference [26] calculated large degassing events associating surface deformations to largest de-pressuring cycles.

Porewater CH_4 can be sampled in situ using modified diffusion cambers called "peepers" [33] . Gas samples collected for dissolved CH_4 analysis using a headspace equilibrium technique presented by reference [61] , a 60-ml plastic syringe adapted with a luer-lock valve containing 30 ml of water sample and 30 ml of pure nitrogen was vigorously shaken during 2 min, transferring the dissolved phase to the nitrogen headspace. The headspace was then injected with a needle in pre-evacuated glass flasks sealed with butyl rubber stoppers for later FID-GC analysis.

3.3. Isotope Analysis

After the GC analysis, the remaining gas contained in the glass flasks acquired from the emitted CH_4 and CO_2 (chambers), CH_4 in the water column and CH_4y/o CO_2 in the sediment bubbles are employed for $\delta^{13}C$-CH_4 and analysis in line with a mass spectrometer (GC-IRMS). Isotope analyses are expressed by $\delta^{13}C$ (see Notation). Water samples are analyzed for the $^{13}C/^{12}C$ ratios of their DIC component by precipitating DIC with $BaCl_2$ ($BaCO_3$).

3.4. Notations

Expression of results and some used formulas.

R defines the ratio between two isotopic species. Example:

$$^{13}R = \frac{^{13}C}{^{12}C}; \quad ^{2}R = \frac{^{2}H^{1}H}{^{1}H_2}$$

α_{A-B} is the fractionation factor between phases or components A and B $\alpha = \frac{R_A}{R_B}$, if > 1, A is enriched in the heavy isotope respect B, conversely if < 1, A is depleted in the heavy isotope respect to B. Also is used the enrichment factor (defined as $\varepsilon = 1000(\alpha - 1)$). Results are expressed in ‰ defined as: $\delta = 1000\frac{R_S - R_R}{R_R}$‰ where S denotes sample, R reference standard And the isotopic deviation in ‰, ^2H or ^{13}C respect the International reference Standard, V-Pee Dee Belemnitella (VPDB) for ^{13}C and V-Standard Mean Ocean Water for (VSMOW) for δ^2H [62] [63] .

Since ^2H and ^{13}C are values actually measured, fractionation factor are calculated from δ by means of the fol-

lowing expression: $\alpha_{A-B} = \frac{1000 + \delta_A}{1000 + \delta_B}$

4. CONCLUSION

Principal processes leading to the production and sinking of CH_4 in wetlands are presented and discussed mainly from the stable isotopes point of view. Isotope fractionation factors associated to production, diffusion, ebullition, bacterial oxidation, etc., allow using the isotopes of carbon and hydrogen as environmental tracers of these pro- cesses. So, it can be applied to study $CH_4$and CO_2 generation, distribution and consume in wetlands. This review presents an updated list of bibliographic references to support its arguments and serve as a guide to researcher in the thematic of CH_4 dynamic from wetland. Also it is presented guidelines of the most reliable field and laboratory methods used at present for the correct sampling and analyzing CH_4 in different stages of occurrence.

REFERENCES

1. IPCC (2015) Climate Change 2014: Synthesis Report. Contribution of Working Groups I, II and III to the Fifth Assessment Report of the Intergovernmental Panel on Climate Change, IPCC, Geneva.
2. Kirschke, S., Bousquet, P., Ciais, P., Saunois, M., Canadell, J.G., Dlugokencky, E.J., et al. (2013) Three Decades of Global Methane Sources and Sinks. Nature Geoscience, 6, 813-823. http://dx.doi.org/10.1038/ngeo1955
3. Stenke, A., Deckert, R. and Gottschaldt, K. (2012) Methane modeling: From Process Modeling to Global Climate Models. In: Schumann, U., Ed., Atmospheric Physics: Background-Methods-Trends, Springer, Berlin, 781-797, http://dx.doi.org/10.1007/978-3-642-30183-4_47
4. IPCC (2007) The Physical Science Basis. Contribution of Working Group I to the Fourth Assessment Report of the intergovernmental Panel on Climate Change. In: Solomon, S., Qin, D., Manning, M., Chen, Z., Marquis, M., Averyt, K.B., Tignor, M., Miller, H.L., Eds., Climate Change 2007, Cambridge University Press, Cambridge and New York, 996. [Citation Time(s):1]
5. Keppler, F., Hamilton, J.T.G., Brab, M. and Rockmann, T. (2006) Methane Emissions from Terrestrial Plants under Aerobic Conditions. Nature, 439, 187-191. http://dx.doi.org/10.1038/nature04420
6. Mikaloff Fletcher, S.E., Tans, P.P., Bruhwiler, L.M., Miller, J.B. and Heimann, M. (2004) CH4 Source Estimated from Atmospheric Observations of CH4 and Its 13C/12C Isotopic Ratios: 1. Inverse Modeling of Source Processes. Global Biogeochemical Cycles, 18, Article ID: GB4004.
7. Mikaloff Fletcher, S.E., Tans, P.P., Bruhwiler, L.M., Miller, J.B. and Heimann, M. (2004) CH4 Source Estimated from Atmospheric Observations of CH4 and Its 13C/12C Isotopic Ratios: 2. Inverse Modeling of CH4 Fluxes from Geographic Regions. Global Biogeochemical Cycles, 18, Article ID: GB4004.
8. Bousquet, P., Ciais, P., Miller, J.B., Dlugokencky, E.J., Hauglustaine, D.A., Prigent, C., et al. (2006) Contribution of Anthropogenic and Natural Sources to Atmospheric Methane Variability. Nature, 443, 439-443. http://dx.doi.org/10.1038/nature05132
9. Chen, Y.-H. and Prinn, R.G. (2006) Estimation of Atmospheric Methane Emissions between 1996 and 2001 Using a Three-Dimensional Global Chemical Transport Model. Journal of Geophysical Research, 111, Article ID: D10307.http://dx.doi. org/10.1029/2005jd006058

10. Neef, L., van Weele, M. and van Velthoven, P. (2010) Optimal Estimation of the Present-Day Global Methane Budget. Global Biogeochemical Cycles, 24, Article ID: GB4024. http://dx.doi.org/10.1029/2009gb003661
11. Whiticar, M.J., Faber, E. and Schoell, M. (1986) Biogenic Methane Formation in Marine and Freshwater Environments: CO2 Reduction vs. Acetate Fermentation—Isotope Evidence. Geochimica et Cosmochimica Acta, 50, 693-709.http://dx.doi.org/10.1016/0016-7037(86)90346-7
12. Conrad, R. (1989) Control of Methane Production in Terrestrial Ecosystems, In: Andreae, M.O. and Schimel, D.S., Ed., Exchange of Trace Gases between Terrestrial Ecosystems and the Atmosphere, John Wiley, New York, 39-58.
13. Whiticar, M.J. (1999) Carbon and Hydrogen Isotope Systematics of Bacterial Formation and Oxidation of Methane. Chemical Geology, 161, 291-314. http://dx.doi.org/10.1016/S0009-2541(99)00092-3
14. Monteil, G., Houweling, S., Dlugockenky, E.J., Maenhout, G., Vaughn, B.H., White, J.W. and Rockmann, T. (2011) Interpreting Methane Variations in the Past Two Decades Using Measurements of CH4 Mixing Ratio and Isotopic Composition. Atmospheric Chemistry and Physics, 11, 9141-9153. http://dx.doi.org/10.5194/acp-11-9141-2011
15. Fisher, R.E., Sriskantharajah, S., Lowry, D., Lanoiselle, M., Fowler, C.M.R., et al. (2011) Arctic Methane Sources: Isotopic Evidence for Atmospheric Inputs. Geophysical Research Letters, 38, Article ID: L21803. http://dx.doi.org/10.1029/2011gl049319
16. Khalil, M. (2000) Atmospheric Methane. Springer, New York. http://dx.doi.org/10.1007/978-3-662-04145-1
17. Wuebbles, D.J. and Hayhoe, K. (2002) Atmospheric Methane and Global Change. Earth-Science Reviews, 57, 177-210. http://dx.doi.org/10.1016/S0012-8252(01)00062-9
18. EPA (2001) Environmental Protection Agency of United States, Office of Water EPA 843-F-01-002b, Oceans and Watersheds (4502T). http://water.epa.gov/ type/ wetlands/ types_index.cfm
19. MacDonald, J.A., Fowler, D., Hargreaves, K.J., Skiba, U., Leith, I.D. and Murray, M.B. (1998) Methane Emission Rates from a Northern Wetland: Response to Temperature, Water Table and Transport. Atmospheric Environment, 32, 3219-3227. http://dx.doi.org/10.1016/s1352-2310(97)00464-0
20. Christensen, T.R., Panikov, N., Mastepanov, M., Joabsson, M.A., Stewart, A., ?quist, M., Sommerkorn, M., Reynaud, S. and Svensson, B. (2003) Biotic Controls on CO2 and CH4 Exchange in Wetlands—A Closed Environment Study. Biogeochemistry, 64, 337-354. http://dx.doi.org/10.1023/A:1024913730848
21. Olson, D.M., Griffis, T.J., Noormets, A., Kolka, R. and Chen, J. (2013) Interannual, Seasonal, and Retrospective Analysis of the Methane and Carbon Dioxide Budgets of a Temperate Peatland. Journal of Geophysical Research: Biogeosciences, 118, 226-238. http://dx.doi.org/10.1002/jgrg.20031
22. Le Mer, J. and Roger, P. (2001) Production, Oxidation, Emission and Consumption of Methane by Soils: A Review. European Journal of Soil Biology, 37, 25-50. http://dx.doi.org/10.1016/S1164-5563(01)01067-6
23. Whalen, S.C. (2005) Biogeochemistry of Methane Exchange between Natural Wetlands and the Atmosphere. Environmental Engineering Science, 22, 73-94. http://dx.doi.org/10.1089/ees.2005.22.73
24. Lai, D.Y. (2009) Methane Dynamics in Northern Peatlands: A Review. Pedosphere, 19, 409-421. http://dx.doi.org/10.1016/S1002-0160(09)00003-4
25. Baird, A., Belyea, L.R., Comas, X., Reeve A.S. and Slater, L.D. (2013) Carbon Cycling in Northern Peatlands. American Geophysical Union (AGU), Washington DC.
26. Glaser, P.H., Chanton, J.P., Morin, P., Rosenberry, D.O., Siegel, D.I., Ruud, O., Chasar, L.I. and Reeve, A.S. (2004) Surface Deformations as Indicators of Deep Ebullition Fluxes in a Large Northern Peatland. Global Biogeochemical Cycles, 18, Article ID: GB1003. http://dx.doi.org/10.1029/2003gb002069
27. Clark, I.D. and Fritz, P. (1997) Environmental Isotopes in Hydrogeology. Lewis Publishers, New York.

28. Coleman, D.D., Liu, C.L., Hackley, K.C. and Benson, L.J. (1993) Identification of Landfill Methane Using Carbon and Hydrogen Isotope Analysis. Proceedings of 16th International Madison Waste Conference, Municipal & Industrial Waste, Department of Engineering Professional Development, University of Wisconsin Madison, Madison, 22-23 September 1993, 303-314.
29. Hackley, K.C., Coleman, D.D. and Liu, C.L. (1996) Environmental Isotope Characteristics of Landfill Leachates and Gases. Groundwater, 34, 827-836. http://dx.doi.org/10.1111/j.1745-6584.1996.tb02077.x
30. Conrad, R. (2005) Quantification of Methanogenic Pathways Using Stable Carbon Isotopic Signatures: A Review and a Proposal. Organic Geochemistry, 36, 739-752. http://dx.doi.org/10.1016/j.orggeochem.2004.09.006
31. Megonigal, J.P., Hines, M.E. and Visscher, P.T. (2004) Anaerobic Metabolism: Linkages to Trace Gases and Aerobic Processes. In: Schlesinger, W.H., Ed., Biogeochemistry, Elsevier-Pergamon, Oxford, 317-424.
32. Hornibrook, E.R.C, Longstae, W.S. and Fyfe, W.S. (1997) Spatial Distribution of Microbial Methane Production Pathways in Temperate Zone Wetland Soils: Stable Carbon and Hydrogen Isotope Evidence. Geochimica et Cosmochimica Acta, 61, 745-753.
33. Dorodnikov, M., Marushchak, M., Biasi, C. and Wilmking, M. (2013) Effect of Microtopography on Isotopic Composition of Methane in Porewater and Efflux at a Boreal Peatland. Boreal Environment Research, 18, 269-279.
34. Walter, K.M., Chanton, J.P., Chapin, F.S., Schuur, E.A. and Zimov, S.A. (2008) Methane Production and Bubble Emissions from Arctic Lakes: Isotopic Implications for Source Pathways and Ages. Journal of Geophysical Research, 113, Article ID: G00A08. http://dx.doi.org/10.1029/2007JG000569
35. Yamamoto, S., Alcauskas, J.B. and Crozier, T.E. (1976) Solubility of Methane in Distilled Water and Seawater. Journal of Chemical & Engineering Data, 21, 78-80. http://dx.doi.org/10.1021/je60068a029
36. Simpkins, W.W. and Parkin T.B. (1993) Hydrogeology and Redox Geochemistry of Methane in a Late Wisconsinan Till and Loess Sequence in Central Iowa. Water Resources Research, 29, 3643-3657. http://dx.doi.org/10.1029/93WR01687
37. Beckwith, C.W. and Baird, A.J. (2001). Effect of Biogenic Gas Bubbles on Water Flow through Poorly Decomposed Blanket Peat. Water Resources Research, 37, 551-558. http://dx.doi.org/10.1029/2000WR900303
38. Baird, A.J. and Waldron, S. (2003). Shallow Horizontal Groundwater Flow in Peatlands Is Reduced by Bacteriogenic Gas Production. Geophysical Research Letters, 30, Article No. 2043. http://dx.doi.org/10.1029/2003GL018233
39. Strack, M., Kellner, E. and Waddington, J.M. (2005) Dynamics of Biogenic Gas Bubbles in Peat and Their Effects on Peatland Biogeochemistry. Global Biogeochemical Cycles, 19, Article ID: GB1003. http://dx.doi.org/10.1029/2004GB002330
40. Kellner, E., Baird, A.J., Oosterwoud, M., Harrison, K. and Waddington, J.M. (2006) Effect of Temperature and Atmospheric Pressure on Methane (CH4) Ebullition from Near-Surface Peats. Geophysical Research Letters, 33, Article ID: L18405. http://dx. doi.org/10.1029/2006GL027509
41. Comas, X., Slater, L. and Reeve, A. (2005). Spatial Variability in Biogenic Gas Accumulations in Peat Soils Is Revealed by Ground Penetrating Radar (GPR). Geophysical Research Letters, 32, Article ID: L08401.http://dx.doi.org/10.1029/2004gl022297
42. Tokida, T., Miyazaki, T., Mizoguchi, M., Nagata, O., Takakai, F., Kagemoto, A. and Hatano, R. (2007) Falling Atmospheric Pressure as a Trigger for Methane Ebullition from a Peatland. Global Biogeochemical Cycles, 21, Article ID: GB2003. http://dx.doi.org/10.1029/2006GB002790
43. Beckmann, M., Sheppard, S.K. and Lloyd, D. (2004) Mass Spectrometric Monitoring of Gas Dynamics in Peat Monoliths: Effects of Temperature and Diurnal Cycles on Emissions. Atmospheric Environment, 38, 6907-6913.http://dx.doi.org/10. 1016/j.atmosenv.2004.08.004

44. Fechner-Levy, E.J. and Hemond, H.F. (1996) Trapped Methane Volume and Potential Effects on Methane Ebullition in a Northern Peatland. Limnology and Oceanography, 41, 1375-1383. http://dx.doi.org/10.4319/lo.1996.41.7.1375
45. Coulthard, T.J., Baird, A.J., Ramirez, J. and Waddington, J.M. (2009) Methane Dynamics in Peat: Importance of Shallow Peats and a Novel Reduced-Complexity Approach for Modeling Ebullition. In: Carbon Cycling in Northern Peatlands. Geophysical Monograph Series, 184, 173-185. http://dx.doi.org/10.1029/2008gm000811
46. Siegel, D.I., Chanton, J.E., Glaser, E.H., Chasar, L.C. and Rosenberry, D.O. (2001) Estimating Methane Production Rates in Bogs and Landfills by Deuterium Enrichment of Pore-Water. Global Biogeochemical Cycles, 15, 967-977.http://dx.doi.org/10.1029/2000GB001329
47. Romanowicz, E.A., Siegel, D.I., Chanton, J.P. and Glaser, P.H. (1995) Temporal Variations in Dissolved Methane Deep in the Lake Agassiz Peatlands, Minnesota. Global Biogeochemical Cycles, 9, 197-212. http://dx.doi.org/10.1029/95GB00634
48. Walter, B.P., Heimann, M. and Matthews, E. (2001) Modeling Modern Methane Emissions from Natural Wetlands: 1. Model Description and Results. Journal of Geophysical Research, 106, 34189-34206. http://dx.doi.org/10.1029/2001JD900165
49. De Visscher, A., De Pourcq, I. and Chanton, J. (2004) Isotope Fractionation Effects by Diffusion and Methane Oxidation in Landfill Cover Soils. Journal of Geophysical Research, 109, Article ID: D18111. http://dx.doi.org/10.1029/2004jd004857
50. Happell, J.D., Chanton, J.P. and Showers, W.J. (1995) Methane Transfer across the Water-Air Interface in Stagnant Wooded Swamps of Florida: Evaluation of Mass-Transfer Coefficients and Isotopic Fractionation. Limnology and Oceanography, 40, 290-298. http://dx.doi.org/10.4319/lo.1995.40.2.0290
51. Chanton, J.P., Martens, C.S., Kelley, C.A., Crill, P.M. and Showers, W.J. (1992) Methane Transport Mechanisms and Isotope Fractionation in Emergent Macrophytes of an Alaskan Tundra Lake. Journal of Geophysical Research, 97, 16681-16688. http://dx.doi.org/10.1029/90JD01542
52. Bhullar, G.S., Iravani, M., Edwards, P.J. and Venterink, H.O. (2013) Methane Transport and Emissions from Soil as Affected by Water Table and Vascular Plants. BMC Ecology, 13, 32. http://dx.doi.org/10.1186/1472-6785-13-32
53. Chanton, J.P. (2005) The Effect of Gas Transport on the Isotope Signature of Methane in Wetlands. Organic Geochemistry, 36, 753-768. http://dx.doi.org/10.1016/ j.orggeochem. 2004.10.007
54. Bergamaschi, P. (1997) Seasonal Variation of Stable Hydrogen and Carbon Isotope Ratios in Methane from a Chinese Rice Paddy. Journal of Geophysical Research, 102, 25383-25393. http://dx.doi.org/10.1029/97JD01664
55. Coleman, D.D., Risatti, J.B. and Schoell, M. (1981) Fractionation of Carbon and Hydrogen Isotopes by Methane-Oxidizing Bacteria. Geochimica et Cosmochimica Acta, 45, 1033-1037. http://dx.doi.org/10.1016/0016-7037(81)90129-0
56. Gebert, J., Streblow, C. and Pfeiffer, E.M. (2011) Impact of Gas Transport on Fractionation of Carbon Stable Isotope Related to the Microbial Oxidation of Methane in Soils. EGU General Assembly. Geophysical Research Abstracts, 13, Article No. 14108.
57. Happell, J., Chanton, J.P. and Showers, W. (1994) The Influence of Methane Oxidation on the Stable Isotope Composition of Methane Emitted from Florida Swamp Forests. Geochimica et Cosmochimica Acta, 58, 4377-4388.http://dx.doi.org/10.1016/0016-7037(94)90341-7
58. Sanci, R., Panarello, H. and Ostera, H. (2009) Assessment of Soil Moisture Influence on CO2 Flux: A Laboratory Experiment. Environmental Geology, 58, 491-497. http://dx.doi.org/10.1007/s00254-008-1522-7
59. Sanci, R., Panarello, H. and Ostera, H. (2012) CO2 Emissions from a Municipal Site for Final Disposal of Solid Waste in Gualeguaychú, Entre Ríos Province, Argentina. Environmental Earth Sciences, 66, 519-528. http://dx.doi.org/10.1007/s12665-011-1260-0
60. Wassmann, R., Neue, H.U., Alberto, M.C.R., Lantin, R.S. Bueno, C., Llenaresas, D., Arah, J.R.M., Papen, H., Seiler, W. and Rennenberg, H. (1996) Fluxes and Pools of Methane in

Wetlands Rices Oils with Varying Organic Inputs. Environmental Monitoring and Assessment, 42, 163-173. http://dx.doi.org/10.1007/BF00394048
61. McAuliffe, C. (1971) Gas Chromatographic Determination of Solutes by Multiple Phase Equilibrium. Chemical Technology, 1, 46-51.
62. Gonfiantini, R. (1978) Standards for Stable Isotope Measurements in Natural Compounds. Nature, 271, 534-536. http://dx.doi.org/10.1038/271534a0
63. Coplen, T. (1994) Reporting of Stable Hydrogen, Carbon, and Oxygen Isotopic Abundances. Pure and Applied Chemistry, 66, 273-276. http://dx.doi. org/10. 1351/pac199466020273

CHAPTER 18

Nitrogen Removal in a Horizontal Subsurface Flow Constructed Wetland Estimated Using the First-Order Kinetic Model

Lijuan Cui [1,2,3,*,†], Wei Li [1,2,3,†], Yaqiong Zhang [1,2,3], Jiaming Wei [1,2,3], Yinru Lei [1,2,3], Manyin Zhang [1,2,3], Xu Pan [1,2,3], Xinsheng Zhao [1,2,3], Kai Li [1,2,3] and Wu Ma [4]

[1]*Institute of Wetland Research, Chinese Academy of Forestry, Beijing 100091, China*

[2]*Beijing Hanshiqiao National Wetland Ecosystem Research Station, Beijing 101399, China*

[3]*The Beijing Key Laboratory of Wetland Ecological Function and Restoration, Beijing 100091, China*

[4]*School of Natural Resources, West Virginia University, Morgantown, WV 26506, USA*

ABSTRACT

We monitored the water quality and hydrological conditions of a horizontal subsurface constructed wetland (HSSF-CW) in Beijing, China, for two years. We simulated the area-based constant and the temperature coefficient with the first-order kinetic model. We examined the relationships between the nitrogen (N) removal rate, N load, seasonal variations in the N removal rate, and environmental factors—such as the area-based constant, temperature, and dissolved oxygen (DO). The effluent ammonia (NH_4^+-N) and nitrate (NO_3^--N) concentrations were significantly lower than the influent concentrations ($p < 0.01$, $n = 38$). The NO_3^--N load was significantly correlated with the removal rate ($R^2 = 0.96$, $p < 0.01$), but the NH_4^+-N load was not correlated with the removal rate ($R^2 = 0.02$, $p > 0.01$). The area-based constants of NO_3^--N and NH_4^+-N at 20 °C were 27 ± 26 (mean ± SD) and 14 ± 10 m·year^{-1}, respectively. The temperature coefficients for NO_3^--N and NH_4^+-N were estimated at 1.004 and 0.960, respectively. The area-based constants for NO_3^--N and NH_4^+-N were not correlated with temperature ($p > 0.01$). The NO_3^--N area-based constant was correlated with the corresponding load ($R^2 = 0.96$, $p < 0.01$). The NH_4^+-N area rate was correlated with DO ($R^2 = 0.69$, $p < 0.01$), suggesting that the factors that influenced the N removal rate in this wetland met Liebig's law of the minimum.

Keywords: horizontal subsurface flow wetland; N removal; first-order kinetics; model

1. INTRODUCTION

The rapid economic development witnessed in recent years in China has led to vast increases in the production of human sewage and other waste effluents, most of which are rich in nitrogen (N) compounds. Nitrogen not only affects the water quality, but also contributes to water eutrophication [1], especially in lakes and rivers [2]. Therefore, N removal from wastewater has become a topic of growing concern, both in China and worldwide [3,4]. Wetlands play an increasingly important role in removing N from aquatic systems via various processes, and can achieve total nitrogen (TN) removals of between 40% and 55% [5]. The two most common forms of N in wetlands are organic and inorganic. Organic N is made up of a variety of compounds, including amino acids, urea and uric acid, and purines and pyrimidines. Inorganic forms of N in wetlands include ammonia (NH_4^+-N), nitrite (NO_2^--N), nitrate (NO_3^--N), nitrous oxide (N_2O), dissolved elemental N, or dinitrogen gas (N_2) [6].

Constructed wetlands (CWs) are rapidly becoming a viable method for wastewater treatment worldwide [7], because of their similarity to natural wetlands and the fact that they cost less to construct, implement, and maintain than other types of treatments [8,9]. Constructed wetlands can be used to prevent water eutrophication [10]. For example, it was reported that, over a period of seven consecutive years, CWs were able to maintain a higher removal rate of N than traditional N removal processes [11]. The

HSSF-CWs have a purification function, in that the sewage water flows to the outlet end in a horizontal direction and transverses the sand medium and plant roots in turn. In HSSF-CWs, N removal is achieved by a combination of physical, chemical, and biological processes, which are affected by a range of factors, such as pH, dissolved oxygen (DO), and temperature [12]. Mathematical models can be used to simulate N removal and the relationships between various parameters and the N removal rates in constructed wetlands to reflect the removal mechanisms in wetlands. This information can then be used to provide decision support and the scientific basis for improving ecosystem service functions of wetlands [13]. Previous studies suggested that the first-order kinetics model could describe N removal in wetland ecosystems [14,15]. However, recent studies that have used the first-order kinetics model for N removal have mainly focused on the application of the area-based constant and volume-based rate constant in HSSF-CWs, and, to date, applications of the area-based constant to HSSF-CWs and assessments of the influence of water pollution loads, temperature, and DO on the area-based constant have been largely neglected [16].

Using as its basis the wetland water balance principle, in which the potential water transfer mechanism inputs and outputs, including runoff, groundwater recharge, and lateral inflow or drainage processes, are balanced, the aim of this study was to examine the removal and seasonal variations of two N forms, namely NO_3^--N and NH_4^+-N, in an HSSF-CW [17,18].

The first-order kinetic model, a commonly used hydrological model, has been frequently used to predict, for example, variations in water chemistry, sudden changes in influent organic loads, NO_3^--N removal, and pharmaceuticals removal [19,20,21,22]. We also used the first-order kinetic model to estimate the area-based constant and temperature coefficient of N removal in this study. Further, to improve the accuracy and consistency of the model parameters, we also examined the relationships between the area-based constant and factors that influenced it, such as temperature, pollution load, and DO.

2. MATERIALS AND METHODS

2.1. Study Site Description

Our study was carried out in a HSSF-CW in the Beijing Wildlife Rescue and Rehabilitation Center, Beijing, China (Latitude: 40°6′14.40″ N, Longitude: 116°42′35.71″ E) (Figure 1). The annual average temperature at the study site is 11.5 °C. The average temperatures in January and July are 4.9 and 25.7 °C, respectively, and the minimum temperature in July 2012 and 2013 was 19.1 °C. The average annual rainfall is 625 mm from 2012 to 2013. The wetland consists of three treatment cells (Table 1), and its purpose is to improve the water quality of the artificial lake in the site.

The main pollution source of the water is the waste water from the bird life in the lake. Gravel is the principal media; there are two layers, one 20 cm deep that comprises gravels with a diameter of between 15 and 30 mm and the other 50 cm deep comprised of finer gravels with diameters of between 5 and 15 mm from bottom to top. The average water depths in the three treatment cells are 0.3, 0.1, and 0.1 m.

Plants in the upper media, planted at a density of two plants per square metre, show vigorous growth. The plant species are mainly *Lythrum salicaria*, *Iris tectorum*, and *Scirpus validus*, with densities of 30, 50, and 60 stem/m², respectively. Birds, such as *Ardea alba*, *Anas platyrhynchos*, and *Anatidae*, are frequently observed at the lake.

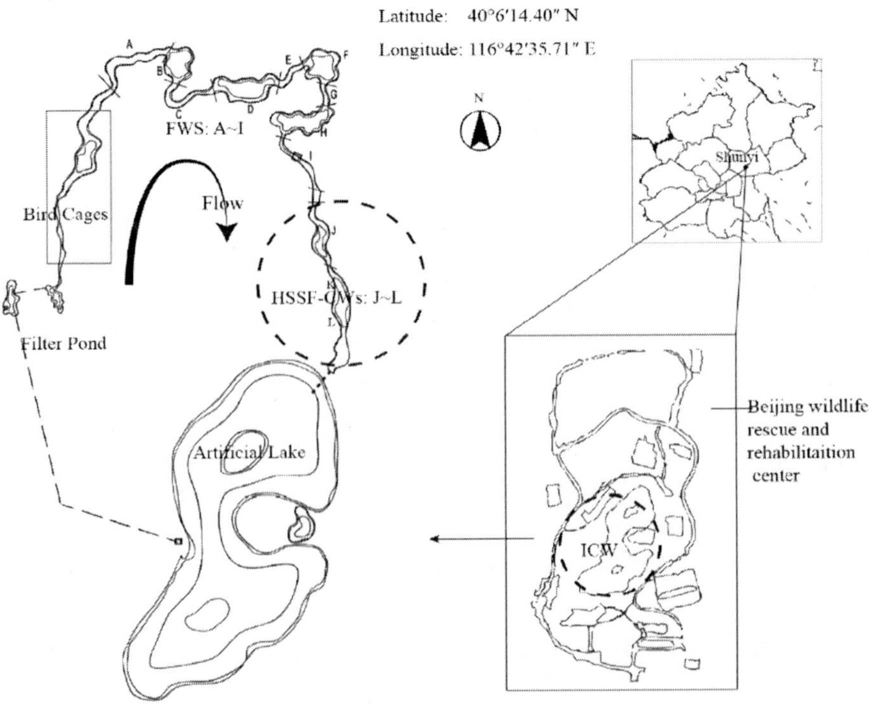

Figure 1. Diagram showing the location of the constructed wetland.

Table 1. Configurations of the different treatment cells of the HSSF-CW.

Configuration	Treatment Cell		
	a	b	c
Length (m)	20	22	25
Area (m^2)	170	160	81
Particle density (g/m^3)	2.43	1.80	1.44
Main Vegetation		*Iris tectorum*	
Plant spacing (m)	2	2	2

2.2. The Water Quality and Hydrology Monitoring

The HSSF-CW was operated with continuous flow from April 2012 to November 2013. The water quality was monitored once a month from April to November in 2012 and 2013. Between December 2012 and March 2013 the low temperatures and frozen water prevented sampling. The sampling points were located in the inflow, middle, and outflow of each cell. Flow-weighted composite water samples were collected from the inlet and outlet of each wetland cell using water samplers. Samples were collected from each point in triplicate. The water samples were collected in 500-mL plastic bottles and stored at 4 °C until analysis. The flow was recorded by a flow meter (SonTek, YSI, Yellow Springs, OH, USA), while meteorological variables, such as rainfall and evaporation, were recorded hourly by a portable weather station (WeatherHawk, CampSci, Logan, Utah, USA). The concentrations of NH_4^+-N, TN, and NO_3^--N were determined using the standard methods for the examination of water and wastewater [23] with a water quality analyzer (SmartChem 200 Chemical Analyzer, West Co., Rome, Italy). A portable water quality probe (EXO, YSI, Yellow Springs, OH, USA) was used to measure DO, temperature, and other In Situ variables.

2.3. Data Analysis and Modeling

Equation (1) represents the wetland water dynamic equilibrium model of the HSSF-CW:

$$dV/dt = Q_{in} - Q_{out} + Q_c + (P - ET - I)\ A \tag{1}$$

where Q_{in} and Q_{out} are the daily volumetric water inflow and outflow rates ($m^3 \cdot day^{-1}$), respectively, Qc is the catchment runoff rate ($m^3 \cdot day^{-1}$), P is the daily precipitation rate ($mm \cdot day^{-1}$), ET is the daily evapotranspiration rate (mm·day^{-1}), I is the daily infiltration rate (mm·day^{-1}), dV/dt is the net change in volume (m^3·day^{-1}), and A is the total area of the wetland (m^2).

2.4. The First-Order Kinetic Model of N Removal Applied to the HSSF-CW

The areal removal rate constant and the temperature coefficient were calculated from the monitoring data and the following first-order model [14]:

$$\ln\left(\frac{C_{out} - C^*}{C_{in} - C^*}\right) = -\frac{K_{(t)}}{q} \quad (2)$$

where C_{in} and C_{out} are the influent and effluent concentrations, respectively, in mg·L^{-1}, C^* is the background concentration (mg·L^{-1}, 0 mg/L), q is the hydraulic loading rate (m·day^{-1}), t is the temperature, and $K_{(t)}$ is the area-based constant of a first-order removal rate on temperature (m·year^{-1}). Equation (3) was used to calculate q (m·year^{-1}):

$$q = \frac{Q}{A} \quad (3)$$

where Q is total water inflow rate (m^3·day^{-1}).

The effect of temperature on the areal removal rate constant was modelled using the following equation:

$$K_{(t)} = K_{(20)}\theta^{(t-20)} \quad (4)$$

$$\ln(K_{(t)}) = \ln\theta(t-20) + \ln(K_{(20)}) \quad (5)$$

The selected variables in this study included concentrations of NH_4^+-N and NO_3^--N, DO, and temperature.

The means of actual measured values are presented. Data fitting was carried out with Origin 8.5 and statistical analyses were performed using SPSS version 18.0. Statistically significant differences were determined at the $\alpha = 0.01$ significance level.

3. RESULTS

3.1. Water Balance

The daily volumetric water inflow and outflow rates of the HSSF-CW were 33 ± 19 and 33 ± 19 $m^3 \cdot day^{-1}$, respectively (Table 2). On the whole, there were no significant differences between the average inflow and outflow rates ($p > 0.01$, $n = 42$), which indicates that rainfall had little influence on the HSSF-CW throughout the experimental period. Water was mainly supplied via injection. The large standard deviations in the daily volumetric water inflow and outflow rates indicate that there were considerable changes in the water volume in the HSSF-CW each month.

Table 2. Flow variations in the HSSF-CW in each treatment cell.

Flow (m^3/Day)	a		b		c		Total	
	Mean	SD	Mean	SD	Mean	SD	Mean	SD
Influent	33	19	33	19	33	19	33	19
Rainfall	0.46	0.50	0.43	0.47	0.22	0.24	1.11	1.21
Effluent	33	19	33	19	33	19	33	19

Note: SD = standard deviation.

Water injection to the HSSF-CW began in April. As time progressed, the inflow gradually increased. Drainage began in October and the flow decreased with drainage, so the annual flow showed an initial increase and then decreased (Figure 2). The maximum rainfall in the experimental period occurred in July 2012, when 337 mm was recorded; the minimum rainfall (0.5 mm) occurred in November 2013. The outflow was higher than the inflow in July because of the more intensive rainfall in this month.

3.2. Influent and Effluent Nitrogen Concentrations

The average annual NO_3^--N concentration was 3.40 ± 3.44 $mg \cdot L^{-1}$, while the annual average NH_4^+-N concentration was 0.36 ± 0.35 $mg \cdot L^{-1}$ (Table 3). Over the monitoring period, the NO_3^--N concentrations were higher than the NH_4^+-N concentrations. The standard deviations of the concentrations of the two N species were very large, which indicates that the concentrations of the two N species changed considerably in different months. When compared to the influent concentrations, we found that the concentrations of NO_3^--N and NH_4^+-N in the effluent had decreased significantly ($p < 0.01$, $n = 38$), and were, on average, 1.22 ± 0.92 and 0.27 ± 0.27 mg·L^{-1}, respectively. The lower concentrations at the outflow of the treatment cell indicate that N was removed from the HSSF-CW in the two years of monitoring. Further, the concentrations of these two N species met Class III of the Surface Water Standards in China. The water standards mainly apply to the centralized drinking water and industrial and agricultural water, and were based on the human health reference values [24].

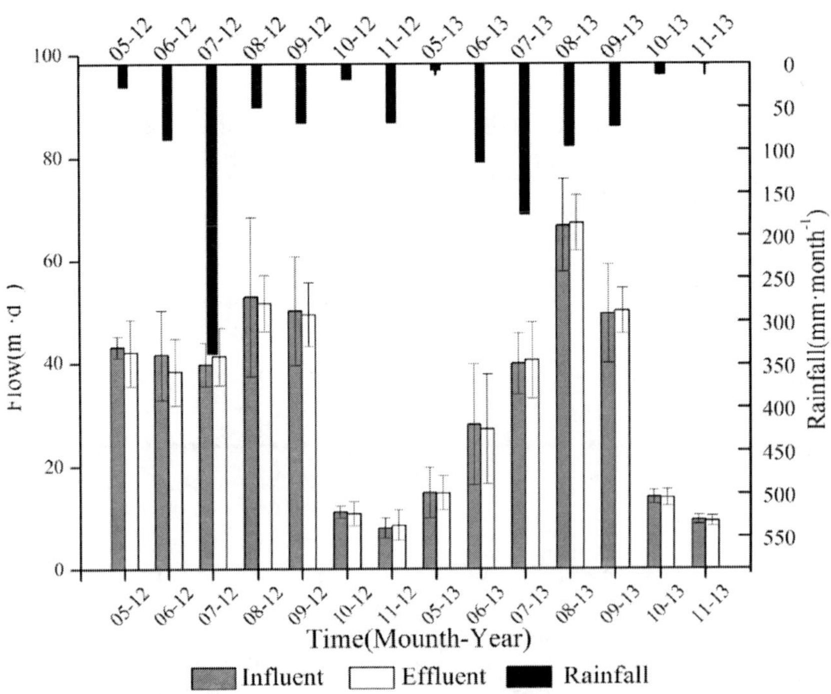

Figure 2. Average water fluxes, including the influent, effluent, and rainfall, in the HSSF-CW in different months.

There were seasonal variations in the N concentrations of the inflow and outflow of the HSSF-CW (Table 4). The inflow concentrations of NH_4^+-N (0.60 ± 0.33 mg·L^{-1}) and NO_3^--N (4.52 ± 3.94 mg·L^{-1}) were highest in spring and summer respectively, while the NO_3^--N (0.21 ± 0.20 mg·L^{-1}) inflow concentration was lowest in spring.

Table 3. Average influent and effluent nitrogen concentrations in the HSSF-CW.

Parameter	Unit	*n*	Influent		Effluent	
			Mean	SD	Mean	SD
NO_3^--N	$mgN{\cdot}L^{-1}$	38	3.40	3.44	1.22	0.92
NH_4^+-N	$mgN{\cdot}L^{-1}$	38	0.36	0.35	0.27	0.27

Notes: *n* = sample number; SD = standard deviation.

Table 4. Average influent and effluent nitrogen concentrations in the HSSF-CW in different months.

Season	Month	*n*	NO_3^--N ($mg{\cdot}L^{-1}$)				NH_4^+-N ($mg{\cdot}L^{-1}$)			
			Influent		Effluent		Influent		Effluent	
			Mean	SD	Mean	SD	Mean	SD	Mean	SD
Spring	4–5	4	0.21	0.20	0.19	0.18	0.60	0.33	0.44	0.25
Summer	6–8	14	4.52	3.94	1.39	1.02	0.34	0.25	0.28	0.23
Autumn	9–11	20	3.25	3.06	1.31	0.82	0.32	0.40	0.23	0.30

Notes: *n* = sample number; SD = standard deviation.

3.3. Nitrogen Loading and Removal Rate

There is a generally a good relationship between the inflow and outflow concentrations and loadings of TN and NH_4^+-N in this research. Also, N removal efficiency in constructed wetlands generally falls within a certain range, regardless of whether the N load is large or small [25].

In this study, we found that the NO_3^--N pollution load and removal rate were strongly correlated ($R^2 = 0.96$, $p < 0.01$) (Figure 3), but that the NH_4^+-N pollution load and removal rate were not correlated ($R^2 = 0.02$, $p > 0.01$) (Figure 4).

During the study period, the NO_3^--N removal rate was higher than that of NH_4^+-N, and was vulnerable to seasonal changes. The highest NO_3^--N and NH_4^+-N removal rates of 44.7% and 37.4% occurred in summer and autumn, respectively, and are consistent with the higher NO_3^--N loads in summer and autumn. The NH_4^+-N removal rate was lowest in summer (19.1%) because of the higher rainfall in summer. An increase in hydraulic loading rates resulted in a decrease in the NH_4^+-N removal rate (Table 5).

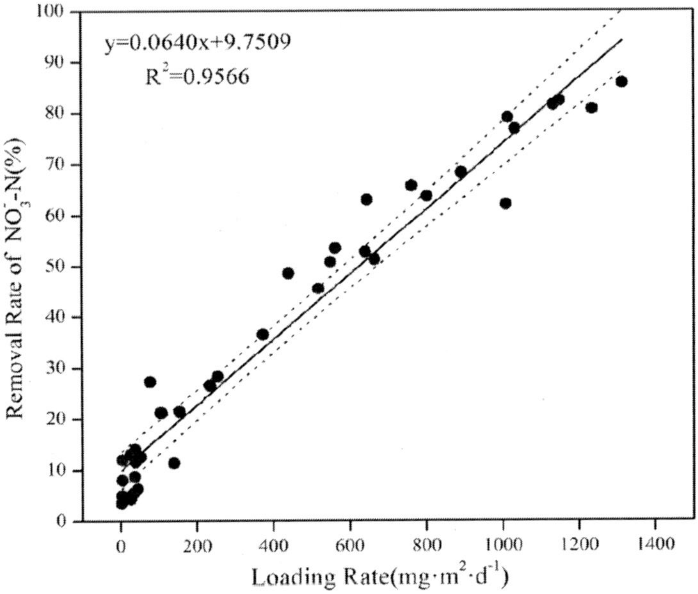

Figure 3. Relationship between the areal loading and removal efficiency for NO_3^--N (the dotted line indicates the 99% confidence limit).

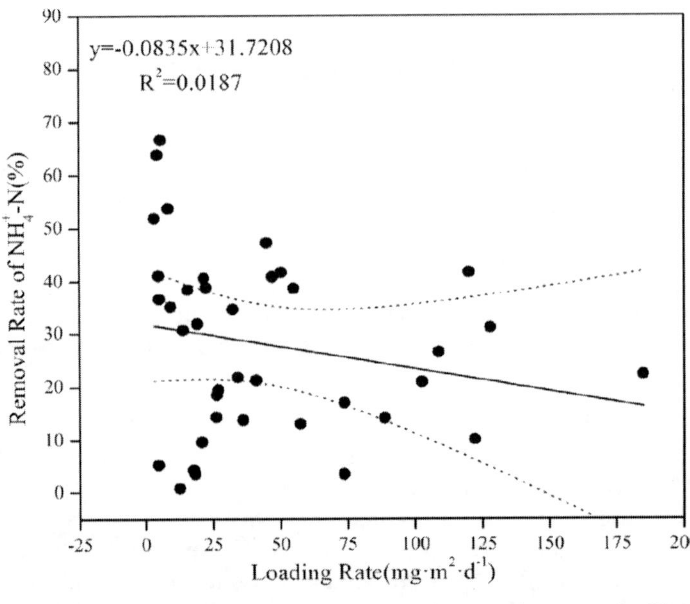

Figure 4. Relationship between the areal loading and removal efficiency for NH4$^+$-N (the dotted line indicates the 99% confidence limit).

Table 5. Areal loading and removal efficiency of the HSSF-CW.

Season	Month	*n*	NO_3^--N Loading Rate ($mg \cdot m^{-2} \cdot Day^{-1}$)		NO_3^--N Removal (%)	NH_4^+-N Loading Rate ($mg \cdot m^{-2} \cdot Day^{-1}$)		NH_4^+-N Removal (%)
			Mean	SD		Mean	SD	
Spring	4–5	4	25.8	25.6	7.4	75.4	41.9	27.2
Summer	6–8	14	564.6	492.0	44.7	42.2	31.8	19.1
Autumn	9–11	20	406.1	382.0	37.4	39.9	49.7	34.4

Notes: *n* = sample number; SD = standard deviation.

3.4. Estimation of the Area-Based Constants and the Temperature Coefficients

Using Equation (1) and the measured values of C_{in} and C_{out}, we calculated area-based constants (K) for NO_3^--N and NH_4^+-N of 27 ± 26 and 16 ± 12 $m \cdot year^{-1}$ (Table 6), respectively. The corrected values of K for a temperature of 20 °C, obtained from Equation (4), were 27 ± 26 and 14 ± 10 $m \cdot year^{-1}$ for NO_3^--N and NH_4^+-N, respectively. When corrected to 20 °C, the $K_{(20)}$ values of both NO_3^--N and NH_4^+-N were slightly lower, indicating that the influence of temperature on N removal was minimal. The $K_{(20)}$ of NO_3^--N and NH_4^+-N were lower and slightly higher, respectively, than those reported by Kadlec and Wallace [6], but were slightly higher than the values obtained by Dzakpasu et al. [26]. The temperature coefficient (θ) values estimated for NO_3^--N and NH_4^+-N were 1.004 and 0.960, respectively, and were lower than those reported by Kadlec and Wallace [6], but were very similar to those reported by Dzakpasu et al. [26].

Table 6. The areal rate constants and temperature coefficients of the HSSF-CW.

Parameter	*n*	θ	K ($m \cdot Year^{-1}$) Mean	K SD	$K_{(20)}$ ($m \cdot Year^{-1}$) Mean	$K_{(20)}$ SD
NO_3^--N	38	1.004	27	26	27	26
NH_4^+-N	38	0.960	16	12	14	10

Notes: *n* = sample number; SD = standard deviation.

The temperature of the inflow to the HSSF-CW ranged from 5 to 30 °C. Curves can be fitted using Equation (4), and are shown in Figure 5 and Figure 6. The figures suggest that, for both NO_3^--N and NH_4^+-N ($p > 0.01$), the area-based constant and temperature were not correlated. At the same time, lower θ values demonstrate that temperature had little or no impact on either the area-based constant or the physical removal process. This contrasts with the conclusion of Huang et al. [27], who reported that the NH_4^+-N and total nitrogen removal rates were significantly influenced by temperature. Moreover, the findings suggest that temperature had a role in promoting the growth of wetland plants. Therefore, appropriate external thermal insulation measures need to be adopted to cover the surface of the wetland, to ensure improved growth of hardy plants all year round.

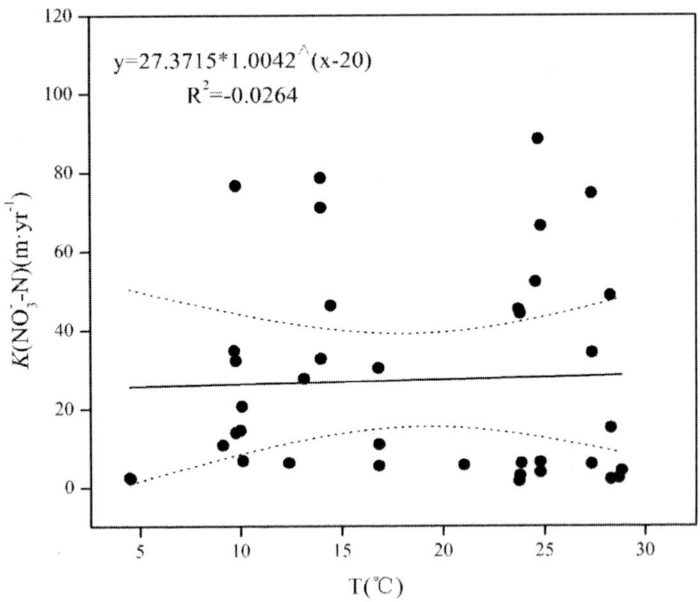

Figure 5. Relationship between the temperature and areal rate constants for NO_3^--N (the dotted line is the 99% confidence limit).

3.5. The Influence of N Loads and DO on the Area-Based Constant

There was a strong linear correlation ($R^2 = 0.96$, $p < 0.01$) between the area-based constants and loads of NO_3^--N (Figure 7). The area-based constants and loads of NH_4^+-N were not correlated ($R^2 = 0.037$, $p > 0.01$) (Figure 8), possibly because the NH_4^+-N loads over the entire study period were lower. This also indicates that the area-based constant can accurately reflect the removal potential of different pollutants in wetland systems that are in receipt of high pollutant loads [28].

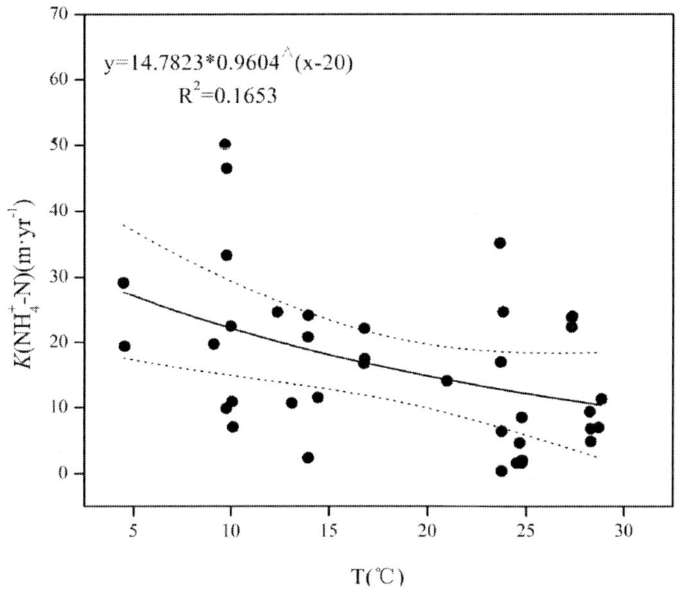

Figure 6. Relationship between the temperature and areal rate constants for NH4⁺-N (the dotted line is the 99% confidence belt).

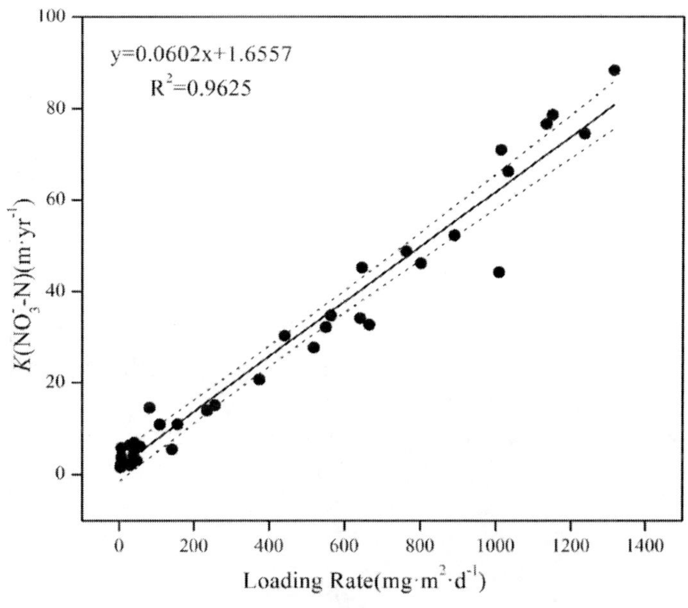

Figure 7. Relationship between the areal loading and areal rate constants for NO_3^--N (the dotted line is the 99% confidence limit).

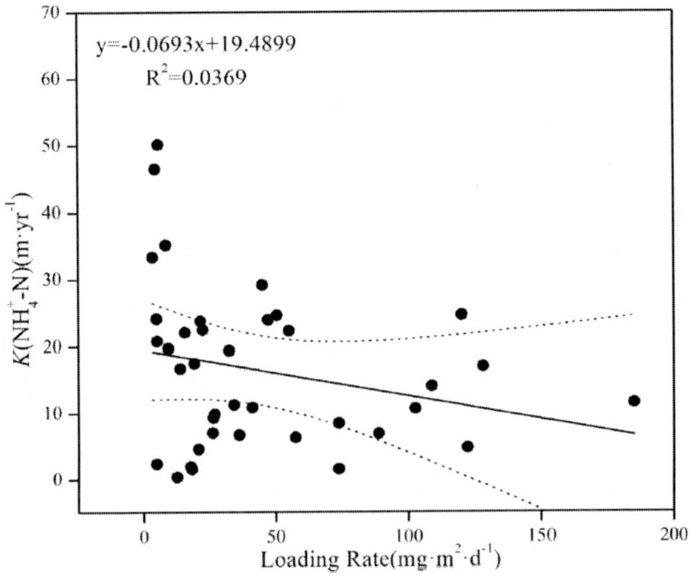

Figure 8. Relationship between the areal loading and areal rate constants for NH_4^+-N (the dotted line is the 99% confidence limit).

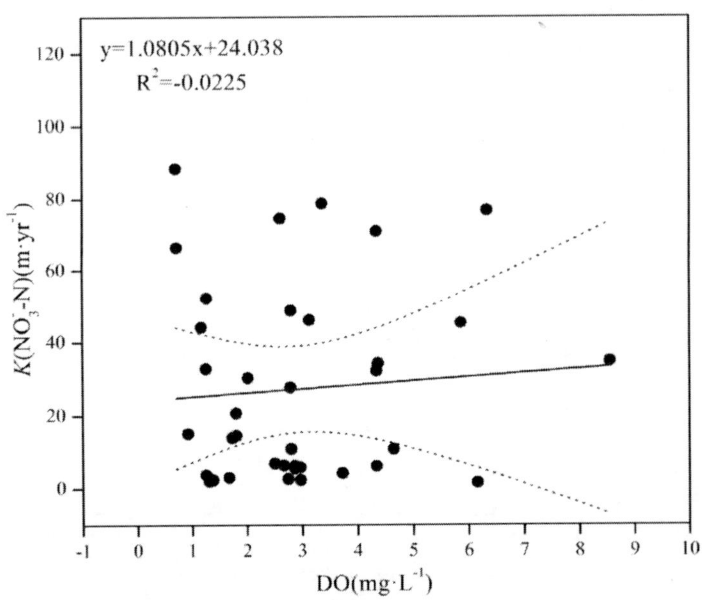

Figure 9. Relationship between DO and areal rate constants for NO_3^--N (the dotted line is the 99% confidence limit).

As shown in Figure 9 and Figure 10, the inflow DO concentrations had a minimal influence on the NO_3^--N area-based constants, and the inflow DO concentration and NH_4^+-N area-based constants were positively correlated ($R^2 = 0.69$, $p < 0.01$). The lower NH_4^+-N removal rate may be the result of the lower DO concentrations; these lower DO concentrations lead to anaerobic conditions in wetlands, increase denitrification, and decrease nitrification. In addition, the low DO concentrations may be related to the lack of temperature effects. The lower DO levels would cause a decrease in biodegradation, which suggests that a significant portion of the N removal is driven by physical processes that are not influenced by temperature.

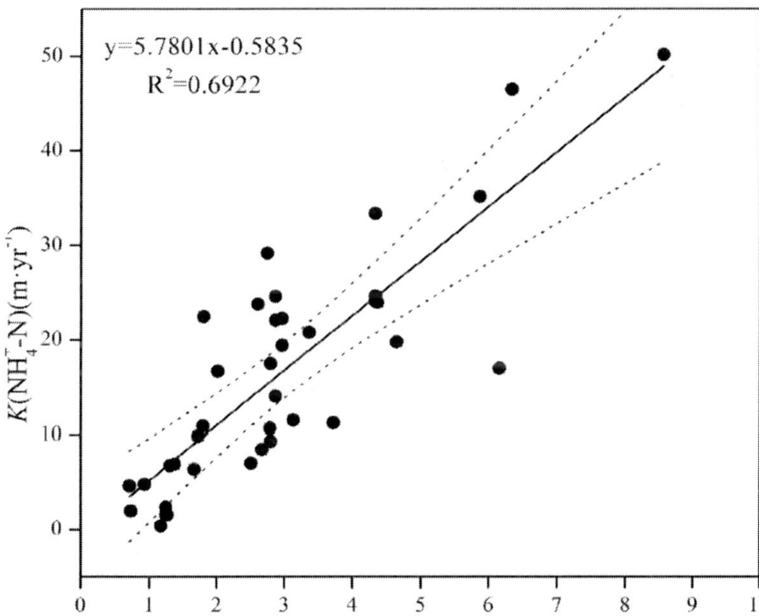

Figure 10. Relationship between DO and areal rate constants for NH_4^+-N (the dotted line is the 99% confidence limit).

4. DISCUSSION

In this study, we examined the wetland hydrology of the HSSF-CW over a period of two years using the wetland water balance principal. The results showed that water injection was the main source of water to the HSSF-CW. The flows into and out of the wetland varied on a monthly basis. In this study, we combined information about the wetland hydrology and N removal, and the results suggest that, when the rainfall is higher, the hydraulic retention time is lower, which would then influence the N removal rate. The water balance model adopted in this study only included hydrological data, and did not consider factors such as soil, topography, and land use [29]; this is a major limitation of the study.

There was obvious removal of both NO_3^--N and NH_4^+-N in this HSSF-CW. The removal rates however varied on a seasonal basis, with higher NH_4^+-N inflow concentrations in spring, because the swimming birds that live in the open water connected to the HSSF-CW were more active in April and May than in other months. The inflow NO_3^--N concentrations were higher in summer, reflecting the weaker denitrification caused by the abundant rainfall (Figure 2) and shorter hydraulic retention time in summer. The NO_3^--N loads were strongly correlated with the removal rate ($R^2 = 0.96$, $p < 0.01$), but the NH_4^+-N load and removal rate ($R^2 = 0.02$, $p > 0.01$) were not correlated. This may reflect the higher and lower inflow loads of NO_3^--N and NH_4^+-N, respectively. Under certain pollutant loads, increased pollutant concentrations are conducive to biofilm formation on the substrate surface, which not only provides good conditions for microbial growth but can also adsorb large quantities of organic matter, thereby contributing to pollutant removal [30].

We examined the relationship between the area-based constant and temperature, pollutant loads, and DO for both NO_3^--N and NH_4^+-N, and also estimated the temperature coefficient θ. The area-based constant and the NO_3^--N load were strongly correlated ($R^2 = 0.96$, $p < 0.01$), but there was no correlation between temperature and DO. The area removal rate constant for NH_4^+-N was correlated with DO ($R^2 = 0.69$, $p < 0.01$), but was not correlated with either the temperature or the NH_4^+-N load. The lower θ value estimated for the two N species illustrates that temperature had little or no impact on nitrification and denitrification.

While many factors influence the N removal efficiency in wetlands, the main controls are nitrification and denitrification [4]. Nitrification needs aerobic conditions and denitrification requires anaerobic conditions. Both processes are affected by the DO concentration, meaning that N removal in constructed wetlands is strongly influenced by the DO concentration. There are three main sources of oxygen in constructed wetlands, including release from plants, atmospheric reaeration, and oxygen transported by water, of which release from plants and atmospheric reaeration dominate [31]. Plants can also partly block the oxygen transfer process from water to the substrate [32], so that the low oxygen concentration in the substrate becomes the main limit on NH_4^+-N removal. In this study, the concentrations of DO in the HSSF-CW varied from 0.5 to 4.5 mg·L^{-1}, and the DO% was as high as 87%. These DO concentrations are much lower than the water-saturated dissolved oxygen concentration [33], indicating that the DO concentrations in this HSSF-CW were sufficiently low to satisfy the conditions for denitrification. The DO concentration did not limit NO_3^--N removal in this HSSF-CW. The NH_4^+-N area-based constant and DO were correlated ($R^2 = 0.69$, $p < 0.01$), suggesting that the low DO concentration became the limiting factor for NH_4^+-N removal. Nitrification of NH_4^+-N is slower than degradation of organics [34], and the conditions for nitrifying bacteria are more stringent than those for fungus that consume organic matter. The oxygen was first used to remove organic matter when the concentrations of organics were high; this was followed by nitrification until the organic concentrations dropped to a low level. The controls on the N removal rate in this wetland met Liebig's law of the minimum, therefore, the main way to improve the pollutant removal in constructed wetlands is to regulate and control the dissolved oxygen content [35].

The flow regime and N migration and transformation processes in constructed wetlands involve many complex physical, chemical, and biological processes. The N removal models are more complicated than those for other conventional wastewater treatments [36]. Also, the background concentrations of pollutants in the same type of wetland will vary [37]. In this study, we considered only three controls on the area-based constants and did not consider the influence of background pollutant concentrations. These other factors will be explored in future studies.

5. CONCLUSIONS

The main water source of this HSSF-CW in the Beijing Wildlife Rescue and Rehabilitation Center was system injection and the main N forms were NH_4^+-N and NO_3^--N. Over a period of two years, the effluent concentrations of both N forms were significantly lower than the influent concentrations ($p < 0.01$, $n = 38$). The NO_3^--N load was significantly correlated with the removal rate and the NO_3^--N removal rate varied seasonally. The inflow concentrations of NH_4^+-N and NO_3^--N were highest in spring and summer, respectively, while the NO_3^--N inflow concentration was lowest in spring. The area-based constant for NO_3^--N was correlated with the pollution load ($p < 0.01$), but was not correlated with either temperature or the DO concentration ($p > 0.01$). The NH_4^+-N area-based constant was correlated with the DO concentration ($p < 0.01$) but was not correlated with either temperature or the pollutant load ($p > 0.01$). The controls on the N removal rate in this wetland meet Liebig's law of the minimum.

ACKNOWLEDGMENTS

This study was supported by the Fundamental Research Funds for the Central Non-Profit Research Institution of CAF "N Removal mechanisms of subsurface constructed wetland in Health wetland" (CAFYBB2014QA029) and "The Lecture and Study Program for Outstanding Scholars from Home and Abroad" (CAFYBB2011007).

REFERENCES

1. Conley, D.J.; Paerl, H.W.; Howarth, R.W.; Boesch, D.F.; Seitzinger, S.P.; Karl, E.; Karl, E.; Lancelot, C.; Gene, E.; Gene, E. Controlling eutrophication: Nitrogen and phosphorus. *Science* 2009, *123*, 1014–1015.
2. Bilgin, M.; Şimşek, İ.; Tulun, Ş. Treatment of domestic wastewater using a lab-scale activated sludge/vertical flow subsurface constructed wetlands by using Cyperus alternifolius. *Ecol. Eng.* 2014, *70*, 362–365.
3. Xiong, J.; Guo, G.; Mahmood, Q.; Yue, M. Nitrogen removal from secondary effluent by using integrated constructed wetland system. *Ecol. Eng.* 2011, *37*, 659–662.
4. Liu, X.; Niu, H.; Yan, H.; Ding, Z.; Lu, F.; Ma, X.; Yang, L.; Liu, Y. Research and application of high-efficiency eco-engineering rural sewage treatment system. *Trans. Chin. Soc. Agric. Eng.* 2013, *29*, 184–191.
5. Vymazal, J. Removal of nutrients in various types of constructed wetlands. *Sci. Total Environ.* 2007, *380*, 48–65.
6. Kadlec, R.H.; Wallace, S. *Treatment Wetlands*; CRC Press: Boca Raton, FL, USA, 2008.
7. Birch, G.F.; Matthai, C.; Fazeli, M.S.; Suh, J.Y. Efficiency of a constructed wetland in removing contaminants from stormwater. *Wetlands* 2004, *24*, 459–466.
8. Bruch, I.; Fritsche, J.; Bänninger, D.; Alewell, U.; Sendelov, M.; Hürlimann, H.; Hasselbach, R.; Alewell, C. Improving the treatment efficiency of constructed wetlands with zeolite-containing filter sands. *Bioresour. Technol.* 2011, *102*, 937–941.
9. Lee, S.; Maniquiz, M.C.; Kim, L.-H. Characteristics of contaminants in water and sediment of a constructed wetland treating piggery wastewater effluent. *J. Environ. Sci.* 2010, *22*, 940–945.
10. Li, L.; Li, Y.; Biswas, D.K.; Nian, Y.; Jiang, G. Potential of constructed wetlands in treating the eutrophic water: Evidence from Taihu Lake of China. *Bioresour. Technol.* 2008, *99*, 1656–1663.
11. Harada, J.; Inoue, T.; Kato, K.; Izumoto, H.; Zhang, X.; Sakuragi, H.; Wu, D.; Ietsugu, H.; Sugawara, Y. Long-term nitrogen compound removal trends of a hybrid subsurface constructed wetland treating milking parlor wastewater throughout its 7 years of operation. *Water Sci. Technol.* 2016, *73*, 1018–1024.
12. Saeed, T.; Sun, G. A review on nitrogen and organics removal mechanisms in subsurface flow constructed wetlands: Dependency on environmental parameters, operating conditions and supporting media. *J. Environ. Manag.* 2012, *112*, 429–448.
13. Wynn, T.M.; Liehr, S.K. Development of a constructed subsurface-flow wetland simulation model. *Ecol. Eng.*2001, *16*, 519–536.
14. Kadlec, R.H. The inadequacy of first-order treatment wetland models. *Ecol. Eng.* 2000, *15*, 105–119.
15. Rousseau, D.P.L.; Vanrolleghem, P.A.; De Pauw, N. Model-based design of horizontal subsurface flow constructed treatment wetlands: A review. *Water Res.* 2004, *38*, 1484–1493.

16. Marsili-Libelli, S.; Checchi, N. Identification of dynamic models for horizontal subsurface constructed wetlands. *Ecol. Model.* 2005, *187*, 201–218.
17. Acreman, M.C.; Miller, F. Hydrological impact assessment of wetlands. In *The Global Importance of Groundwater in the 21st Century*; Ragone, S., de la Hera, A., Hernandez-Mora, N., Eds.; NGWA Press: Alicante, Spain, 2007; pp. 89–92.
18. Szporak-Wasilewska, S.; Piniewski, M.; Kubrak, J.; Okruszko, T. What we can learn from a wetland water balance? Narew National Park case study. *Ecohydrol. Hydrobiol.* 2015, *15*, 136–149.
19. Zhang, Y.; Lv, T.; Carvalho, P.N.; Zhang, L.; Arias, C.A.; Chen, Z.; Brix, H. Ibuprofen and iohexol removal in saturated constructed wetland mesocosms. *Ecol. Eng.* 2016, in press. Availableonline: http://www.sciencedirect.com/science/article/pii/S0925857416303123 (accessed on 4 November 2016).
20. Darwiche-Criado, N.; Comín, F.A.; Masip, A.; García, M.; Eismann, S.G.; Sorando, R. Effects of wetland restoration on nitrate removal in an irrigated agricultural area: The role of in-stream and off-stream wetlands. *Ecol. Eng.* 2016, in press. Available online: http://www.sciencedirect.com/science/article/pii/S0925857416301653 (accessed on 4 November 2016).
21. Rizzo, A.; Langergraber, G. Novel insights on the response of horizontal flow constructed wetlands to sudden changes of influent organic load: A modeling study. *Ecol. Eng.* 2016, *93*, 242–249.
22. McCabe, A.J.; Arnold, W.A. Seasonal and spatial variabilities in the water chemistry of prairie pothole wetlands influence the photoproduction of reactive intermediates. *Chemosphere* 2016, *155*, 640–647.
23. Rice, E.; Baird, R.; Eaton, A.; Clesceri, L. *Standard Methods for the Examination of Water and Wastewater*; American Public Health Association: Washington, DC, USA, 2012.
24. Environmental Protection Administration of China. *Environmental Quality Standards for Surface Water*; GB3838-2002; China Environmental Science Press: Beijing, China, 2002.
25. Kim, H.W. Evaluation of Hydrology in an Agricultural Watershed and Nitrogen Removal by Constructed Wetlands. Ph.D. Thesis, North Carolina State University, Raleigh, NC, USA, 2009.
26. Dzakpasu, M.; Hofmann, O.; Scholz, M.; Harrington, R.; Jordan, S.N.; McCarthy, V. Nitrogen removal in an integrated constructed wetland treating domestic wastewater. *J. Environ. Sci. Health A* 2011, *46*, 742–750.
27. Huang, J.; Cai, W.; Zhong, Q.; Wang, S. Influence of temperature on micro-environment, plant eco-physiology and nitrogen removal effect in subsurface flow constructed wetland. *Ecol. Eng.* 2013, *60*, 242–248.
28. Öövel, M.; Tooming, A.; Mauring, T.; Mander, Ü. Schoolhouse wastewater purification in a LWA-filled hybrid constructed wetland in Estonia. *Ecol. Eng.* 2007, *29*, 17–26.
29. Toscano, A.; Langergraber, G.; Consoli, S.; Cirelli, G.L. Modelling pollutant removal in a pilot-scale two-stage subsurface flow constructed wetlands. *Ecol. Eng.* 2009, *35*, 281–289.
30. Faulwetter, J.L.; Gagnon, V.; Sundberg, C.; Chazarenc, F.; Burr, M.D.; Brisson, J.; Camper, A.K.; Stein, O.R. Microbial processes influencing performance of treatment wetlands: A review. *Ecol. Eng.* 2009, *35*, 987–1004.
31. Zhang, D.; Wu, S.; Song, Y.; Liu, Q.-Q.; Pang, C.-L.; Dong, R.-J. Wastewater purification experiment study of a lab-scale tidal flow constructed wetland. *J. China Agric. Univ.* 2011, *16*, 110–116.
32. Wu, H.M.; Zhang, J.; Li, W.J.; Zhang, B. Relationship between oxygen release from plants in constructed wetland and oxygen demand for pollutant degradation. *Chin. J. Environ. Eng.* 2010, *4*, 1973–1977.
33. Zhang, C. Study on Calculation Method of Saturation Values of Dissolved Oxygen in Waters. *Res. Environ. Sci.*1999, *12*, 54–55.
34. Sun, G.; Austin, D. Completely autotrophic nitrogen-removal over nitrite in lab-scale constructed wetlands: Evidence from a mass balance study. *Chemosphere* 2007, *68*, 1120–1128.

35. Nivala, J.; Wallace, S.; Headley, T.; Kassa, K.; Brix, H.; van Afferden, M.; Müller, R. Oxygen transfer and consumption in subsurface flow treatment wetlands. *Ecol. Eng.* 2013, *61*, 544–554.
36. Marsili-Libelli, S. Modelling and automation of water and wastewater treatment processes. *Environ. Model. Softw.* 2010, *25*, 613–615.
37. Van de Moortel, A.M.; Rousseau, D.P.; Tack, F.M.; De Pauw, N. A comparative study of surface and subsurface flow constructed wetlands for treatment of combined sewer overflows: A greenhouse experiment. *Ecol. Eng.* 2009, *35*, 175–183.

CHAPTER 19

Large Constructed Wetlands for Phosphorus Control: A Review

Robert H. Kadlec

Wetland Management Services, 6995 Westbourne Drive, Chelsea, MI 48118, USA

ABSTRACT

This paper reviews aspects of the performance of large (>40 ha) constructed treatment wetlands intended for phosphorus control. Thirty-seven such wetlands have been built and have good data records, with a median size of 754 ha. All are successfully removing phosphorus from a variety of waters. Period of record median concentration reductions were 71%, load reductions 0.77 $gP \cdot m^{-2} \cdot year^{-1}$, and rate coefficients 12.5 $m \cdot year^{-1}$. Large wetlands have a narrower performance spectrum than the larger group of all sizes. Some systems display startup trends, ranging to several years, likely resulting from antecedent soil and vegetation conditions. There are internal longitudinal gradients in concentration, which vary with lateral position and flow conditions. Accretion in inlet zones may require attention. Concentrations are reduced to plateau values, in the range of about 10–50 $mgP \cdot m^{-3}$. Vegetation type has an effect upon performance measures, and its presence facilitates performance. Trends in the performance measures over the history of individual systems display only small changes, with both increases and decreases occurring. Such trends remove little of the variance in behavior. Seasonality is typically weak for steady flow systems, and most variability appears to be stochastic. Stormwater systems display differences between wet and dry season behavior, which appear to be flow-driven. Several models of system performance have been developed, both steady and dynamic.

Keywords: treatment wetlands; free water surface wetlands; phosphorus; performance; variability; models

1. INTRODUCTION

In past decades, wastewater treatment plants sometimes discharged to existing marshes. Phosphorus removal was not historically part of pretreatment, and hence considerable P loadings reached those marshes. In other situations, agricultural drainage made its way into neighboring wetlands. For instance, Brillion Marsh in Wisconsin received wastewaters from the town of the same name. A 156 ha natural wetland received incoming wastewater, beginning in 1923. No long-term studies were done, but an intensive study was conducted over the two-year period 1974–1975 [1,2]. The marsh was a uniform stand of *Typha latifolia* and *Typha angustifolia*, which received a channel input of treated wastewater. Outflow was also via a channel, and additionally there were small tributary streams to the marsh conveying agricultural runoff. During the two-year study, 16,300 $m^3 \cdot day^{-1}$ of influent carried 3.3 ± 2.2 $gP \cdot m^{-3}$ of phosphorus into the marsh. The hydraulic loading rate of one centimeter per day carried a load of 12.6 $gP \cdot m^{-2} \cdot year^{-1}$. The outgoing phosphorus was 2.6 $gP \cdot m^{-3}$, a concentration reduction of 22%, and 2.8 $gP \cdot m^{-2} \cdot year^{-1}$ was removed by the marsh. This marsh continued to remove phosphorus after fifty years of operation. Similar results were obtained at the 170-ha Cootes Paradise (*Glyceria grandis*) Marsh near Dundas Ontario, which was found to be removing phosphorus after 60 years of wastewater discharge [3].

Observations at natural wetlands, such as Brillion and Dundas, led to engineered attempts to use natural wetlands for phosphorus reduction. Projects sited in natural wetlands in Michigan and Florida, in the late 1970s, further explored the concept [4,5]. The construction of large wetlands for the purpose of phosphorus control commenced in 1976, with Boney Marsh in Florida [6,7]. Following in 1980 were the Delta Marsh cells in Manitoba [8].

More large constructed P-control wetlands were added in the 1980s, with the development of the Orlando Easterly Wetland (OEW) [9] and Orange County Eastern Service Area Wetland (OCESA) [10], both in 1987. The Balaton system in Hungary began in 1985 [11]. Numerous other small wetlands were constructed during the same time period, many with the expectation of a phosphorus removal benefit. Many other constructed P-control wetlands were added in the 1990s and later.

From these early studies, it was found that constructed treatment wetlands have a demonstrated ability to remove total phosphorus (TP) from various waters, but at relatively low areal efficiencies [12]. Concentration and flow measurements for surface waters entering and leaving the wetland provide the majority of data. A number of mechanisms contribute to the observed P reductions. Particulate forms may settle and become trapped in the litter and floc layers. Dissolved forms may be convected to those bottom regions. P partitions to the soil and sediment surfaces to form storage exchangeable with porewater phosphorus. Biological processes at several scales utilize and convert phosphorus, ranging from microorganisms and algae to macrophytes. New sediments and soils are formed as residuals from the biogeochemical pathways, a process termed bio-accretion. Such new solid accretions are a long-term sink for phosphorus in the wetland. This combination of these physical, chemical and biological mechanisms promotes P removal to new sediments, but generally requires continuously wet conditions for stability.

It is the purpose of this paper to examine and summarize the performance of large constructed free water surface wetlands for phosphorus removal. Herein, "large" is defined as more than 40 ha. The term "constructed" is more ambiguous, but is taken to mean that there are some number of structures and levees that define the wetland. Information from small and natural wetlands is incorporated to some small extent, when useful. The emphasis is on total phosphorus, and upon performance measures, rather than on mechanisms or processes.

2. PERFORMANCE MEASURES

There are several different metrics that may be used to evaluate performance. The main goal of the systems under consideration here is the reduction of TP concentrations. Simple measures include:

- Concentration reduction percentage. For large constructed P-control wetlands, it is always the case that, on average, lesser TP concentrations depart from the wetland than enter it. An increase in percent TP concentration reduction would be viewed as an improvement, while a decrease would be interpreted as degradation of performance. This measure does not reflect the flow rate the wetland receives, or the magnitude of the TP concentrations.

- Areal P load reduction. Depending upon many factors, differing P load removals may be achieved. Whatever the level that prevails, long-term trends in the mass of P removed per hectare of treatment wetland are a sign of improvement or degradation as the case may be. This measure reflects the inlet TP concentration and the inlet flow that the wetland receives, but does not distinguish between them.
- Steady-state removal rate coefficients. Simple first-order TP-removal models utilize annual average removal rate coefficients. These vary from system to system, and from year to year. These (partially) compensate for the inlet TP concentrations and hydraulic loadings.

2.1. Wetland Definition

Wetlands have been constructed with a number of cells, arranged in series and parallel. Performance measures may be applied to individual cells, or to combinations of cells, provided that appropriate data are collected. In this overview, an individual wetland is considered to comprise all cells in series in an independent flow path. Parallel flow paths that are structurally independent are considered to be separate wetlands, even though they may be adjacent to one another, and receive the same input water.

2.2. Defining Equations

The design of a wetland system may be for either concentration reduction or for mass removal, but these goals conflict. Interestingly, this distinction has not been well elucidated for P-removal wetlands, but has been described repeatedly for N-removal wetlands [12,13,14]. All large P-removal wetlands have been designed for concentration reduction. Wetland P-removal performance has been documented to follow a first-order rule, according to which P removal is greater at higher P concentrations. This model (Equation (1)) assumes a near-direct proportionality: doubling the concentration above background doubles the removal rate. For a wetland without significant water gains or losses, a simple extension is the PkC* model [15]. Large constructed P-control wetlands typically fit this description to a reasonable approximation [15].

$$\frac{C_o - C*}{C_i - C*} = \left(1 + \frac{k}{Nq}\right)^{-N} \qquad (1)$$

where C_i: inlet TP concentration, $gP \cdot m^{-3}$; C_o: outlet TP concentration, $gP \cdot m^{-3}$; C^*: background concentration, $gP \cdot m^{-3}$; h: water depth, m; k: apparent rate coefficient, $m \cdot year^{-1}$; q: inlet hydraulic loading, $m \cdot year^{-1}$ ($= cm \cdot day^{-1} \cdot 3.65$); and N: number of tanks in series (TIS). It is important to note that the first-order model is imperfect, and rate coefficients may display a dependence on hydraulic loading [16].

TP concentrations decrease as water passes through the treatment wetland, and removal rates also decrease. However, the actual mass of phosphorus that is removed increases with increasing hydraulic loading. Thus, increasing hydraulic loads results in more tons removed, but at the expense of a lesser concentration reduction. This trade-off between concentration reduction efficiency and load reduction may be quantified for the situation of no significant water gains or losses:

$$CR = \frac{C_i - C_o}{C_i} \tag{2}$$

$$PLR = q_i C_i - q_o C_o \tag{3}$$

where CR: fractional P concentration reduction; PLR: P load reduction, $gP \cdot m^{-2} \cdot year^{-1}$; q_i: inlet hydraulic loading, $m \cdot year^{-1}$; and q_o: outlet hydraulic loading, $m \cdot year^{-1}$.

The concentration reductions increase with decreasing hydraulic loading (increasing detention time), but not quite in an exponential manner, due to the PkC* model. The maximum load of phosphorus that can be removed results from a hydraulic load so high that little or no concentration reduction is achieved. It is, thus, apparent that there is no such thing as a "typical" or "average" load reduction, because the effects of hydraulic loading and inlet TP concentration create a full spectrum of potential load reductions.

3. CONSTRUCTED P-CONTROL WETLANDS INVENTORY

The size spectrum of constructed wetlands may be separated into three major groups: mesocosms, small, and large systems. Size matters, because of issues of scale-up. Microcosms and mesocosms are useful for investigating component processes, and, in general, are easily replicated to bolster statistical confidence. But, mesocosms have edge effects, may not have representative vegetation densities, and are not exposed to a realistic set of meteorological conditions. They may contain only a few of the components of a large system, such as soil without plants, litter, or animals. The plants, algae, and microbes in a mesocosm ecosystem are exposed to a time-changing water chemistry, which is different from the spatially distributed patterns of these along the flow direction [17]. Thus, there is a possibility that mesocosm studies produce replicated, well-controlled results of artifacts, rather than processes that actually occur in the field [18,19,20]. Therefore, the database summarized here excludes studies of systems smaller than pilot-scale, generally defined as systems of several tens of square meters extent, located in the out-of-doors.

3.1. Large Systems

There are 94 treatment wetlands of 40 ha or more that were found in the published literature or on the Internet, with a median size of 167 ha. Of these 94 wetlands, 66 are constructed wetlands, with a median size of 210 ha. The remaining 28 are natural wetlands that have been or are being used for water quality improvement, and 20 of these natural systems are forested.

Most of the large constructed wetlands target the reduction of phosphorus (P), and these number 43 of the 66 (Table 1). Some are in a startup condition, and some have only limited amounts of operating data. There are 37 wetlands with extensive phosphorus data, including two that were not specifically intended to remove phosphorus (Columbia and Lakeland). These 37 wetlands total 27,937 ha, with a median size of 754 ha.

Most P-removal wetlands are intended to reduce low P concentrations to even lower levels. The median inlet total phosphorus (TP) for the 37 P-target systems is 114 $mgP{\cdot}m^{-3}$; the median outlet TP is 38 $mgP{\cdot}m^{-3}$ (Table 2). This 71% concentration reduction (CR) is achieved at a low median hydraulic loading of 2.55 $cm{\cdot}day^{-1}$. The amount of phosphorus stored (phosphorus load removed, PLR) in these systems is low, with a median of 0.77 $gP{\cdot}m^{-2}{\cdot}year^{-1}$.

First-order rate coefficients (k) have a median of 12.5 $m{\cdot}year^{-1}$. This is somewhat higher than reported medians of 11.0 $m{\cdot}year^{-1}$ ([34], $N = 20$), and 10.0 $m{\cdot}year^{-1}$ ([15], $N = 282$).

The 37-wetland P-removal dataset under consideration contains information totaling 357 wetland-years, out of a total of 586 wetland-years of operation. The average operating age is 15 years. The median number of cells in a wetland is 2.0. All but three of the 37 are in the state of Florida, although 26 of the 66 total large constructed wetlands are outside Florida. Reasons for this Floridian concentration have been discussed elsewhere (e.g., [35,36]).

Table 1. Large constructed wetlands treating phosphorus. Columbia and Lakeland do not specifically target phosphorus.

Site	State/Country	System	Area	Cells	Start	Operating	Data	Ref.
			ha		Year	Years	Years	
Columbia	MO	Columbia	53	23	1994	23	14	[21]
Lakeland	FL	Lakeland	354	7	1987	30	12	[22]
Balaton	Hungary	Feneki Pond	1600	2	1992	25		[11]
Venice	Italy	Fusina	100	6	2011	6		[23]
Portage la Prairie	Manitoba	Delta	50	10	1980	37	10	[8]
Putrajaya	Malaysia	Putrajaya	197	23	1998	19	1	[24]
Seagoville	TX	East Fork	745	24	2009	8	1	[25]
Melbourne	FL	Viera	81	5	2002	15		[26]
Okeechobee Co.	FL	Nubbin Slough	313	2	2013	4	1	[27]
Martin Co.	FL	Lakeside Ranch	691	8	2013	4		[27]
Okeechobee Co.	FL	Taylor Creek	48	2	2010	7	3	[27]
Tarrant	TX	Shannon	171	4	2002	15	6	[28]
Balaton	Hungary	Hidvegi Pond	1800	2	1985	32	5	[11]
Orange Co.	FL	OCESA	56	4	1987	30	5	[10]
Orlando	FL	OEW	481	17	1987	30	13	[29]
Boney Marsh	FL	West	49	1	1976	41	11	[7]
Apopka	FL	B1	73.1	1	1990	27	10	[30]
Apopka	FL	B2	49.2	1	1990	27	10	[30]
Apopka	FL	C1	77.9	1	1990	27	10	[30]
Apopka	FL	C2	76.2	1	1990	27	10	[30]
Titusville	FL	Blue Heron	101	6	1997	20	7	[31]
Palm Beach Co.	FL	ENRP	1508	5	1994	5	5	[32]
Palm Beach Co.	FL	STA1E East	436	2	2005	12	2	[33]
Palm Beach Co.	FL	STA1E Central	782	3	2005	12	10	[33]
Palm Beach Co.	FL	STA1E West	796	3	2005	12	10	[33]
Palm Beach Co.	FL	STA1W East	875	3	2000	17	14	[33]
Palm Beach Co.	FL	STA1W West	552	3	2000	17	12	[33]
Palm Beach Co.	FL	STA1W North	1211	2	2000	17	16	[33]
Palm Beach Co.	FL	STA2 FW1	742	1	2000	17	14	[33]
Palm Beach Co.	FL	STA2 FW2	957	1	2000	17	15	[33]
Palm Beach Co.	FL	STA2 FW3	926	1	2000	17	15	[33]
Palm Beach Co.	FL	STA2 FW4	2415	3	2009	8	7	[33]
Palm Beach Co.	FL	STA2 FW5	1208	2	2013	4	3	[33]
Palm Beach Co.	FL	STA3/4 East	2611	2	2004	13	11	[33]
Palm Beach Co.	FL	STA3/4 Central	2157	2	2004	13	11	[33]
Palm Beach Co.	FL	STA3/4 West	1815	2	2004	13	11	[33]
Palm Beach Co.	FL	STA3/4 PSTA	40	1	2008	9	8	[33]
Hendry Co.	FL	STA5/6 FWY 1	975	2	1999	18	16	[33]
Hendry Co.	FL	STA5/6 FWY 2	834	2	1999	18	16	[33]
Hendry Co.	FL	STA5/6 FWY 3	775	2	2008	9	6	[33]
Hendry Co.	FL	STA5/6 FWY 4	754	2	2012	5	3	[33]
Hendry Co.	FL	STA5/6 FWY 5	1065	2	2012	5	3	[33]
Hendry Co.	FL	STA5/6 FWY 6	766	2	2013	4	6	[33]
Hendry Co.	FL	STA5/6 FWY 7	250	1	1997	20	13	[33]
Hendry Co.	FL	STA5/6 FWY 8	98	1	1997	20	14	[33]
Median				2.0		15	10	

Many of the Florida systems are designated as stormwater treatment areas (STAs). Twenty-eight constructed large stormwater wetlands were identified, and 26 of these are in Florida. The remaining 38 constructed large systems receive a relatively steady flow, usually via pumping. Stormwater systems receive water on an episodic basis, as determined by rainfall events. Rainfall runoff is pumped or flows by gravity into these wetlands. As a consequence, flows occur mostly in the wet season, and to a much lesser extent in the dry season. Seventy percent of Florida's rain occurs June–October.

Some systems receive highly treated municipal wastewater (e.g., OEW, OCESA), and others treat river water that has a high content of such treated effluent (e.g., Shannon, Blue Heron). Only Columbia and Lakeland receive high phosphorus municipal effluent, and their design goals do not include P removal.

Table 2. Performance of large constructed wetlands treating phosphorus.

System	HLR	TP In	TP Out	CR	PLR	k
	$cm \cdot day^{-1}$	$gP \cdot m^{-3}$	$gP \cdot m^{-3}$	%	$gP \cdot m^{-2} \cdot year^{-1}$	$m \cdot year^{-1}$
Columbia	13.30	2.550	2.410	0.05	6.67	2.76
Lakeland	1.02	5.080	3.030	0.40	7.63	1.99
Taylor Creek	9.00	0.321	0.241	25%	10.10	9.80
Shannon	5.45	0.888	0.539	39%	6.94	10.3
Hidvegi Pond	2.76	0.544	0.232	0.57	3.14	9.2
OCESA	0.97	0.210	0.040	0.81	0.40	6.9
OEW	5.61	0.280	0.106	0.62	3.56	21.7
Boney West	2.16	0.051	0.019	0.63	0.25	9.7
Apopka B1	9.88	0.110	0.076	0.31	1.40	15.6
Apopka B2	11.76	0.095	0.071	0.26	1.40	14.5
Apopka C1	10.20	0.109	0.088	0.19	1.16	8.8
Apopka C2	9.34	0.102	0.095	0.07	0.77	2.5
Blue Heron	0.79	0.288	0.049	0.83	0.69	6.0
ENRP	3.11	0.108	0.022	0.80	0.97	23.6
STA1E East	2.52	0.091	0.013	0.86	0.63	27.50
STA1E Central	2.55	0.114	0.030	0.81	0.81	21.03
STA1E West	1.90	0.173	0.086	0.64	0.58	7.92
STA1W East	2.65	0.164	0.041	0.81	1.07	16.01
STA1W West	3.61	0.156	0.037	0.77	1.08	23.73
STA1W North	3.16	0.186	0.060	0.84	1.48	23.66
STA2 FW1	3.15	0.093	0.012	0.89	0.94	61.90
STA2 FW2	4.32	0.107	0.027	0.76	1.12	36.66
STA2 FW3	4.06	0.095	0.018	0.82	1.13	33.31
STA2 FW4	1.47	0.093	0.022	0.71	0.43	14.71
STA2 FW5	0.51	0.045	0.020	0.53	0.05	2.98
STA3/4 East	2.42	0.085	0.016	0.81	0.51	20.87
STA3/4 Central	2.22	0.069	0.019	0.71	0.33	14.84
STA3/4 West	3.07	0.066	0.016	0.77	0.50	29.62
STA3/4 PSTA	4.63	0.019	0.011	0.40	0.12	16.05
STA5/6 FWY 1	2.55	0.153	0.070	0.53	0.55	8.22
STA5/6 FWY 2	2.25	0.196	0.079	0.66	0.73	12.15
STA5/6 FWY 3	0.57	0.278	0.038	0.85	0.48	6.30
STA5/6 FWY 4	0.60	0.257	0.033	0.87	0.49	6.30
STA5/6 FWY 5	0.36	0.190	0.032	0.83	0.23	3.09
STA5/6 FWY 6	2.20	0.177	0.052	0.70	0.78	12.46
STA5/6 FWY 7	1.94	0.107	0.033	0.72	0.36	14.98
STA5/6 FWY 8	2.26	0.082	0.035	0.55	0.30	8.45
Median	2.55	0.114	0.038	0.71	0.77	12.5

3.2. Small Systems

A total of 87 small constructed marsh systems were identified as having reasonable data quality and quantity. This means that subsurface flow systems and natural wetland systems have been excluded. The 87 have a median size of 0.25 ha, but range up to 33 ha. Because most of these are pilot systems they have relatively short operating lives and short periods of data record.

4. PERIOD OF RECORD RESULTS

The performance of large P removal wetlands may be compared to a broader population of small wetlands. Phosphorus removal data for large

(>40 ha) wetlands is restricted to a smaller number, 37 of the total of 45 listed in Table 1. However, those 37 have an average period of record (POR) of about fourteen years, and data record periods with an average of ten years. The median size of the 37 large data wetlands is 754 ha. In the analyses that follow in this section, annual averages for the entire data record period are averaged, and performance measures computed (Table 2).

4.1. Concentration Reduction

Of the 282 total wetlands reported by Kadlec and Wallace (2009) [15], 274 (97%) lower the phosphorus concentration. The median reduction is 41%. Of the 37 large wetlands, all lower the P concentration, with a median of 71%. The distribution of concentration reductions is rather narrow for large wetlands, spanning 26%–85% for the 10th–90th percentiles (Figure 1). The better performance of large wetlands is at least partly due to their lower hydraulic loading rates, 2.6 cm·day^{-1} for large *versus* 3.4 cm·day^{-1} for the set of small wetlands.

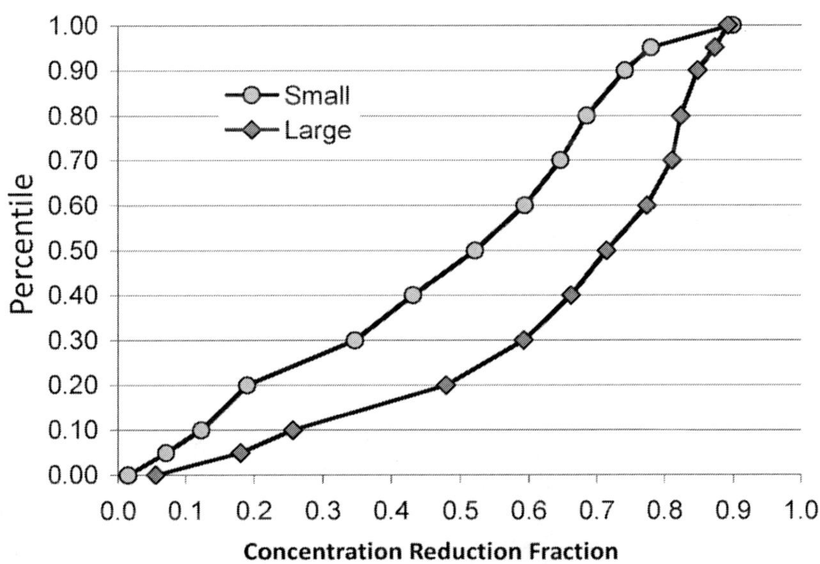

Figure 1. Distributions of P concentration reductions for large (N = 43) and small (N = 87) constructed wetlands.

4.2. Load Reduction

There is a large difference between the P load reductions achieved by large wetlands and that of small wetlands: median 0.75 gP·m^{-2}·year^{-1} for the former, and 2.96 gP·m^{-2}·year^{-1} for the latter (Figure 2). This may be partially due to the young age of small systems, during which soils have not been P-saturated and vegetation is still establishing. The distribution of load reductions is rather narrow for large wetlands, spanning 0.24–1.32 gP·m^{-2}·year^{-1} for the 10th–80th percentiles (Figure 2). The small dataset has a median incoming P loading of 10.4 gP·m^{-2}·year^{-1}, whereas the large systems have a median incoming P loading of 1.22 gP·m^{-2}·year^{-1}. In turn, this is partly due to the global median inlet concentration of 0.870 gP·m^{-3}, *versus* 0.114 gP·m^{-3} for the large systems.

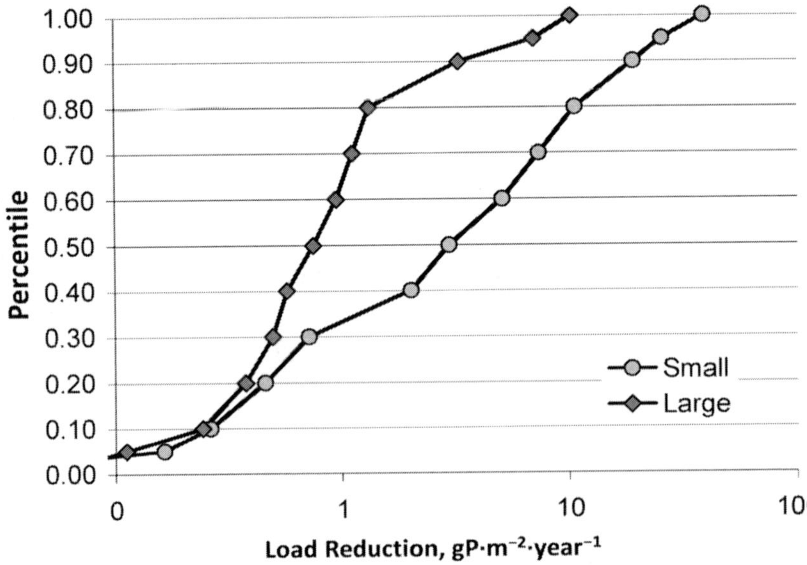

Figure 2. Distributions of P load reductions for large (N = 43) and small (N = 87) constructed wetlands.

4.3. Rate Coefficients

There is not much difference between the P median removal rate coefficients for large wetlands (14.6 m·year^{-1}) and small (12.3 m·year^{-1}) (Figure 3). The distribution of rate coefficients is somewhat narrower for large wetlands, spanning 6.1–28.8 m·year^{-1} for the 10th–90th percentiles (Figure 3). The narrow distribution may be in part due to the narrower range of inlet concentrations and hydraulic loadings that characterize large wetlands.

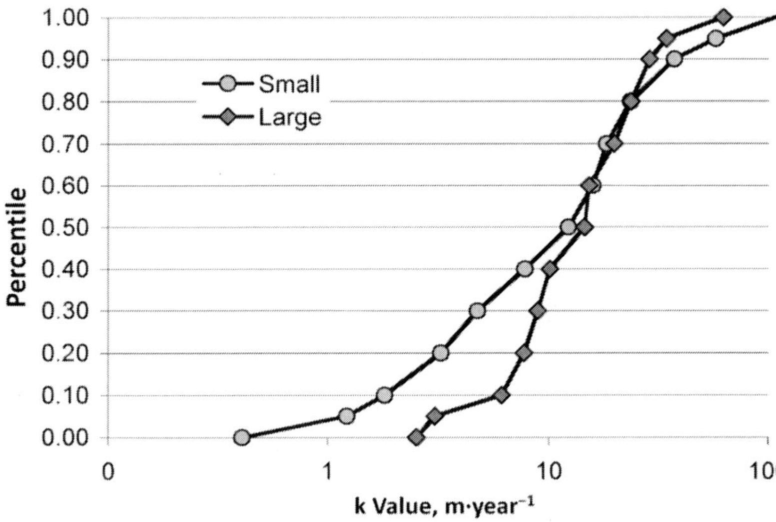

Figure 3. Distributions of P removal rate coefficients for large (N = 43) and small (N = 87) constructed wetlands. A plateau concentration of 6 mgP·m^{-3} has been used. Large wetlands are assumed to have a distribution of detention times represented by eight tanks in series, and for the small dataset three tanks in series.

5. WATER MOVEMENT

5.1. Flows

5.1.1. Timing

There is a notable distinction between steady flow and event-flow (stormwater) wetlands. In the latter case, large flows occur for only short periods of time. A two-year example of event-driven inflows for STA1W shows that about two-thirds of the time there was no flow into the system (Figure 4). Therefore, the P contents of the incoming and outgoing water were measured using flow-proportional samplers, which provide a flow-weighted mean concentration.

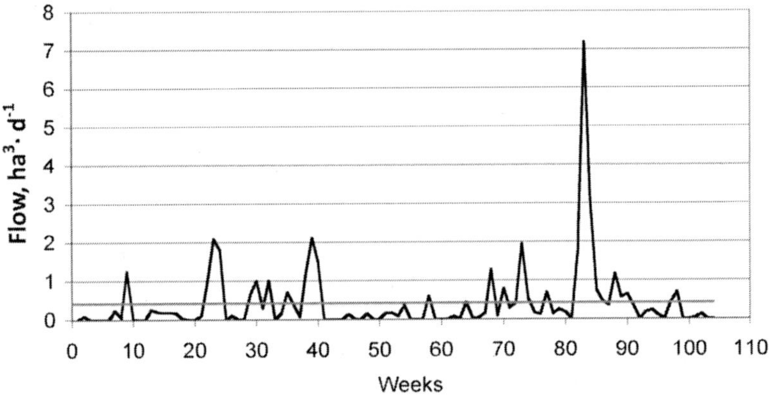

Figure 4. Weekly average flow entering STA1West over a two year POR (period of record) (2011–2013). The mean daily flow was 0.42 ha^3·day^{-1}. Flow occurred on 30% of the days, at a mean rate of 1.38 ha^3·day^{-1}.

In contrast, the steady flow systems can be grab sampled for P content. For instance, the Apopka Wetland was sampled weekly [37]. Comparable procedures were followed at other systems.

5.1.2. Distribution

In general, water is brought into the STAs via gated structures and spreader canals. This arrangement is believed to help avoid short circuits at the system inlet. Collection is also into a canal at the downstream end of the flow path, and thence to a control structure or pump. Most of the other P-control wetlands also attempt to provide uniform distribution across the flow direction at the inlet, such as Apopka, OCESA and Shannon (Table 1). However, some large P-control systems, such as OEW, utilize point distribution of water.

5.2. Depths

Constructed wetland water depths and flow rates are controlled by either or both the outlet structure and resistance to flow within the wetland. In small wetlands, depth control is generally via the settings of the outflow structure. In contrast, the depths in large wetlands are determined by vegetation density, topography, and flow rate, not by an outflow structure setting [15,38]. However, at low flow, or no flow, an outlet structure can control the depth of the residual water in a stormwater wetland.

Examples of depth profiles through the course of the year are shown in Figure 5. A steady flow system, the Everglades Nutrient Removal Project (ENRP), also operated at relatively steady depths. In contrast, STA56 flow way 7 received no water in the dry season, resulting in complete dry-out to negative depths (below ground). Such complete dry-out has performance implications, because stored organic sediments can oxidize under dry conditions, releasing stored phosphorus [39]. Emergent wetland plants can, in general, withstand dry conditions, but submerged aquatic vegetation (SAV) cannot. Therefore, SAV wetlands, like STA2 C3, are kept hydrated to preserve the desired vegetative cover (Figure 5). Supplemental water is required for this purpose.

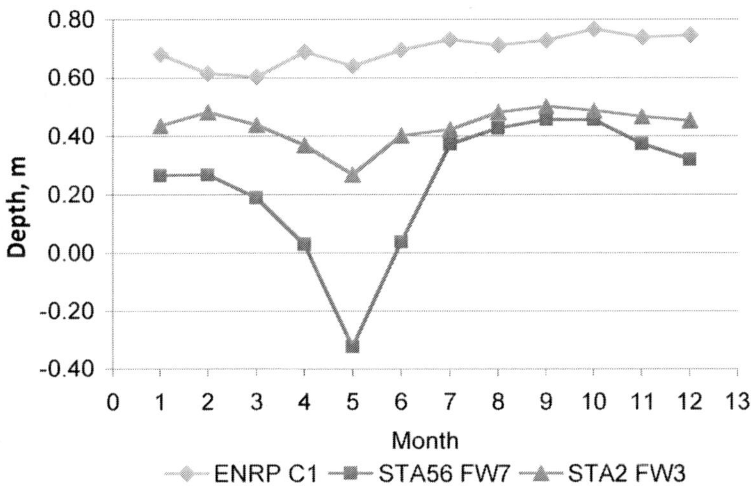

Figure 5. Average monthly water depths for three large P-control wetlands. ENRP C1 (*N* = 5 years) contained emergent vegetation and was operated at nearly constant flow and depth. STA2 C1 (*N* = 8 years) fosters submerged aquatic vegetation and, therefore, was kept hydrated in the dry season. STA56 FW7 (*N* = 8 years) contains emergent vegetation, but was not kept hydrated in the dry season.

5.3. Internal Patterns

The patterns of water movement through the wetland have a significant influence on the efficiency and extent of phosphorus movement. Many of the important biogeochemical reactions rely on contact time between water being treated and biological components and their associated substrate. Any short-circuiting or dead zones will, therefore, impact treatment efficiency. Non-ideal flow patterns in wetland treatment systems have received considerable attention in the literature [15].

Three types of hydraulic inefficiencies may occur in treatment wetlands: one caused by topographical features, a second due to preferential flow channels caused by vegetative non-uniformity, and a third caused by mixing effects, in both the vertical and transverse directions. Large wetlands are particularly susceptible to these effects. Large size makes land leveling prohibitively expensive, and existing topography may be quite non-uniform. Antecedent land uses may mean that remnant roads and ditches may be included within the footprint of a large system. Vegetation establishment in large systems is most often by natural colonization and growth, because of the expense of planting a large area. Such natural regrowth is not uniform, and reflects factors such as water depth, which in turn depends upon topography. Wind fetches in large systems can be quite large. If there exists open water or SAV, then surface water is driven in the wind direction, with opposing return flows created at other depths.

As a consequence, tracer testing has been utilized to assess non-ideal flow effects in large wetlands [40,41,42,43]. Results are extremely variable, with the tanks-in-series (TIS) characterization ranging from 1 to 9. Poor internal hydraulics corresponded to preferential flow routes, while better performance was found for multiple cells in the flow path. The large wetland TIS determinations bracket the central tendency for smaller wetlands, reported as TIS = 3–4 [15].

5.4. Seepage

It is often not technically feasible to provide impervious liners for large wetlands, although lining was done at the Columbia MO Wetland to protect regional potable water supply. Depending upon the transmissivity of the substrates, vertical flows may occur, either as leakage from or into the wetland. If outward leakage constitutes a potential threat to adjacent properties, then the wetland may be rimmed with a seepage collection canal, which collects seepage and redirects it into the wetland. Seepage collection has been done at several STAs.

6. PHOSPHORUS SPECIATION

The regulatory measure for treatment wetlands is almost invariably total phosphorus. Total phosphorus in water entering a treatment wetland is comprised of several groups of constituents. The analytical categories of particulate (PP), and dissolved organic (DOP) and soluble reactive (SRP) forms comprise TP. These constituents are also mixtures. Particulate P exists in a spectrum of particle sizes and types. Dissolved organic P is comprised of a suite of organic molecules of varying types and sizes. SRP contains mono- and di-basic phosphates, polyphosphates, and other compounds.

The performances of P-removal constructed wetlands can be quite different, depending upon which forms of phosphorus are sent into them. Further, there are interconversions that occur among the various P species. For instance, SRP may be consumed and converted to PP by algae, and PP may decompose, releasing SRP and DOP.

Wetlands receiving flows dominated by PP have been found to remove P by settling and filtration, which occur at rates that are considerably higher than those for total phosphorus ([30,44,45]). For example, the median annual TP removal rate coefficient for the Apopka, FL systems was 15 m·year^{-1}, while the apparent rate coefficient for PP was 32 m·year^{-1}.

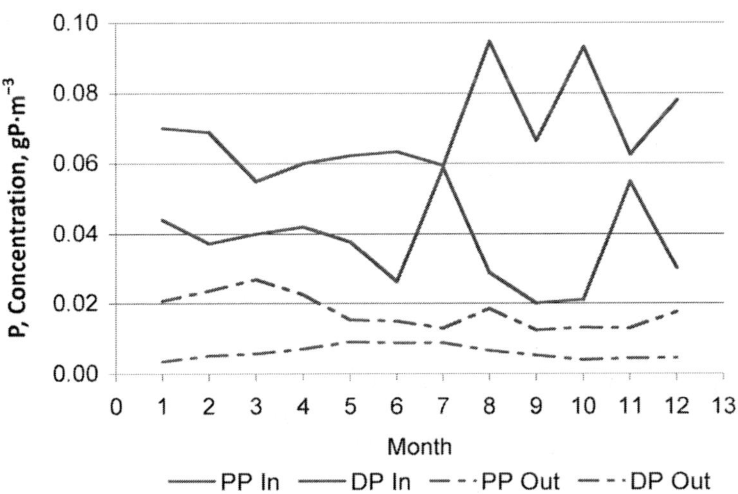

Figure 6. Monthly performance of the ENRP over a four-year POR. The average hydraulic loading was 2.4 cm/day. PP (particulate) and Dissolved phosphorus (DP) are reduced in all months. Overall TP reduction is slightly less in the first six months of the year (73%) and slightly greater in the second six months (82%).

The removed PP is not necessarily inactive, but may release some of the particulate phosphorus, which may appear as soluble phosphorus. If the wetland has a low hydraulic loading, then released soluble P can be taken up by wetland processes. However, if the system receives a high hydraulic loading, then the mechanisms that utilize soluble P may not be able to utilize all of the released P. An example of a low-loaded system is the ENRP, which demonstrated reductions of all forms of P without any large seasonal variation (Figure 6). The Apopka project was loaded at a much higher rate, but at similar concentration levels, and behaved much differently (Figure 7). PP decreased in all months, but Dissolved phosphorus (DP) increased markedly in all months. As a result, TP was reduced in all months, but at much higher percentages in the first six months of the year.

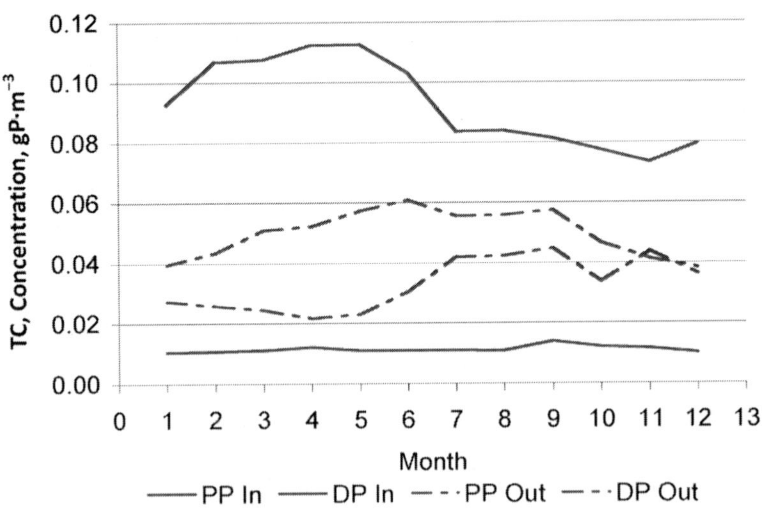

Figure 7. Monthly performance of the Apopka Wetlands over a ten-year POR. The average hydraulic loading was 10.3 cm/day. PP is reduced in all months, but dissolved P is increased. Overall TP reduction is greater in the first six months of the year (34%, and much less in the second six months (2%).

Soluble reactive phosphorus pulses are rapidly consumed in wetlands [46,47]. Juston and DeBusk (2011) [48] measured P species on multiple-point transects over multiple years in STA2 Cell 3. The internal P speciation profile showed that SRP was reduced to the analytical detection limit more quickly than DOP or PP, and that SRP was reduced to lower levels. As a result, SRP exiting the STAs is frequently at the minimum detection level (Table 3).

Table 3. Phosphorus speciation in water entering and leaving the stormwater treatment areas (STAs) during a POR ending in 2007 ($gP \cdot m^{-3}$).

System	Location	TP	SRP	DOP	PP	DP
STA1E	In	0.143	0.059	0.013	0.070	0.073
STA1W	In	0.158	0.085	0.009	0.064	0.094
STA2	In	0.097	0.068	0.006	0.022	0.075
STA34	In	0.088	0.043	0.008	0.036	0.052
STA56	In	0.059	0.011	0.007	0.041	0.019
Median		**0.097**	**0.059**	**0.008**	**0.041**	**0.073**
STA1E	Out	0.023	0.005	0.010	0.007	0.015
STA1W	Out	0.074	0.043	0.008	0.023	0.051
STA2	Out	0.021	0.011	0.004	0.006	0.015
STA34	Out	0.019	0.004	0.005	0.010	0.009
STA56	Out	0.017	0.006	0.004	0.006	0.011
Median		**0.021**	**0.006**	**0.005**	**0.007**	**0.015**

7. VEGETATION

Surprisingly, it is not yet clear whether vegetation type is an important determinant of the performance of large P treatment wetlands. This is in large part because of the difficulty of maintaining any particular species, or any particular vegetative community type, on a large scale. There are no system-wide macrophyte monocultures present in any of the large P treatment wetlands. Intersystem comparisons based on vegetation are compromised by differences in driving variables of flow amounts and timing, inlet concentrations, and climatic variations.

The broad categories of vegetation include emergent aquatic vegetation (EAV), submerged aquatic vegetation (SAV) and algal mats (periphyton). However, floating aquatic vegetation (FAV) has been found to volunteer in patches in some systems, including *Eichhornia*, *Pistia*, and *Ludwidgia*. EAV, typically, is dominated by *Typha*. SAV is often comprised of *Chara*, *Ceratopyllum*, *Najas*, *and Hydrilla*.

The vegetation patterns in these large systems are typically quite complex and change with time during operational history. Some changes are spontaneous; others are engendered by management activities. The reader interested in the evolutionary details of species present, and their spatial distribution and abundance should consult the parent literature on the individual systems.

In no case has vegetation been harvested in the large systems, except as it occurs as a management strategy, such the intentional replacement of EAV with SAV.

7.1. Florida Stormwater Systems

Early analyses of the Florida STA projects, including size scales ranging from mesocosms to large field-scale systems, indicated that SAV could provide treatment advantages [49].

Accordingly, the flow paths of the STAs were built or retrofitted to contain significant areal percentages of SAV. Performance data now exist for these constructed wetlands, as well as surveys of vegetation percentages for PORs of several years [50]. A bivariate analysis of that data is shown in Figure 8, which indicates no correlation between the rate coefficient and the fraction of EAV (R^2 = 0.0005, N = 125 wetland-years).

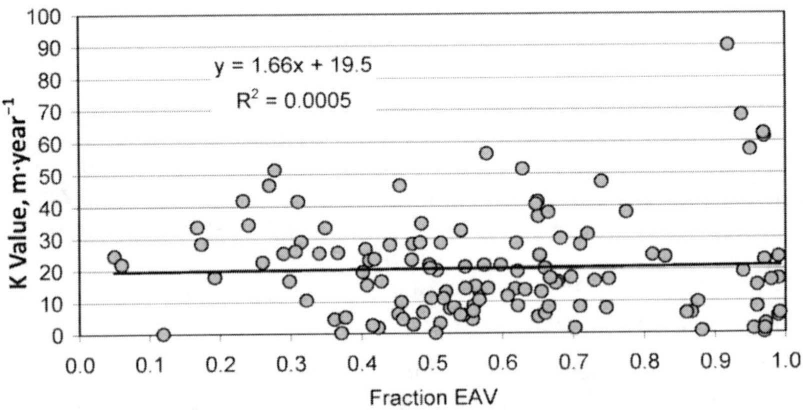

Figure 8. The dependence of the rate coefficient on the fraction EAV in SAV–EAV (submerged aquatic vegetation–emergent aquatic vegetation) stormwater wetlands.

7.2. Periphyton Systems

There are two strategies that have been explored to facilitate the growth of periphyton in wetlands that contain only very sparse macrophytes (Periphyton STA = PSTA). One technique is to add a layer of calcareous substrate on top of antecedent peat soils; the second is to remove organic soils down to below layers of calcareous bedrock. There is at present one large treatment wetland (STA34 PSTA), which has been in operation for eight years, implemented via method two (scrapedown). The hypothesis has long been advanced that such systems should be capable of achieving very low outflow P concentrations. Low P has been demonstrated at the mesocosm level [51] and at field scale for very low incoming P concentrations [52]. However, other mesocosm, pilot, and field-scale work was not particularly successful [53].

The 40 ha STA34 PSTA project was established by scraping a thin layer of organic soil from the underlying bedrock. It has achieved an eight-year average outlet P concentration reduction from 19 to 11 $mgP{\cdot}m^{-3}$, at a hydraulic loading rate of 4.6 $cm{\cdot}day^{-1}$ (Table 2). This corresponds to a low P load reduction, and to rate coefficient (16 $m{\cdot}year^{-1}$), about equal to the average for other large-scale wetlands.

7.3. Vegetative Transitions

Constructed wetlands usually begin existence with an antecedent cover crop that is not compatible with the hydrologic regime imposed by the treatment function. Over time, a mix of vegetation will develop, primarily determined by hydropattern and water quality conditions.

Most large P treatment wetlands are not monotypic communities, but rather contain a patchwork of open water, SAV, EAV, and FAV. This may in part be due to the system topography, or to intentionally utilizing successive cells with different bathymetries.

Vegetation contributes to treatment, and so the amount contributes to treatment efficacy. Phosphorus removal has been strongly linked to the fractional coverage of different community types ([15,54]). Deeper systems, without macrophytes, display lower removal rate coefficients, in the range of 3–9 $m{\cdot}year^{-1}$ [55].

Vegetative transitions can be slow to occur, either due to processes of growth or to species competition. The dynamics of transitions have been documented in several of the South Florida Environmental Reports [50]. The example in Figure 9 shows that several years may be needed to achieve vegetative grow-in.

In the South Florida STAs, vegetation management is an ongoing process, designed to maintain community health and coverage and to avoid species replacements regarded as undesirable. In other systems species management has typically not been practiced.

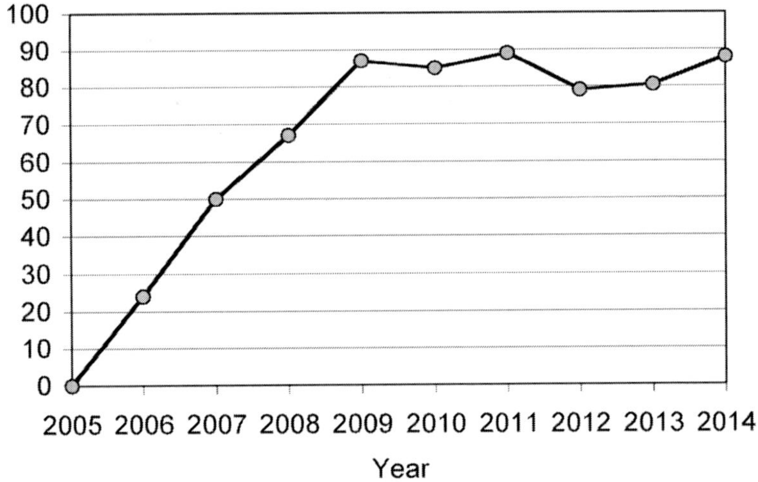

Figure 9. The development of cattail cover in STA1E Cell 5 (231 ha). Flow-through began in October 2005. The site was a citrus grove prior to wetland construction.

7.4. Management Issues

There are conditions in which vegetation is perceived to require management. For large wetlands, vegetation management can be costly and can interfere with the water quality goals of the system. Some of the problems that have occurred include the following:

- Floating cattail islands. It is not uncommon that *Typha* forms floating mats [15]. In fact, a contiguous mat with 100% cover is conducive to reasonable P removal [56]. The difficulty arises if the mat is not complete, but consists of "tussocks", with sizes of a few meters to several hectares. These floating islands can move with the wind and can wreak havoc with bottom sediments and water control structures. Removal was undertaken in STA1W Cell 2.

- Dense surface covers of rooted or floating plants. Some macrophytes have growth habits that can include the formation of thick foliage at the water surface, and to a small extent, above water. *Hydrilla verticillata* and *Panicum repens* (torpedo grass) are examples that have influenced operations in large-scale P-removal wetlands. Dense growths smother below-water activity. Several STA cells have experienced this phenomenon and have been treated with herbicides as a cure.

- Deep water damage. A number of wetland plants are susceptible to damage by episodic deep water and/or high water velocities. Brief periods of high inundation can be tolerated, as can periods of dryout of varying intensities. However, prolonged deep water can lead to plant uprooting or fatal oxygen deprivation. An extreme example was the loss of vegetation in STA1E Cell 6, in which vegetation export clogged the outflow and stopped operation [57]. Large stormwater wetlands are particularly susceptible to deep water in their inlet zones, because the hydraulic gradients in large systems are controlled by the vegetation density and not by structure settings. The pulse flows that reach a stormwater wetland can and do occur at rates that are many times the annual average flow and are limited only by the capacity of the inlet pumps (if any) and by flow management strategies.
- Wind damage. The fetches of large wetlands can reach many kilometers, creating conditions for large waves and wind-driven water pileup. In the southeastern USA, primarily in Florida, hurricanes and tropical storms often bring very high winds. Some vegetative communities can withstand these events, including firmly-rooted, sturdy emergents. However, softer less well-rooted plants, including SAV of various kinds, cannot. Instances of severe SAV wind damage have occurred, such as in STA2 Cell 3 due to the 2005 hurricane Wilma [57].

8. SOILS

8.1. Floc

Investigators have often observed a flocculent detrital layer, or "floc", overlaying the soil surface in both natural and constructed wetlands, either unimpacted or impacted by excess phosphorus [58,59]. The bulk density of floc is low, *ca.* 0.05 g/cm^3, and, hence, the thicknesses of the floc layers ranges between a few centimeters up to more than 20 cm.

Large amounts of floc have developed in the STAs. For example, the mean floc depth in 13 STA cells in 2003 was 10.9 ± 5.5 cm, weighing 8.1 ± 5.5 kgDW·m^{-2}. The P content was 6.4 ± 3.8 gP·m^{-2}. These accumulations are very much larger than the sum of all TSS inputs, which were approximately 0.1 kgDW·m^{-2}·year^{-1}. This leads to the conclusion that the STAs, and other P-treatment wetlands, manufacture particulate matter and floc.

Floc materials can quickly respond to nutrient enrichment [60], and thus have the most immediate effect on phosphorus in the water column. Floc is suspendable, via shear forces or bioturbation [61]. However, suspended solids concentrations in the flowing water in the STAs is typically very low, *ca.* 3 mg/L. Therefore, some researchers have hypothesized that particulate P, in the form of floc, is transported downstream via a spiraling mechanism [60].

8.2. Accretion

The long-term storage of phosphorus is dominated by the formation and accretion of the new soils via the processes of sedimentation and bio-accretion. Wetlands with large incoming sediment loads can contribute to solids buildup in the inlet region via the process of sedimentation. Bio-accretion, created by internal generation processes, is likely to produce one or two centimeters per year of new solids [15] and over a period of many years this may influence the wetland hydraulics. This range of rates will not, in general, threaten vegetation, because root and rhizome development can keep pace on the timescale of years.

Early research on impacted natural wetlands in Florida utilized deposited radioisotope tracers to establish accretion rates [62,63]. A newer technique utilizes change points in soil chemistry along a vertical gradient downward from the soil surface [64]. Applied to the Florida stormwater treatment wetlands, both the vertical soil accretion and the P accretion were determined for the cells in eight of the STA wetlands (Table 4). The P accumulated should correspond to the P removed from the water. The soils data show accumulations about double the removed P, but the variability in both measures is quite high.

Accretion may require maintenance activity after a decade or two of operation. The accretions are likely to be spatially non-uniform, with gradient effects in the flow direction, and patchy development. Sedimentation typically creates a delta effect, and bio-accretion is greater in regions of greater water nutrient content. In turn, this can cause preferential flow path development, due to local channelization of new soils, particularly in the inlet zone. These hydraulic changes may occur without compromising the freeboard of the containment levees. Nevertheless, it is possible that sediment removal will be necessary to restore optimal hydraulics. The only report to date of this maintenance activity is from the Orlando Easterly Wetland, where about 20 cm of sediment was removed from a 59 ha inlet zone of the 481 ha system [29]. In that case, the accumulation was due entirely to bio-accretion, because the incoming water contained no suspended sediment.

Table 4. Phosphorus accumulation in soils and removed from water in STAs. Accretion and P accumulation were measured at a small number of locations in each wetland for soil cores taken in 2009 [64]. The P removed is calculated from annual performance data for the water entering and leaving the wetlands, which typically spanned the majority of the operating history.

System	Wetland	Accretion		P Accumulation		P Removed	
		$cm \cdot year^{-1}$		$gP \cdot m^{-2} \cdot year^{-1}$		$gP \cdot m^{-2} \cdot year^{-1}$	
		mean	stdev	mean	stdev	mean	stdev
STA1W	East	0.97	0.31	1.20	0.61	1.21	0.65
STA1W	West	0.90	0.22	1.47	0.36	1.64	1.01
STA1W	North	1.10	0.54	2.05	1.24	1.15	0.97
STA2	FW1	1.00	0.26	1.00	0.13	0.99	0.36
STA2	FW2	1.00	0.26	1.60	0.41	1.34	0.52
STA2	FW3	1.20	0.40	2.50	1.00	1.14	0.49
STA3/4	East	1.70	0.66	2.55	0.94	0.90	0.44
STA3/4	Central	1.75	0.78	4.65	2.85	0.69	0.58
mean		**1.20**		**2.13**		**1.13**	
stdev		**0.34**		**1.17**		**0.29**	

9. VARIABILITY

9.1. Temporal Variability

9.1.1. System Startup

The basins for a constructed treatment wetlands can start from differing initial conditions soil type, and vegetation type and density. When these new wetlands are exposed to a long hydroperiod, it is obvious that some period of adaptation will ensue, during which soils and vegetation undergo changes. It is expected that the terrestrial basins constructed to contain treatment wetlands do not have the full suite of ecosystem processes that are capable of P removal. Sustainable P removal requires and utilizes the full biogeochemical cycle, with all of the size scales, ranging from macrophytes downward through the microflora and microfauna.

Water quality regulatory requirements do not always allow for startup periods. As an example, a large (41 ha) natural wetland near Orlando released phosphorus for several initial months in 1988–1989, engendering enforcement actions and system termination [34]. Current Floridian stormwater wetland operating permits recognize the possibility that constructed wetlands require a period of development. Two initial periods are considered: (1) attainment of concentration reduction, termed net benefit; and (2) attainment of a relatively stable operating mode, termed stabilization.

Most large-scale P treatment wetlands have not had long startup transients, in part due to the strategy of holding the wetland in a no-flow condition until sorption and grow-in processes have stabilized. The STAs were required to achieve concentration reduction before commencing flow-through.

Antecedent Phosphorus

Wetland P-removal performance may be initially influenced by transition from the past history of the soils and vegetation on the site. The large P-control wetlands are often built on agricultural soils, and may have large antecedent stores of soil P. In turn, that may temporarily result in high concentrations in floodwaters. As an illustration, the adaptation period for the northern flow path of STA1W in Florida lasted about one and a half years. During that time, by regulation, it was held in a no-discharge mode to accommodate the transition (Figure 10) [15]. This wetland clearly contained mobile legacy phosphorus, because it was built on a former agricultural field.

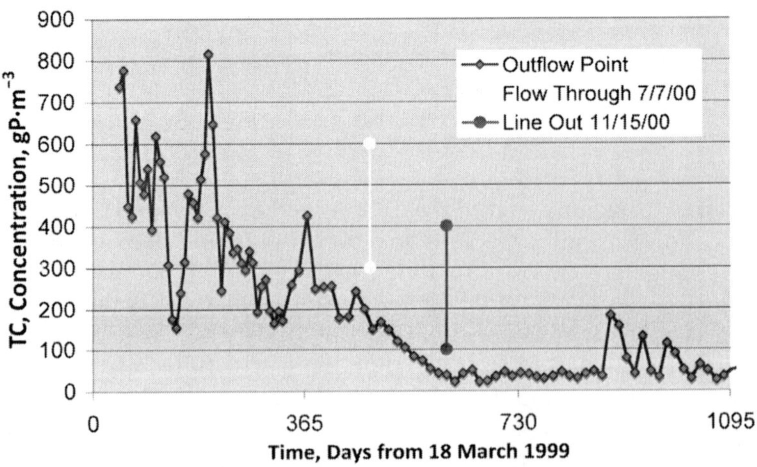

Figure 10. Startup of STA1W Cell 5, beginning18 March 1999.

In contrast, prior land uses may have involved low levels of phosphorus. When P-rich floodwater is added, inorganic forms of phosphorus partition to soils and sediments. The process is well understood [65]. This initial sorptive capacity for phosphorus will persist only until the new P-loading establishes a new sorbed P equilibrium [56]. As a case in counterpoint, the Shannon Wetland in Texas [28] displayed no startup pattern, neither an enhancement nor decrement (Figure 11). This 98 ha wetland showed no time trend in rate coefficients over the first five years. Vegetation alterations took place during this period, including sparse plantings.

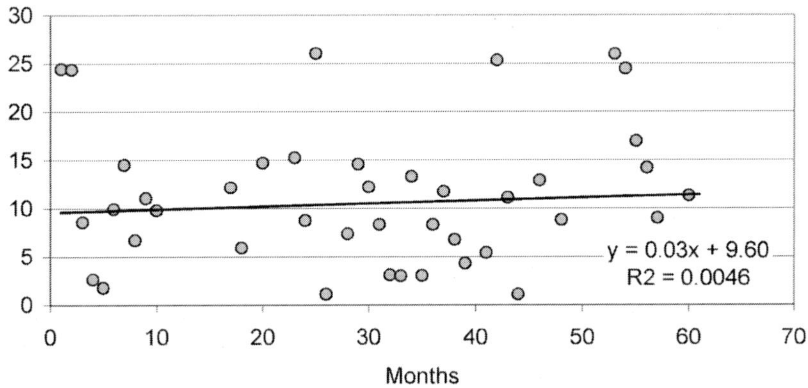

Figure 11. Startup rate coefficients for the Shannon Wetland. Operations began in June 2003.

Thus, the initial period of performance may show P release, or lesser or greater P removal than in the long-term. Some systems have started close to that long-term performance. Antecedent soils and their P loading is an important factor in the initial startup performance of the soils. Short-term uptake or release should be anticipated in the operational strategy, and discounted in forecasts of long-term P removal.

Vegetation Grow-In and Rearrangement

Large wetlands are more expensive to plant with candidate vegetation than smaller systems. The large quantities of propagules needed are not necessarily available. Nonetheless, a few large P wetlands have been at least partially planted, e.g., Orlando Easterly and Shannon.

The growth of the full complement of wetland vegetation requires a supply of nutrients that ultimately resides in the stable, fully developed system biomass, both living and dead. Although this new storage of phosphorus can vary depending on vegetation type and density, it is not trivial compared to typical P loadings to treatment wetlands. As a hypothetical conceptual reference, suppose the marsh develops 1000 $gDW{\cdot}m^{-2}$ of above- and belowground biomass, with a P content of 0.2%. The grow-in will then require 2.0 $gP{\cdot}m^{-2}$. Supposing that the wetland receives 2.0 $gP{\cdot}m^{-2}{\cdot}year^{-1}$ (e.g., 20 $m{\cdot}year^{-1}$ of water at 0.1 $gP{\cdot}m^{-3}$), and removes half, it would require two full year's addition of water to supply this growth requirement.

The establishment of the wetland standing crop of phosphorus is not to be counted as a sustainable P-removal function. Nonetheless, it will be reflected in the input/output mass balance for the wetland during its startup period, if any. This uptake will produce an over-estimate of wetland capacity during startup, and should be discounted in forecasts of long-term P removal.

9.1.2. Seasonal Variation

Three factors are likely candidates for modifying phosphorus uptake over the course of a year: (1) water temperature; (2) vegetation growth patterns; and (3) patterns of water addition. Water temperature may modify microbial processes, which are involved in rapid P uptake and in P release during the decomposition of detritus. Substantial amounts of phosphorus are withdrawn from porewaters and surface waters to support plant and algal growth. Temperate wetlands have a period of plant dormancy occasioned by winter conditions, which differs markedly from warm climate wetlands. Most of the large P-control wetlands are ion-warm climates. These systems may also have periods of relative low vegetative activity, but they do not experience the complete growth stoppage seen in temperate systems. Plankton and periphyton, which depend on light and hence on season, are a variable part of the overall biology, and microbial processes are also known to be important. Stormwater wetlands, such as the South Florida STAs, experience incoming flows and concentrations driven by seasonal patterns of precipitation. Accordingly, it is reasonable to expect temperature or seasonal effects in steady flow systems, but those could be masked by seasonal flow effects in stormwater systems.

Large P wetland performance data were analyzed for potential seasonal effects. Several years' monthly concentration reduction data were folded into monthly means, and examined for patterns during the course of the year. Three types of behavior were identified.

Steady flow systems showed essentially no seasonal variability in concentration reduction (Figure 12). Temperature effects on rate coefficients (k) are often presumed to follow a modified Arrhenius equation:

$$k_T = k_{20}\,\theta^{(T-20)} \quad (4)$$

where k_T: rate coefficient at temperature (T) = T °C, m·year^{-1}; k_{20}: rate coefficient at 20 °C, m·year^{-1}; and q: temperature coefficient. Water temperature in the five Floridian systems typically varies only about 10°C during the course of the year [32], but the Shannon Wetland experiences a 20°C annual variation [28]. Only minimal temperature trends have been found for P removal for emergent marshes in general [15]. For six steady flow large P systems, the median q = 1.012, which implies a 13% increase in k for a 10 °C temperature increase. However, Equation (4) removes only 3% of the variability in k values.

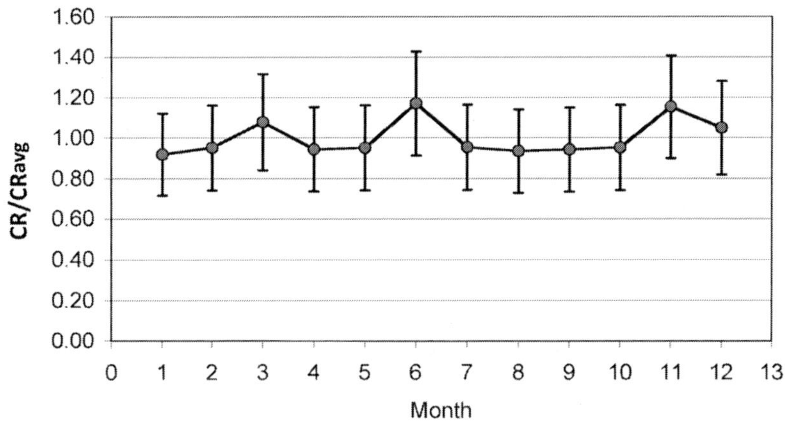

Figure 12. Relative concentration reductions (CR) over the course of the year for six steady flow large P wetlands (OEW, Boney, Titusville, ENRP, Lakeland, Shannon). Data have been normalized by the long-term mean CR for each wetland. Error bars are the average standard deviation for all months.

A second seasonal pattern was displayed in the performance of the four Apopka wetlands (Figure 13) [30]. Concentration reductions during the first half of the calendar year were 34% ± 8%; and during the second half of the year were 2% ± 9%. A distinguishing characteristic of these wetlands was that most P removal was particulate P. Dunne *et al.* (2015) [30] speculate that during the second-half periods, dissolved P was released to the water column, possible from decaying algal and plant biomass.

The STAs display a third type of seasonal variation (Figure 14). STAs show decreased rate coefficients during the winter–spring months, which is also the dry season. Relative rate coefficients during the seven dry months were 39% ± 16% of average and during the remaining five wet months were 185% ± 47% of average. Three operational features characterize the dry season months: lower incoming flows, slightly lower incoming concentrations, and slightly shallower depths in the wetlands.

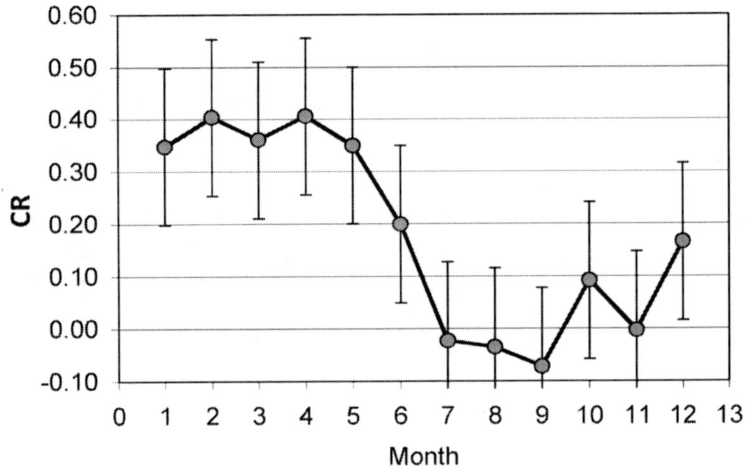

Figure 13. Concentration reductions (CR) over the course of the year for the four Apopka steady flow large P wetlands. Error bars are the average standard deviation for all months.

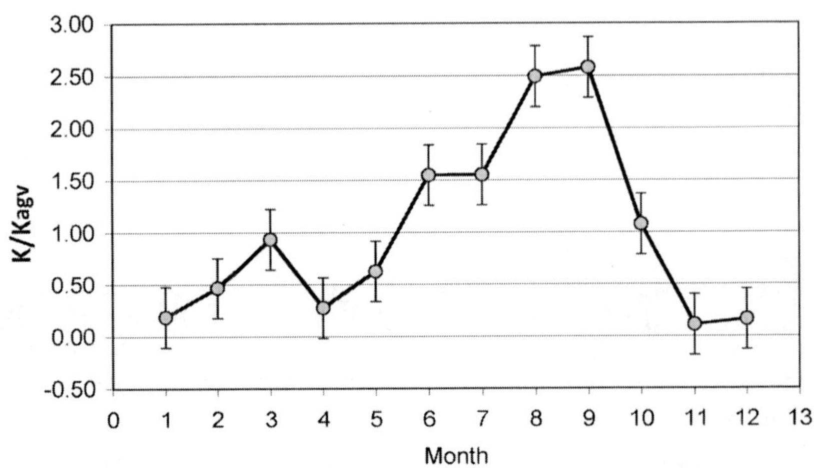

Figure 14. Relative k-values over the course of the year for five STA large P wetlands (STA2C1, STA2C2, STA2C3, STA34 East, STA34 West). Data have been normalized by the long-term mean K-value for each wetland. Error bars are the average standard deviation for all months.

9.1.3. Stochastic Behavior

A sizeable fraction of the variability in outlet concentrations cannot be ascribed to the major driving forces of flow per unit area and inlet concentration. Among other potential causes are actions of animals and wind, and the status of the vegetative and soil components of the ecosystem. This unascribable variability is deemed to be stochastic.

Steady Flow Systems

The framework for interpreting treatment wetland performance, in general, is comprised of a deterministic component, characterized by appropriate equations, together with a stochastic component, characterized by one or more probability distributions [66]. Performance is typically seasonal, as shown in Figure 15.

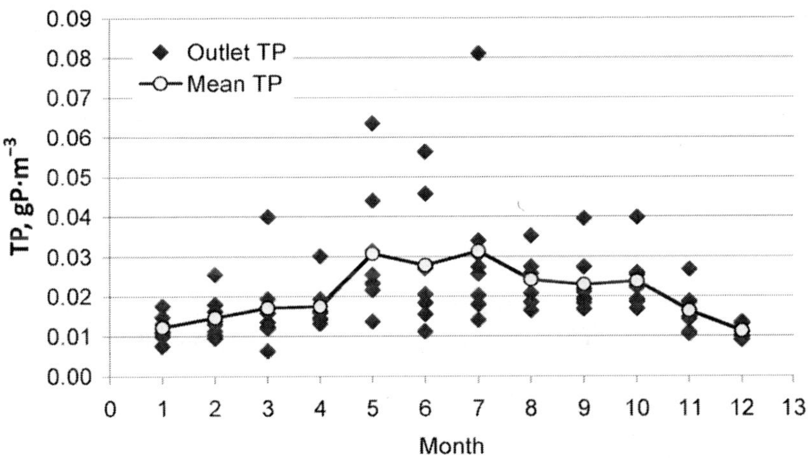

Figure 15. Variability of outlet TP concentrations at Boney Marsh. The mean TP values are means for the designated month. (Source: unpublished data from South Florida Water Management District (SFWMD)).

Therefore, quantification of stochastic variability requires that the performance be seasonally detrended:

$$C=C_{trend}+E \quad (5)$$

where C: outlet concentration, $gP{\cdot}m^{-3}$; C_{trend}: trend component of outlet concentration, $gP{\cdot}m^{-3}$; and E: stochastic component of outlet concentration, gP·m^{-3}. The random variation is captured via a multiplier on the trend value, given by (Y):

$$\Psi = \frac{C_{trend} + E}{C_{trend}} \tag{6}$$

The data from Boney Marsh in Florida provides an illustration of the variability to be expected for mean long-term performance for P removal in large wetlands. This wetland received an average inlet concentration of 0.051 gP·m^{-3} for 1979–1987, at an average hydraulic loading rate of 2.16 cm·day^{-1}, and produced a mean outlet concentration of 0.019 gP·m^{-3}. Figure 14 shows a considerable scatter band around the mean cyclic trend performance for the monthly means of weekly TP measurements.

The upper percentile points of that distribution are of interest for large steady flow P-control wetlands because of regulatory requirements. For instance, the 90th percentile represents the fractional addition to the monthly trend that may be expected to occur one time out of ten during the period of record. Steady flow systems are often subject to "not to exceed" requirements for monthly effluent values.

Table 5 lists percentiles of the monthly Ψ-distributions for phosphorus for representative large P-removal wetlands. These data indicate that the median of the 95th percentile is an additional 79% above the trend, and that the median of the 90th percentile is an additional 59%. For the larger population of P-removal wetlands of all sizes, these medians are higher [15].

Table 5. Multipliers for monthly trends in total phosphorus exiting large steady flow wetlands.

Wetlands	Years	Percentile			
		50%	80%	90%	95%
ENRP	4	1.00	1.24	1.35	1.44
Lakeland	7	1.00	1.23	1.46	1.51
Boney	8	0.92	1.15	1.55	1.73
Titusville	7	0.89	1.39	1.77	2.00
OEW	10	0.72	1.39	2.07	2.50
Apopka	9	0.82	1.26	1.64	1.86
Median		0.91	1.25	1.59	1.79

Stormwater Systems

The regulatory criteria for the large STA systems are based upon annual performance, and upon multi-year averages as well. Such criteria are founded on the statistical analysis of the variability of annual average TP in the wetland outlet. The 90th percentile of the distributions of annual outlet concentrations for the STAs is 1.67 times the mean over the POR (N = 236 flow path years).

9.2. Spatial Variability

It is very difficult to view a very large wetland as behaving as a "black box." The size of large P wetlands invites examination of the internal profiles of P concentrations. Two distinctly different procedures have been followed. If the wetland is compartmentalized, with structures conveying flow from one cell to the next, then those intercell flows can be sampled for TP. If flow rates are also measured, then the TP in the intercell flow can be flow-weighted. If the wetland is not compartmentalized, then a grid of interior points may be sampled for TP and other parameters. Often, a spatially uniform grid is sampled, with cross-flow points representing a specific distance from the inlet. In general, it is not possible to measure the flow rate that may be representative of a given sample point. Flow weighting is not possible.

9.2.1. Compartmentalization

Both the P-load reduction and the rate coefficient depend upon the hydraulic loading, which is the water flow per unit of wetland area. A common, but not universal, feature of large P-removal wetlands is compartmentalization, meaning they contain a number of (relatively) independent sequential cells. In the absence of such compartmentalization, inlet regions are, nonetheless, likely to perform differently from outlet regions. Concentrations, and often flow rates, are measured at the inter-cell connection points. As a consequence of the sequential removal of P through such systems, the inlet cells bear most of the P burden. The inlet cells see the highest P concentrations, and are the recipient of solids deposits originating with the incoming water. Downstream cells see less phosphorus, and accrete solids generated by internal wetland cycling.

The number of cells in the large P-constructed dataset range from one to two dozen, but the median is two (Table 1). When performance is analyzed for cells in series, a decreasing trend is observed from cell to cell along the flow path (e.g., Shannon [28]; Lakeland [34]; ENRP [67]). Cell-wise performance information for the STAs was reported through

2012, at which time intermediate cell data acquisition was terminated [68].

9.2.2. Internal Measurements

If the wetland is evaluated in its entirety, internal trends may be subsume, and whole system trends may not be observed. Time-changing zones can be entirely within the wetland as a whole. In general terms, the zone nearest the inlet may undergo alteration in plant speciation, while zones further downstream may retain their species and performance, and at long distances a background ecosystem will prevail [69].

A case in point is the natural Houghton Lake treatment wetland. Although not compartmentalized, it was analyzed at different size scales. The overall wetland comprises 700 ha, and input–output data were collected at that scale. In the initial few years of the project, which has spanned 39 years to date, attention was focused upon an inlet area of some 20 ha [70,71]. The P removal in this small area declined with time over the initial five years. At the opposite extreme, the "performance" of the entire 700 ha wetland did not change over a 30-year history; that is, there was no change in the effluent P concentration [56]. In recognition of this disparity, monitoring of a 100-ha inlet region commenced in year four of operation. This intermediate-sized zone displayed time trends of decreasing k-values over a period of about five years, but there was no k-value time trend over the ensuing 25 years [56].

9.2.3. Internal Behavior of the STAs

Periodically, many of the STAs have been sampled internally for water quality, on grids that extend both parallel to flow and cross flow ([48,72,73,74]). This information is vitally useful in understanding STA performance.

Transect information can be used to determine rate coefficients. These are snap-shots of performance, linked to the instantaneous flows and concentration at the time of the transect. Therefore, transect rate coefficients must be concatenated over the annual flow distribution to be compared to annual rate coefficients, which are determined from flow-weighted averages of inlet and outlet information. However, transect coefficients allow insights into the P-uptake patterns, as these are related to the instantaneous flow rates.

Areal rate coefficients allow forecasts of the longitudinal variation in P concentrations, as these relate to the hydraulic loading up to that distance (Equation (1)). As one moves downstream from the inlet, the upstream hydraulic loading becomes lower, until at the outlet it becomes the loading for the entire wetland. The kC* model fits transect data quite well; for instance, for twenty transects in STA2C3, the average fit was $R^2 = 0.89 \pm 0.09$. These instantaneous transect rate coefficients are strongly dependent on flow rate. When the transect data are flow-weighted, the concentration reductions show the expected pattern of decreases with increasing hydraulic loading rate (Equation (1)) (Figure 16). Furthermore, the monthly rate coefficients, determined from input/output data, cluster at the low end of the loading spectrum, and are consistent with the flow-weighted transect information. The mean transect rate coefficient for the five-year period was 44 m·year^{-1}, while for the same five years, the mean annual system rate coefficient was 38 m·year^{-1}.

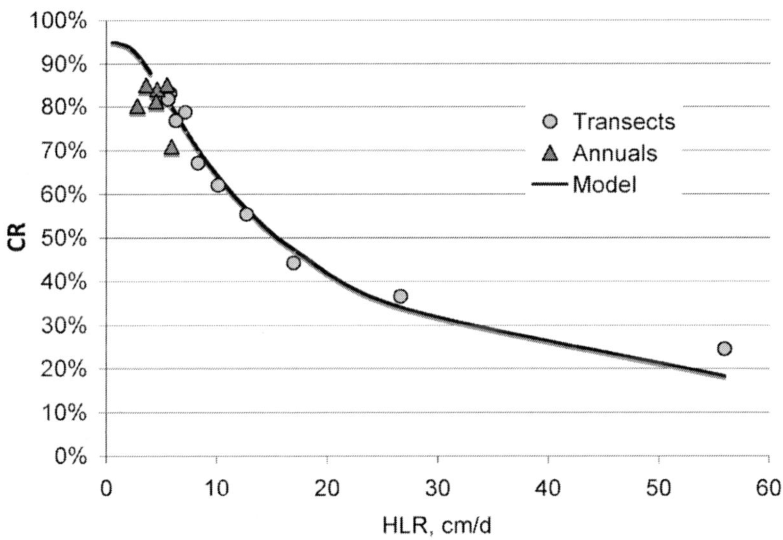

Figure 16. Transect data and the kC* model for 28 grid transects for STA2C3 taken over five years. The data were flow-weighted. The hydraulic loading rates decrease as the location moves along the transect from inflow to outflow. Shown for comparison are the annual input/output flow-weighted concentration reductions for the five-year period. The model parameters are: K = 44.4 m/year, and C* = 6 µg/L, as co-determined by nonlinear regression. $R^2 = 0.977$.

The instantaneous transect coefficients change with flow, as do the monthly system I/O coefficients. When flows are high, the coefficients are also high (Figure 17). The instantaneous P-uptake coefficients are very nearly proportional to flow. This behavior is distinctly different from that presumed up till now in STA calculations, which has been a rate coefficient invariant with flow.

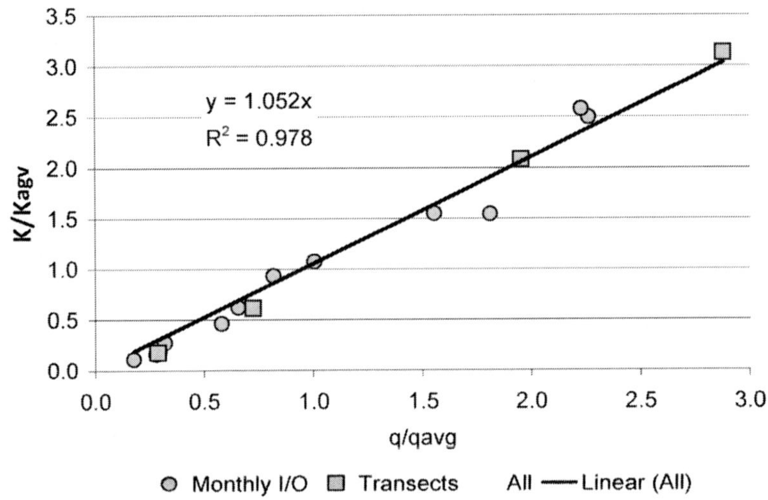

Figure 17. Relative monthly rate coefficients (K/Kavg) *versus* the relative hydraulic loading (q/qavg) for five STA wetlands. Also shown are four flow groups of transect rate coefficients for STA2C3. (q = hydraulic loading, cm/day).

There is also spatial variability between parallel transects across the flow direction. As an example, TP was measured on transects on 20 occasions over a span of four years, along five parallel paths in STA2 [74]. The average coefficient of variation perpendicular to flow was 25%. The spatial average concentration, across the width of a wetland, is typically lower than the flow-weighted concentration at that same distance. This is due to the fact that slower speed zones occupy the majority of the wetland width, and so spatial sampling emphasizes slow speeds, with better treatment [75]. The difference between inlet and outlet flow-weighted concentrations and the internal spatial averages for STA2 Cell 3 were about 10 μgP/L near the inlet, and about 4 μgP/L near the outlet.

10. SUSTAINABILITY AND TIME TRENDS

A recurring theme in older wetland literature is that when a wetland receives new inputs of phosphorus it will only remove some portion of the incoming load until the system becomes "saturated" with phosphorus. For instance, Richardson (1985) [70] stated: "Wetlands tested as wastewater filtration systems became phosphorus-saturated in a few years, with the export of excessive quantities of phosphate." Such early misconceptions ignored the soil building that occurs in treatment wetlands, and were subsequently supplanted by acknowledgement of the sustainable P removal to newly created soils [65]. Based on studies of the storage of P in marshes removing phosphorus in Florida marshes, Craft and Richardson (1993) [63] later recognized that "The efficiency of any wetland to store P on a long-term basis is thus determined by the peat or soil accretion rate..."

Although most wetland science literature now correctly identifies accretion as a principal long-term storage for phosphorus (e.g., [62,63]), there still persists literature that does not directly address the soil-building mechanism as a primary route of phosphorus immobilization. For instance, USEPA (1999) [76] states: "New constructed and natural wetlands are capable of adsorbing phosphorus (P) loadings until the capacity of the soils and new plant growth are saturated." Mitsch and Gosselink (2000) [77] identify only chemical precipitation, adsorption, and plant uptake as removal mechanisms. Crites *et al* (2006) [78] state: "Adsorption and precipitation reactions are the major pathways for phosphorus removal..." Interestingly, USEPA (2000) [79] correctly identifies the sustainable mechanism as "accretion and burial." However, identification of accretion removal does not address the question of whether there may be continuing deterioration of the P-removal mechanisms.

At this point in the history of treatment wetland technology, there now exist some monitoring histories of modestly long duration. Those that pertain to large constructed P-control wetlands are of interest here. Time trends allow confirmation of the long-term sustainability of P removal in these marshes.

10.1. Trends for Steady Flow Wetlands

Time trends for steady flow systems are shown in the top part of Table 6. An example is the Orlando Easterly Wetland (Figure 18). This wetland went into operation in 1988, and underwent a period of vegetative grow-in until 1990, and of reducing inlet TP for another two years. Here, the marsh portion (318 ha) is considered. For the ensuing twelve years, phosphorus loads to the system were fairly stable, with modest interannual variability, until excavation management was undertaken in 2002. That twelve-year POR is considered here. Only slight trends were present in performance, as quantified by linear regression (Table 6).

Additionally, these trends account for only a small amount of the variability in the three measures: R^2 = 0.032, 0.050, and 0.061 for concentration reduction, P load reduction, and rate coefficient, respectively. Excavation was done because this steady flow system had monthly performance criterion, and monthly excursions were becoming more pronounced (Figure 18).

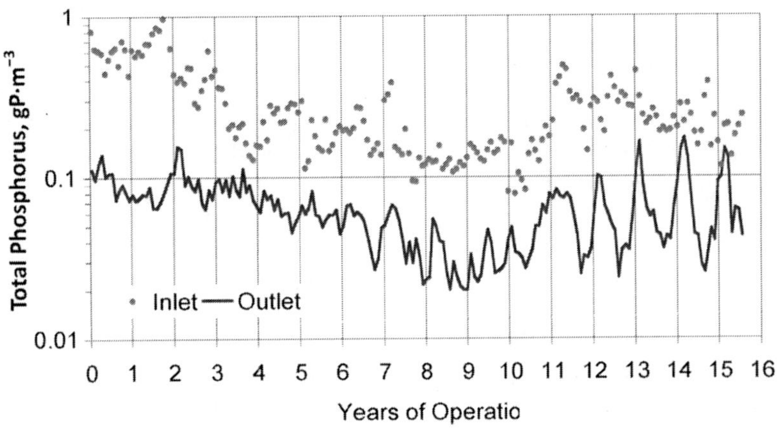

Figure 18. Time progression of inlet and outlet TP concentrations for the OEW system.

Data was analyzed for nine large P steady flow treatment wetlands, plus two large natural wetlands. Linear regression was used to establish the time-trend slope of the three performance measures, expressed as percent change per year (Table 6). Median rates of change were 0.2, 2.7, and 2.8% $year^{-1}$, for CR, k, and PLR, respectively. These median rates of change are all slightly positive. However, the linear trends account for only a very small fraction of the annual variability, with median R^2 values of 0.087, 0.014, and 0.0.050 for CR, PLR, and k values, respectively. In other words, the interannual data scatters, and displays almost no trends.

10.2. Trends for Stormwater Wetlands

The south Florida stormwater treatment areas form a well-studied subset (N = 19) of large wetlands designed to remove phosphorus. The systems receive pulses of flow on an intermittent basis. Linear regression was also used to establish the time-trend slope of the three performance measures for the stormwater wetlands (Table 6). Median rates of change were 1.0, −0.2, and −7.6% year^{-1}, for CR, k, and PLR, respectively. The median rate of change for PLR is negative, which may reflect lower incoming P loads and/or persistent startup conditions. Nonetheless, the linear trends account for only a small fraction of the annual variability, with median R^2 values of 0.093, 0.074, and 0.113 for CR, k, and PLR values, respectively. In other words, the interannual data scatters, and displays almost no trends for CR and k.

Table 6. Time trends in annual concentration reduction (CR), rate coefficients (k), and P load removed (PLR) for large P wetlands. Two natural wetlands are included because of their long history. Slopes are in percent per year.

Site	System	Years	CR		k		PLR	
		Trend	Slope	R^2	Slope	R^2	Slope	R^2
Boney	West	11	1.8	0.114	2.4	0.094	3.9	0.193
OEW	Stratum 4	12	−1.0	0.032	4.7	0.061	3.1	0.050
Titusville		7	0.4	0.003	4.5	0.014	−12.0	0.167
Lakeland	Cells 1–4	12	−1.1	0.019	2.7	0.041	−4.4	0.130
Shannon		6	0.2	0.190	7.5	0.347	8.1	0.670
Bayou Marcus	Natural	14	−3.0	0.289	0.1	0.000	12.9	0.697
Houghton Lake	Natural	26	−0.7	0.334	−0.2	0.004	−0.2	0.002
Apopka	B1	9	8.6	0.563	4.0	0.115	2.8	0.036
Apopka	B2	9	2.6	0.026	−0.7	0.001	−0.7	0.000
Apopka	C1	9	4.1	0.001	−2.6	0.003	−2.9	0.003
Apopka	C2	7	−16.4	0.087	4.4	0.003	9.0	0.013
Median Steady		9	**0.2**	**0.087**	**2.7**	**0.014**	**2.8**	**0.050**
STA1E	Central	10	−2.7	0.102	−7.3	0.076	−11.4	0.217
	West	10	4.1	0.014	4.5	0.021	−17.1	0.162
STA1W	East	14	2.9	0.333	−2.4	0.041	−4.0	0.143
	West	12	1.5	0.317	0.8	0.012	0.3	0.000
	North	16	5.2	0.500	6.1	0.323	2.3	0.039
STA2	FW1	12	0.5	0.072	4.8	0.095	0.7	0.004
	FW2	13	0.3	0.019	1.2	0.030	−0.3	0.001
	FW3	13	1.3	0.185	−0.7	0.010	0.5	0.003
	FW4	7	2.2	0.034	−0.2	0.000	−13.4	0.200
STA34	East	11	−0.6	0.093	−3.4	0.074	−15.4	0.497
	Central	11	−1.3	0.092	−9.1	0.198	−23.4	0.391
	West	11	0.0	0.000	5.6	0.254	0.0	0.000
	PSTA	8	−7.5	0.507	−2.4	0.021	−1.4	0.007
STA56	FW1	16	8.7	0.534	1.9	0.014	−7.6	0.113
	FW2	16	5.0	0.166	−0.1	0.000	−12.4	0.247
	FW3	6	0.2	0.005	10.7	0.142	0.1	0.000
	FW6	6	5.7	0.549	−14.3	0.274	−16.9	0.772
	FW7	13	−1.0	0.066	−11.8	0.528	−8.7	0.081
	FW8	14	1.0	0.010	−19.8	0.312	−9.4	0.134
Median Storm		12	**1.0**	**0.093**	**−0.2**	**0.074**	−7.6	**0.113**
Global Median		11	0.5	0.0925	0.5	0.041	−1.1	0.097

11. ANCILLARY BENEFITS

The primary objective of the large P-control wetlands is water quality improvement. Secondary benefits that can accrue in treatment wetlands include (1) vegetative biodiversity; (2) protection and production of fauna; and (3) aesthetic, recreational, commercial, and educational human uses [15]. The first of these is not typical of the large P wetlands, because of perceptions of the desirability, or undesirability, of certain vegetation species for water quality treatment. However, the other two ancillary benefits, animal habitat and recreation, are very important in many of the large P wetlands. The large size of these systems makes them especially important as wildlife habitats, because species richness goes up with size [80]. In turn, wildlife attracts fishermen, bird watchers, and hunters.

11.1. Wildlife Use

Large treatment wetlands are popular with fish, birds, and reptiles. Many thousands of waterfowl frequent the Florida STAs and other P-treatment wetlands, although many are American coots (*Fulica americana*). These make the systems a stopover on winter migration, but also forage year-round on wetland vegetation. Wading birds are also very prevalent, including uncommon species, such as roseate spoonbills (*Platalea ajaja*), white pelicans (*Pelecanus erythrorhynchos*), wood storks (*Mycteria americana*), and flamingoes (*Phoenicopterus ruber*). Herons and egrets abound.

Fish occur in large wetlands as well. The southerly location of many of the systems creates conditions conducive to the non-native tilapia (*Oreochromis aureus*). In general, fishing is not allowed in the Florida STAs.

Alligators (*Alligator mississippiensis*) have been observed at quite high densities in wetland treatment systems with open water areas and high fish populations. Both the Florida and Texas large P wetlands are home to alligators.

11.2. Wildlife Damage

Nutria (Myocastor coypus) are an invasive species from South America, but they now range across southern and northwestern USA. Nutria feed on a broad range of plants in treatment wetlands, including cattails, grasses, water hyacinth, and duckweed. Nutria cut vegetation to build feeding platforms and nest mounds. They are prolific and achieve damaging population densities unless they are controlled by trapping or shooting Nutria have occurred in problematic numbers in many treatment wetlands, including the East Fork Wetland in Texas. From January through December 2010, almost 13,000 nutria were eliminated from the East Fork Wetland. Another exotic, invasive species, feral hogs

(*Sus scrofa*), also caused limited impacts to the wetland vegetation and levees [25].

American coots (*Fulica americana*) have caused severe damage to the SAV in Florida STAs, They eat the SAV, causing destruction of the vegetative component of the system.

Black-necked stilts (*Himantopus mexicanus*), burrowing owls (*Athene cunicularia*), and snail kites (*Rostrhamus sociabilis*) have all caused interruptions of the operations of Floridian treatment wetlands [33].

11.3. Human Use

The diversity and abundance of birds in and around wetlands attracts bird watchers, who observe that species lists are longer and counts higher when they include large P-treatment wetlands in their counting areas. All the Florida STAs have public access. For instance, STA34 has two public access points, featuring information kiosks, paved parking, concrete sidewalks, and restrooms for hikers, bicyclers, and bird-watchers. A public dual-lane boat ramp is open seven days a week and offers access to 27 miles of perimeter canals. The STA-3/4 East entrance provides parking and toilet facilities, along with informational signage and access to levee trails. The Audubon Society offers escorted driving trips in the STAs on weekends.

The other large P wetlands, such as the OEW and Shannon Wetlands, also have much public use. Thousands of birds visit the Shannon Wetlands, including several types of egrets, great blue herons, wood storks, white ibises, and American white pelicans. In winter, bird numbers soar, because thousands of migratory birds descend on the area. Gators, deer, and feral hogs also frequent the Shannon Wetlands.

A number of wetland bird species are game species and are hunted. Treatment wetlands frequently provide good waterfowl habitat. In a few cases, these treatment wetlands have been used for duck hunting as a recognized secondary benefit. The stormwater treatment constructed wetlands in Florida are open to duck hunting, and are deemed the best hunting spots in the entire state. Alligator densities in the Florida stormwater treatment constructed wetlands have reached proportions sufficient to allow hunting. On designated weekends in specific areas of the STA, waterfowl and alligator hunts are managed by the Florida Fish and Wildlife Conservation Commission.

12. MODELS

Almost all models of large wetlands, with one exception, address total phosphorus, a lumped measure. As a result, they do not deal with solids movement, related to particulate phosphorus. Soluble reactive phosphorus is known to behave differently from dissolved organic P and from particulate P. Models do not do a good job of describing the very low P concentrations that occur in some wetlands. Yet, a very large effort has been expended on modeling of large P wetlands, which is briefly summarized here.

12.1. Steady State Models

Early models of P removal in marshes were predicated on steady state, or long-term average conditions. This was a necessity created by the nature of the available databases. Historically, hydraulic conditions were presumed to be either a single well-mixed unit (lakes), or plug flow (rivers, wetlands). Removal was presumed to be a first-order process, with rates varying in proportion to the water concentration ([34,81,82,83]). The calibration parameter is the removal rate coefficient, k. Hydraulic models can account for flow patterns intermediate to plug flow or perfect mixing, through the use of the tanks in series (TIS) model. Several large-scale wetlands have been tracer tested, and produced a central tendency of about N = 4–6 for cells or flow paths of a large wetland (see, for instance, [40,41,42]).

12.1.1. Speciation

Total phosphorus (TP) is, in fact, a mixture comprised of particulate (PP), and dissolved organic (DOP) and soluble reactive (SRP) forms. Each fraction of the lumped TP possesses its own net removal rate coefficient. As the mixture passes through the wetland, its composition changes due to weathering.

As shown in Kadlec (2003) [84], and in Kadlec and Wallace (2009) [15], it is possible to account for speciation in modeling without resorting to multi-species models. In the wetland environment, hydraulics and weathering interact to produce the overall observed reduction in a lumped category of pollutants. Weathering behavior in wetland situations may also be represented by a TIS model, where the hydraulic parameter (N) is relaxed to become a fitting parameter (see Equation (1)).

Applications of this model for large constructed P wetlands may be found in, for instance, Kadlec *et al.* (2011) [28] and Juston *et al.* (2013) [74], and it is used in this work.

12.1.2. Including Solid Compartments

The lower bound in Equation (1), labeled C*, is the result of internal recycle processes in the wetland, with small contributions from atmospheric deposition. Model fitting research has typically determined C* levels in the range 10–20 mgP·m^{-3} [48]. The contribution from atmospheric deposition is estimated at about 2–4 mgP·m^{-3}, based on rainfall TP and dry deposition measurements. The contribution of SAV in STA2C3 has been calibrated at about one-third of the total C* [74]. The submerged aquatic vegetation phosphorus model (SAVPM) is a steady state model patterned after the PkC* model, but additionally containing interacting compartments for biomass, recently accreted sediments, and underlying soils [74]. The model was calibrated and exercised for STA2C3, an SAV wetland.

12.2. Dynamic Models

When the wetland is presented with time-varying flows, there can be large effects of the time sequence of those flows. Proper accounting then requires that system dynamics be considered in modeling. This greatly increases the complexity of calculations. It also brings into consideration the various timescales of wetland processes, including the response times of the water sheet and floc (rapid) and those of the sediments, soils, and biota (slow). The stormwater treatment wetlands manifest the collective effects of these dynamic processes.

As a consequence, spatial and temporal models have been explored for describing P processing in the South Florida STAs [85,86]. These involve two or more compartments and account for spatial effects; they are briefly described here. In general, these models of large wetlands get the hydrology correct, including water depths and stages, without excessive parameterization. They also are all moderately successful in describing the high end of the concentration range, *i.e.*, from a few hundred mgP·m^{-3} down to 20–25 mgP·m^{-3}, on the timescale of a year. However, none are particularly successful at describing wetland performance at the monthly or faster timescale, or in the low concentration range, *i.e.*, less than about 20 mgP·m^{-3}.

12.2.1. Wetland Water Quality Model (WWQM)

This model was developed to describe phosphorus removal in the STAs [87,88,89]. The WWQM is a dynamic model that uses two spatial dimensions. Hydrodynamics are calculated first, followed by chemical interactions between water and four stationary compartments: periphyton, macrophytes, aerobic sediments, and anaerobic sediments. Chemical species include calcium, iron, carbon, nitrogen, and sulfur. As a consequence, the code solves over 100 equations requiring over 200 coefficients. The attendant computational times were found to be quite large. To date, this model has not been successfully implemented. It was originally calibrated to data from cell 4 of the ENRP. This model is not in general use, and has been supplanted by the LOEM-CW model described below.

12.2.2. Exploratory Dynamic Modeling

Researchers working on various aspects of the STAs developed dynamic models of less complexity as an adjunct to investigating specific ecosystem types. One such was the Process Model for SAV (PMSAV). This model was developed to describe the P-removal processes in wetlands dominated by submerged aquatic vegetation [90,91]. It allows P removal from water by two routes: a biologically mediated pathway and a sedimentation pathway. It also splits storage into a biomass compartment, representing the SAV, and an active sediment compartment. It also utilizes a combination of Monod and first-order processes to interconnect the water column and the biomass components. There are seven adjustable parameters. It can be run easily on a desktop computer, and there are minimal calibration issues. This model was used to conduct screening level computations for the SAV alternative for P removal from Everglades Agricultural Area (EAA) runoff. It has been calibrated to phosphorus and biological data from wetlands at several scales, up to 200 ha, in South Florida. This model is not in general use.

A second study-specific model was the PSTA Forecast Model. This model was developed to describe the P-removal processes in wetlands dominated by periphyton [92]. It utilizes a combination of Monod and first-order processes to interconnect the water column and the biomass components. It can be run easily on a desktop computer, and there are minimal calibration issues. This model was used to conduct screening level computations for the PSTA alternative for P removal from EAA runoff. The model considers hydrologic factors (rain, evaporation), but not hydrodynamic factors. It is not a spatial model, and utilizes a single well-mixed unit. Accordingly, hydraulic efficiencies were very low, and the resulting wetland areas needed for acceptable treatment were very large. At the present time, this model is not in general use.

A third dynamic forecast model was used to explore design and operational strategies for the whole set of STAs, the Everglades

Phosphorus and Hydrology (EPH) model. This model spatially discretized an STA into plug-flow cells in series, selected by the user to reflect landscape and water quality regions. It was simplified by consideration of only a few transfers, and by the selection of a monthly time step. Phosphorus removal was calculated using a first-order areal uptake equation, and return i taken to be zero order, and operative only during dry periods [93]. EPH was used to theoretically investigate the results of different operational and design strategies for the STAs. This model is no longer in general use.

12.2.3. The Dynamic Model for Stormwater Treatment Areas (DMSTA)

The STA treatment wetlands operate in a dynamic mode. The principal parts of DMSTA include hydraulics, water TP, and stored P. Storage in solid compartments occurs during periods of high phosphorus availability, and release back to the water occurs in periods of P famine. Permanent removal is to residual sediments containing unavailable P. DMSTA [55,94] is an unsteady state model that removes phosphorus to permanent burial in proportion to the amount of labile P in storage. There is a generic labile pool of phosphorus, in addition to the permanently buried P. That labile pool is presumed to be drawn down by a return flux, or bleed-back. Temporal variations in the water budget (flows, rainfall, evapotranspiration, seepage, storage) are also simulated. DMSTA allows for spatial variability via cells in series or parallel, and accommodates PTIS for each wetland cell.

DMSTA does not attempt to describe the startup of a wetland. For wetlands built on agricultural soils, there is generally a period of startup, during which internal and outlet P concentrations exceed input values. The length of this period is dependent on the size of the antecedent P load contained in the soils. During startup, the wetland undergoes the initial stages of development, such as increases in biomass and the extent of areal coverage. However, the occurrence of a switch to a condition of net P removal does not necessarily signal the end of the ecosystem transition, which may continue over an extended period of stabilization. Data show that startup and stabilization may take two or more years for a newly converted wetland [50]. DMSTA avoids startup by seeking a stationary state solution for repetitive time series.

Numerous calibration datasets have been used, collected from operating wetlands over the years prior to 2005, and at a wide range of size scales [55]). These include individual cells of the STAs, shown in Table 1 and Table 2. This model has seen extensive use in connection with the STAs.

12.2.4. Transport and Reaction Simulation Engine (TARSE)

This model solves the advection–dispersion equation on an unstructured triangular mesh and incorporates a wide range of user-selectable mechanisms for phosphorus uptake and release parameters [95]. In general, the phosphorus model contains transfers between stores; examples of stores that can be included are soil, water column (solutes), pore water, macrophytes, suspended solids (plankton), and biofilms. Examples of transfers are growth, senescence, settling, and diffusion, described with first-order, second-order, and Monod types of transformations. Local water depths and velocities are determined from an existing two-dimensional, overland-flow hydrologic model, the South Florida Water Management District Regional Simulation Model.

There are nine state variables, including particulate and soluble P in groundwater, porewater, and surface water, together with plants and microbiota [95]. There about 17 transfers of P among compartments, each requiring a rate model and at least one adjustable parameter. Calibration is therefore difficult, and subject to the estimating ability of the user.

The model was applied to one cell 4 of the ENRP [96]. Inflow total phosphorus concentrations and flow rates for the period between 1995 and 2000 were used to simulate cell 4 phosphorus removal. The long-term mean TP at the outlet, 22 $mgP \cdot m^{-3}$, was matched well, but the root mean square error between model and data over six years was 11 $mgP \cdot m^{-3}$ for outlet TP. Flows to this cell were relatively steady during this period, and thus the pulse response of the model was not severely stressed. This model has not been used in design, or in the evaluation.

12.2.5. Lake Okeechobee Environment Model for Constructed Wetlands (LOEM-CW)

This dynamic model attempts to describe movement of water, nutrients, and solids in three dimensions [97,98]. The data demands are very high: for example, hourly wind, solar radiation, air temperature, relative humidity, evapotranspiration, and cloud coverage as the external forcings. There is extreme complexity in both the water quality section (130 model parameters) and in the sediment section 27 state variables [99]. As a result, the model requires large amounts of computer execution time. LOEM-CW has been applied to the western flow path of STA34. The root mean square error between model and data over six years was 19 $mgP \cdot m^{-3}$ for outlet TP, which had an observed mean of 23 $mgP \cdot m^{-3}$ [98].

13. ECONOMICS

The large P-control wetlands are costly to build and operate, because of their size. However, they are not usually planted, nor is a liner typically used, and those two costs comprise about 40% of the total capital cost for smaller surface flow [15]. The large wetlands are not located in areas of excessive land values. As for any treatment works, there is an economy of scale that lowers the per hectare cost [15]. Delivery of water to a large wetland may require very large pumps, which may partially offset an economy of scale.

The median capital cost for 16 large P-control wetlands was $69,000 per hectare, of 53–6742 ha, median 630 ha. In contrast, the median cost for 39 small constructed wetlands, of 1–40 ha, median 6.3 ha, was $132,000 per hectare. These 2016 capital costs do not, in general, include the cost of land acquisition.

The median operation and maintenance (O&M) cost for nine large systems, median 2,210 ha, was $770 per hectare per year. The median operation and maintenance cost for 17 small systems, median 6.0 ha, was $3,620 per hectare per year. These 2016 O&M costs do not, in general, include the cost of water quality monitoring.

14. CLOSURE

Most of the large P-control constructed wetlands are in the USA, with the earliest finished in 1976. These marshes total about 28,000 ha, and cost about $2 billion to construct. On an areal basis, these large systems comprise 99% of all constructed P-control wetlands. The large systems also have long data records, in contrast to many smaller wetlands that have only a few years' data. Due to practicality, the large system data does not include replicated, side-by-side studies. However, the economic importance of scientific understanding has resulted in large investments in monitoring and research studies, and much of the knowledge of phosphorus processing in marshes has been gained from the large systems.

This review shows that large constructed wetlands all remove phosphorus. They do so more efficiently than the population of smaller counterparts, as measured by concentration reduction (median 71%) or removal rate coefficients (median 12.5 $m \cdot year^{-1}$) for the entire period of record. However, large systems display lesser P load reductions (median 0.77 $gP \cdot m^{-2} \cdot year^{-1}$) than the larger general population of wetlands, in part because the large systems typically operate at lower incoming P loads (median 1.22 $gP \cdot m^{-2} \cdot year^{-1}$). In common with smaller systems, P concentrations are not reduced to zero, but to a plateau, typically less than 20 $mgP \cdot m^{-3}$, but more than the rainfall P limit of 2–4 $mgP \cdot m^{-3}$. Inlet cells or zones receive the brunt of the P-removal effects, and consequently may require maintenance more than downstream portions, notably to remove solids buildups.

Neither large nor small constructed wetlands have a complete array of removal mechanisms from the moment of flooding. Legacy phosphorus in the soils may be released before benefits are exhibited. Startup transitions occur, with duration depending upon the antecedent condition of the landscape in which the constructed wetland is placed.

Intra-annual behavior displays seasonal rather than temperature-driven behavior. Temperature trends remove almost none of the variability in wetland outflow concentrations. Stormwater wetland performance varies strongly with the flow patterns that accompany the wet and dry seasons. There remains a good proportion of variability in outflow concentrations, which appears to be stochastic, with the 90th percentile of outlet concentrations at 67% higher than the mean.

The type of vegetative community affects the performance measures. Additionally, differences in the lowest achievable outlet concentration may well depend upon vegetation, or upon the soil type that helps determine the plant communities. The existence of vegetation is important, both as a factor in biogeochemical cycling and as an influence on other portions of the suite of removal processes.

The hydraulics of the large systems differs importantly from those of the smaller counterparts, in that water levels are not controllable with outlet structure settings. Vegetation resistance controls depths and levels at medium and high flows. Water flow, and to a lesser extent water depth, have important effects on the removal rate coefficient (Figure 16). Reasons for this dependence have not been determined, nor has the dependence been considered in modeling.

Phosphorus speciation is an issue, because of the differences in removal rates of SRP, DOP, and PP. It is probable that the PP is generated within the wetland, especially in downstream sections. PP, and the related floc, are mobile and can move with water or by spiraling. Yet, no studies of PP movement have been undertaken.

Internal measurements along the flow path have yielded insights not provided by inlet-outlet measurements. Figure 15 shows that I/O data cluster, and scatter, in a small group at the end of transect data. In turn, examination of this cluster alone caused a conclusion that "bivariate analysis...showed a lack of relationship in predicting outflow TP concentrations using inflow HLR..." [100]. In contrast, transect data extend the range of Hydraulic Loading Rates (HLRs) under consideration, and show a very strong relation, corresponding to the PkC* model used herein (R^2 = 0.977).

It has sometimes been asserted that the phosphorus treatment function of a wetland wears out [70,101]. The track records of the various constructed large P systems does not substantiate those early evaluations. Changes do occur over the operating histories of most systems, but there is no commonality for the direction of such trends. Further, the rates of change are limited to small percentages of the various performance measures, and the global medians for concentration reductions and removal rate coefficients are positive, meaning the wetlands improve slightly over time. Furthermore, the amount of variability contained in the trends is typically very low, less than 10% of the total variability for both steady flow and stormwater systems. These observations must be tempered by the data periods involved, which are about a dozen years.

If the wetland is adequately maintained, by taking the steps necessary to accommodate bio-accretion, then there is no apparent limit to wetland lifetime for phosphorus removal. This is borne out in the performance of natural wetlands that have been receiving nutrient-laden discharges for long periods of time, as well as by the performance of constructed wetlands that have been constructed and operated for up to three decades.

The socio-economic aspects of wetland construction are greatly magnified at large-scale. Large system establishment usually means land is taken out of agriculture, which in turn affects farmers, growers, and ranchers, and ultimately has an effect on food and water supply. On the other hand, environmental interests like the restoration of more natural landscapes, albeit not as pristine as many would desire. Wildlife values are greatly enhanced by these large wetlands, resulting in ancillary benefits of human use for birding and hunting. As a consequence, associated political conflicts reverberate at the state and national level.

A number of mathematical models have been advanced that have been calibrated to data from large P-treatment wetlands. A necessary requirement is the inclusion of some form of P return from soils, sediments, and vegetation. The simplest models, such as the PkC* model used herein, add a fitting parameter for the low plateau in achievable P concentrations. More complicated models attempt to calculate P return from solid compartments, but there is generally no available data on the P content of those solid compartments with which to calibrate that part of a model. Most fit short-term datasets well at moderate concentrations, but no model has demonstrated ability to forecast accurately at concentrations below about 20 mgP·m^{-3}. It may be that stochastic behavior dominates in this low P region.

Models with a small number of adjustable parameters have seen the most usage, but because of their high degree of aggregation of processes and compartments these have questionable attributes and accuracy. More complexity does not mean a better model, however [86]. It has been suggested that needed improvements include floc description, vegetation cycling description, and integration of data acquisition with modeling efforts [85].

Transferability is an issue, because model parameters have been found to strongly depend upon several system features. Vegetation type is important, but not the sole determinant of model parameters. Soil types, and their antecedent condition, are also important, as is the form of phosphorus entering the wetland. Thus, intersystem variability has been found to be large, even for attempted replicate systems in a small geographical region, which makes design model forecasting uncertain.

ACKNOWLEDGMENTS

The author is grateful to the several organizations that have shared data from the projects discussed herein. The acquisition of that data was the result of massive efforts and expenditures.

REFERENCES

1. Fetter, C.W.; Sloey, W.E.; Spangler, F.L. Phosphorus accumulation—Discharge cycles in marshes. *Water Resour. Bull.* **1977**, *13*, 1191–1201.
2. Fetter, C.W.; Sloey, W.; Spangler, F. Use of a natural marsh for wastewater polishing. *J. Water Pollut. Control Fed.* **1978**, *50*, 290–307.
3. Mudroch, A.; Capobianco, J.A. Effects of treated effluent on a natural marsh. *J. Water Pollut. Control Fed.* **1979**, *51*, 2243–2256.
4. Tilton, D.L.; Kadlec, R.H. The utilization of fresh-water wetland for nutrient removal from secondarily treated waste water effluent. *J. Environ. Qual.* **1979**, *8*, 328–334.
5. Ewel, K.C., Odum, H.T., Eds.; *Cypress Swamps*; University of Florida Press: Gainesville, FL, USA, 1984.
6. Mierau, R.; Trimble, P. *Hydrological Characteristics of the Kissimmee River Floodplain Boney Marsh Experimental Area*; Technical Memorandum. South Florida Water Management District: West Palm Beach, FL, USA, 1988.
7. Moustafa, M.Z.; Chimney, M.J.; Fontaine, T.D.; Shih, G.; Davis, S. The response of a freshwater wetland to long-term "low level" nutrient loads—Marsh efficiency. *Ecol. Eng.* **1996**, *7*, 15–33.
8. Murkin, H.R., van der Valk, A.G., Clark, W.R., Eds.; *Prairie Wetland Ecology*; Iowa State University Press: Ames, IA, USA, 2000.

9. USEPA (United Sates Environmental Protection Agency). Orlando, FL (Orlando Easterly Wetlands Reclamation Project). *Constructed Wetlands for Wastewater Treatment and Wildlife Habitat: 17 Case Studies*; EPA 832/R-93/005. U.S. EPA Office of Research and Development: Washington, DC, USA, 1993; pp. 95–106.
10. Schwartz, L.N.; Wallace, P.M.; Gale, P.M.; Smith, W.F.; Wittig, J.L.; McCarty, S.L. Orange County Eastern Service Area reclaimed water reuse system. *Water Sci. Technol.* **1994**, *29*, 273–282.
11. Pomogi, P. Nutrient retention by the Kis-Balaton Water Protection System. *Hydrobiologia* **1993**, *251*, 309–320.
12. Kadlec, R.H. Phosphorus removal in emergent free surface wetlands. *J. Environ. Sci. Health A* **2005**, *40*, 1293–1306.
13. Trepel, M.; Palmeri, L. Quantifying nitrogen retention in surface flow wetlands for environmental planning at the landscape scale. *Ecol. Eng.* **2002**, *19*, 127–140.
14. Kadlec, R.H. Nitrate removal in treatment marshes. *Crit. Rev. Environ. Sci. Technol.* **2012**, *42*, 934–1005.
15. Kadlec, R.H.; Wallace, S.D. *Treatment Wetlands*, 2nd ed.; CRC Press: Boca Raton, FL, USA, 2009.
16. Kadlec, R.H. The inadequacy of first order treatment wetland models. *Ecol. Eng.* **1999**, *15*, 105–120.
17. Stein, O.R.; Hook, P.B.; Beiderman, J.A.; Allen, W.C.; Borden, D.J. Does batch operation enhance oxidation in subsurface flow constructed wetlands? *Water Sci. Technol.* **2003**, *48*, 149–156.
18. Eriksson, P.G.; Weisner, S.E.B. Nitrogen removal in a wastewater reservoir: The importance of denitrification by epiphytic biofilms on submersed vegetation. *J. Environ. Qual.* **1997**, *26*, 905–910.
19. Ahn, C.; Mitsch, W.J. Scaling considerations of mesocosm wetlands in simulating large created freshwater marshes. *Ecol. Eng.* **2002**, *18*, 327–342.
20. Juston, J.; DeBusk, T.A. Comment on "Protecting the Florida Everglades wetlands with wetlands: Can stormwater phosphorus be reduced to oligotrophic conditions?" by W. Mitsch *et al. Ecol. Eng.* **2015**.
21. Kadlec, R.H.; Cuvellier, C.; Stober, T. Performance of the Columbia, Missouri treatment wetland. *Ecol. Eng.* **2010**, *36*, 672–684.
22. USEPA (United Sates Environmental Protection Agency). The Lakeland Wetland Treatment System. *Constructed Wetlands for Wastewater Treatment and Wildlife Habitat: 17 Case Studies*; EPA 832/R-93/005. U.S. EPA Office of Research and Development: Washington, DC, USA, 1993; pp. 107–114.
23. Frank, P.; Albano, C.; Bays, J.; Turco, N.L.; de Nat, L.; Rosetto, P.; Zanovello, G. *The Fusina Treatment Wetlands: From Concept through Construction*; IWA Specialist Group on Use of Macrophytes in Water Pollution Control; Newsletter No. 39; IWA Publishing: London, UK, 2011; pp. 14–29.
24. Sim, C.H.; Yusoff, M.K.; Shutes, B.; Ho, S.C.; Mansor, M. Nutrient removal in a pilot and full scale constructed wetland, Putrajaya city, Malaysia. *J. Environ. Manag.* **2008**, *88*, 307–317.
25. Mokry, L.E. North Texas Municipal Water District's East Fork wetland: Initial operational issues and performance evaluation. In Proceedings of the Texas Water Conference, Dallas, TX, USA, 13–16 April 2010; Texas American Water Works Association: Austin, TX, USA, 2010.
26. FDEP (Florida Department of Environmental Protection). *Domestic Wastewater to Wetlands Program BCUD's Viera Wetlands*; FDEP: Tallahassee, FL, USA, 2011. Available online: http://www.dep.state.fl.us/water/wastewater/dom/wetviera.htm (accessed on 30 April 2016).
27. Hazen and Sawyer, PC. Compilation of Benefits and Costs of STA and Reservoir Projects in the South Florida Water Management District. Prepared for the World Wildlife Fund: Hollywood, FL, USA, 2011. Available online: http://www.fresp.org/pdfs/ Compilation%20of%20STA%20and%20REZ%20Benefits%20Costs%20HandS%2011_2011.pdf (accessed on 30 April 2016).

28. Kadlec, R.H.; Bays, J.S.; Mokry, L.E.; Andrews, D.; Ernst, M.R. Performance analysis of the Richland-Chambers treatment wetlands. *Ecol. Eng.* **2011**, *37*, 176–190.
29. Wang, H.; Jawitz, J.W.; White, J.R.; Martinez, C.J.; Sees, M.D. Rejuvenating the largest municipal treatment wetland in Florida. *Ecol. Eng.* **2006**, *26*, 132–146.
30. Dunne, E.J.; Coveney, M.F.; Hoge, V.R.; Conrow, R.; Naleway, R.; Lowe, E.F.; Battoe, L.E.; Wang, Y. Phosphorus removal performance of a large-scale constructed treatment wetland receiving eutrophic lake water. *Ecol. Eng.* **2015**, *79*, 132–142.
31. Ecotech Consultants, Inc. Blue Heron Wetland Treatment System. 2011 Monitoring Report. Report to Titusville City Council, Titusville, FL, 2012. Available online: http://www.titusville.com/SIB/files/13-P-017%20bhwts%202011%20annual%20report.pdf (accessed on 29 April 2016).
32. Gu, B.; Chimney, M.J.; Newman, J.; Nungesser, M.K. Limnological characteristics of a subtropical constructed wetland in south Florida (USA). *Ecol. Eng.* **2006**, *27*, 331–344.
33. SFWMD (South Florida Water Management District); FDEP (Florida Department of Environmental Protection). *2016 South Florida Environmental Report*; South Florida Water Management District: West Palm Beach, FL, USA, 2016.
34. Kadlec, R.H.; Knight, R.L. *Treatment Wetlands*; CRC Press: Boca Raton, FL, USA, 1996.
35. Chimney, M.J.; Goforth, G. History and description of the Everglades Nutrient Removal Project, a subtropical constructed wetland in south Florida (USA). *Ecol. Eng.* **2006**, *27*, 268–278.
36. Kadlec, R.H. Everglades and Agriculture in South Florida. In *The Wetlands Handbook*; Maltby, E., Barker, T., Eds.; Wiley-Blackwell: Chichester, UK, 2009; Chapter 42; pp. 961–982.
37. Dunne, E.J.; Coveney, M.F.; Marzolf, E.R.; Hoge, V.R.; Conrow, R.; Naleway, R.; Lowe, E.F.; Battoe, L.E. Efficacy of a large-scale constructed wetland to remove phosphorus and suspended solids from Lake Apopka, Florida. *Ecol. Eng.***2012**, *42*, 90–100.
38. Wilsnack, M.M.; Welter, D.E.; Montoya, A.M.; Restrepo, J.I.; Obeysekara, J. Simulating flow in regional wetlands with the MODFLOW wetlands package. *JAWRA* **2001**, *37*, 655–674.
39. Dierberg, F.E.; DeBusk, T.A.; Henry, J.L.; Jackson, S.D.; Galloway, S.; Gabriel, M.C. Temporal and Spatial Patterns of Internal Phosphorus Recycling in a South Florida (USA) Stormwater Treatment Area. *J. Environ. Qual.* **2012**, *41*, 1661–1673.
40. Dierberg, F.E.; Juston, J.J.; DeBusk, T.A.; Pietro, K.; Gu, B. Relationship between hydraulic efficiency and phosphorus removal in a submerged aquatic vegetation dominated treatment wetland. *Ecol. Eng.* **2005**, *25*, 9–23.
41. DBE (DB Environmental, Inc. and Milian Swain and Associates). *Baseline Tracer Study: STA-2, Cell 3 Final Report*; Appendix 4–9, 2006 South Florida Environmental Report. South Florida Water Management District: West Palm Beach, FL, USA, 2005.
42. Martinez, C.J.; Wise, W.R. Hydraulic analysis of the Orlando Easterly Wetland. *J. Environ. Eng.-ASCE* **2003**, *129*, 553–560.
43. Keller, C.H.; Bays, J.S. Tracer studies for treatment wetlands. In *Treatment Wetlands for Water Quality Improvement: Quebec 2000 Conference Proceedings*; Pries, J., Ed.; CH2M Hill: Waterloo, ON, Canada, 2002; pp. 173–182.
44. Braskerud, B.C. Factors affecting phosphorus retention in small constructed wetlands treating agricultural non-point source pollution. *Ecol. Eng.* **2002**, *19*, 41–62.
45. Johannesson, K.M.; Kynkäänniemi, P.; Ulén, B.; Weisner, S.E.B.; Tonderski, K.S. Phosphorus and particle retention in constructed wetlands—A catchment comparison. *Ecol. Eng.* **2015**, *80*, 20–31.
46. McCormick, P.V.; O'Dell, M.B.; Shuford, R.B.E., III; Backus, J.G.; Kennedy, W.C. Periphyton responses to experimental phosphorus enrichment in a subtropical wetland. *Aquat. Bot.* **2001**, *71*, 119–139.
47. Pietro, K.; Chimney, M.J.; Steinman, A.D. Phosphorus removal by the Ceratophyllum/periphyton complex in a south Florida (USA) freshwater marsh. *Ecol. Eng.* **2006**, *27*, 290–300.
48. Juston, J.M.; DeBusk, T.A. Evidence and implications of the background phosphorus concentration of submerged aquatic vegetation wetlands in Stormwater Treatment Areas for Everglades restoration. *Water Resour. Res.* **2011**, *47*, W01511.

49. Dierberg, F.E.; DeBusk, T.A.; Jackson, S.D.; Chimney, M.J.; Pietro, K. Submerged aquatic vegetation-based treatment wetlands for removing phosphorus from agricultural runoff: Response to hydraulic and nutrient loading. *Water Res.* **2002**, *36*, 1409–1422.
50. South Florida Water Management District (SFWMD) and Florida Department of Environmental Protection (FDEP). *Everglades Consolidated Reports and South Florida Environmental Reports 2002–2016*; SFWMD: West Palm Beach, FL, USA, 2016.
51. DeBusk, T.A.; Grace, K.A.; Dierberg, F.E.; Jackson, S.D.; Chimney, M.J.; Gu, B. An investigation of the limits of phosphorus removal in wetlands: A mesocosm study of a shallow periphyton-dominated treatment system. *Ecol. Eng.* **2004**, *23*, 1–14.
52. Wetland Solutions, Inc. (WSI); ANAMAR. *STA-1E Periphyton Stormwater Treatment Area (PSTA) Final Report (W912-EP-09-E-0013)*; Prepared for the U.S. Army Corps of Engineers; U.S. Army Corps of Engineers: Washington, DC, USA, 2011.
53. CH2M Hill. *PSTA Research and Demonstration Project, Phases 1, 2 and 3 Summary Report*; South Florida Water Management District: West Palm Beach, FL, USA, 2003.
54. Lakhsman, G. *A Demonstration Project at Humboldt to Provide Tertiary Treatment to the Municipal Effluent Using Aquatic Plants*; SRC Publication No. E-820-11-B-82; Saskatchewan Research Council: Saskatoon, SK, Canada, 1982.
55. Walker, W.W.; Kadlec, R.H. Dynamic Model for Stormwater Treatment Areas, Model Version 2, Documentation Update. 2008. Available online: http://www.wwwalker.net/dmsta/ (accessed on 30 April 2016).
56. Kadlec, R.H. Wastewater treatment at the Houghton Lake wetland: Hydrology and water quality. *Ecol. Eng.* **2009**, *35*, 1287–1311.
57. Pietro, K. *Synopsis of the Everglades Stormwater Treatment Areas, Water Year 1996–2012*; Technical Publication ASB-WQTT-12-001. South Florida Water Management District: West Palm Beach, FL, USA, 2012.
58. Droppo, I.G. Rethinking what constitutes suspended sediment. *Hydrol. Process.* **2001**, *15*, 1551–1564.
59. Noe, G.B.; Childers, D.L. Phosphorus budgets in Everglades wetland ecosystems: The effects of hydrology and nutrient enrichment. *Wetl. Ecol. Manag.* **2007**, *15*, 189–205.
60. Noe, G.B.; Childers, D.L.; Edwards, A.L.; Gaiser, E.; Jayachandran, K.; Lee, D.; Meeder, J.; Richards, J.; Scinto, L.J.; Trexler, J.C.; *et al.* Short-term changes in phosphorus storage in an oligotrophic Everglades wetland ecosystem receiving experimental nutrient enrichment. *Biogeochemistry* **2002**, *59*, 239–267.
61. Larsen, L.G.; Harvey, J.W.; Gregory, B.; Noe, G.B.; Crimaldi, J.P. Predicting organic floc transport dynamics in shallow aquatic ecosystems: Insights from the field, the laboratory, and numerical modeling. *Water Resour. Res.* **2009**, *45*, W01411.
62. Reddy, K.R.; Delaune, R.D.; Debusk, W.F.; Koch, M.S. Long-term nutrient accumulation rates in the Everglades. *Soil Sci. Soc. Am. J.* **1993**, *57*, 1147–1155.
63. Craft, C.B.; Richardson, C.J. Peat accretion and phosphorus accumulation along a eutrophication gradient in the northern Everglades. *Biogeochemistry* **1993**, *22*, 133–156.
64. Bhomia, R.K.; Inglett, P.W.; Reddy, K.R. Soil and phosphorus accretion rates in sub-tropical wetlands: Everglades Stormwater Treatment Areas as a case example. *Sci. Total Environ.* **2015**, *533*, 297–306.
65. Reddy, K.R.; DeLaune, R. *Biogeochemistry of Wetlands: Science and Applications*; CRC Press: Boca Raton, FL, USA, 2008.
66. Kadlec, R.H. Deterministic and stochastic aspects of constructed wetland performance and design. *Water Sci. Technol.* **1997**, *35*, 149–156.
67. Kadlec, R.H. Free surface wetlands for phosphorus removal: The position of the Everglades Nutrient Removal Project. *Ecol. Eng.* **2006**, *27*, 361–379.
68. 2013 South Florida Environmental Report. Available online: http://www.sfwmd.gov/portal/page/portal/pg_grp_sfwmd_sfer/portlet_prevreport/2013_sfer/v1/vol1_table_of_contents.html (accessed on 30 April 2016).
69. Lowe, E.F.; Keenan, L.W. Managing phosphorus-based cultural eutrophication in wetlands: A conceptual approach. *Ecol. Eng.* **1997**, *9*, 109–118.
70. Richardson, C.J. Mechanisms controlling phosphorus retention capacity in freshwater wetlands. *Science* **1985**, *228*, 1424–1427.

71. Richardson, C.J.; Marshall, P.E. Processes controlling movement, storage, and export of phosphorus in a fen peatland. *Ecol. Monogr.* **1986**, *56*, 279–302.
72. 2014 South Florida Environmental Report. Available online: https://issuu.com/southfloridawatermanagement/docs/2014_sfer_at_a_glance (accessed on 30 April 2016).
73. 2015 South Florida Environmental Report. Available online: https://issuu.com/southfloridawatermanagement/docs/sfer_glance2015 (accessed on 30 April 2016).
74. Juston, J.M.; DeBusk, T.A.; Grace, K.A.; Jackson, S.D. A model of phosphorus cycling to explore the role of biomass turnover in submerged aquatic vegetation wetlands for Everglades restoration. *Ecol. Model.* **2013**, *251*, 135–149.
75. Kadlec, R.H. Transport of Phosphorus in Wetlands. In *Phosphorus Biogeochemistry in Subtropical Ecosystems*; Reddy, K.R., O'Connor, G.A., Schelske, C.L., Eds.; Lewis Publishers: Boca Raton, FL, USA, 1999; pp. 355–376.
76. USEPA (United Sates Environmental Protection Agency). *Free Water Surface Wetlands for Wastewater Treatment: A Technology Assessment*; USEPA 832-S-99-002. Office of Water: Washington, DC, USA, 1999.
77. Mitsch, W.J.; Gosselink, J.G. *Wetlands*, 3rd ed.; Wiley: New York, NY, USA, 2000.
78. Crites, R.W.; Middlebrooks, E.J.; Reed, S.C. *Natural Wastewater Treatment Systems*; CRC Press: Boca Raton, FL, USA; New York, NY, USA, 2006.
79. USEPA (United Sates Environmental Protection Agency). *Manual: Constructed Wetlands Treatment of Municipal Wastewaters*; EPA/625/R-99/010. USEPA Office of Research and Development: Cincinnati, OH, USA, 2000.
80. Chimney, M.J.; Gawlik, D.E. The avifauna of constructed treatment wetlands in south Florida used for Everglades restoration. *Fla. Field Nat.* **2007**, *35*, 9–21.
81. Vollenweider, R.A. Input-output models with special reference to the phosphorus loading concept in limnology. *Schweiz. Zeit. Hydrol.* **1975**, *37*, 53–84.
82. Canfield, D.E.; Bachman, R.W. Prediction of total phosphorus concentrations, chlorophyll-a, and Secchi depths in natural and artificial lakes. *Can. J. Fish. Aquat. Sci.* **1981**, *30*, 414–423.
83. Walker, W.W. Design basis for Everglades stormwater treatment areas. *Water Resour. Bull.* **1995**, *31*, 671–685.
84. Kadlec, R.H. Effects of pollutant speciation in treatment wetland design. *Ecol. Eng.* **2003**, *20*, 1–16.
85. Min, J.; Paudel, R.; Jawitz, J.W. Mechanistic biogeochemical model applications for Everglades restoration: A review of case studies and suggestions for future modeling needs. *Crit. Rev. Environ. Sci. Technol.* **2011**, *41*, 489–516.
86. Paudel, R.; Jawitz, J.W. Does increased model complexity improve description of phosphorus dynamics in a large treatment wetland? *Ecol. Eng.* **2012**, *42*, 283–294.
87. HydroQual, Inc. *SFWMD Wetlands Model: Calibration of the Coupled Periphyton/Vegetation Model to WCA2A*; Project SFWD0105; Report to South Florida Water Management District. SFWMD: West Palm Beach, FL, USA, 1997.
88. HydroQual, Inc. *SFWMD Wetlands Model: Calibration of the Coupled Periphyton/Vegetation Model to the ENR*; Report to South Florida Water Management District, Project SFWD0105. SFWMD: West Palm Beach, FL, USA, 1998.
89. Moustafa, M.Z.; Hamrick, J.M. Calibration of the wetland hydrodynamic model to the Everglades Nutrient Removal Project. *Water Qual. Ecosyst. Model.* **2000**, *1*, 141–167.
90. DBE (DB Environmental, Inc.). *Supplemental Technology Standard of Comparison (STSOC) Analysis for Submerged Aquatic Vegetation/Limerock Technology*; Final Report to South Florida Water Management District. SFWMD: West Palm Beach, FL, USA, 2002.
91. SFWMD (South Florida Water Management District). *2003 Everglades Consolidated Report*; SFWMD: West Palm Beach, FL, USA, 2003. Available online: http://my.sfwmd.gov/portal/page/portal/xweb%20about%20us/agency%20reports (accessed on 9 September 2016).
92. CH2M Hill. *PSTA Research and Demonstration Project Phase 1 and 2 Summary Report*; Prepared for South Florida Water Management District. SFWMD: West Palm Beach, FL, USA, 2001. Revised 2002.

93. Munson, R.K.; Roy, S.B.; Gherini, S.A.; MacNeill, A.L.; Hudson, R.J.M.; Blette, V.L. Model prediction of the effects of changing phosphorus loads on the Everglades Protection Area. *Water Air Soil Poll.* **2002**, *134*, 255–273.
94. Walker, W.W., Jr.; Kadlec, R.H. Modeling phosphorus dynamics in Everglades wetlands and stormwater treatment areas. *Crit. Rev. Environ. Sci. Technol.* **2011**, *41*, 430–446.
95. Jawitz, J.W.; Muñoz-Carpena, R.; Muller, S.; Grace, K.A.; James, A.I. *Development, Testing, and Sensitivity and Uncertainty Analyses of a Transport and Reaction Simulation Engine (TaRSE) for Spatially Distributed Modeling of Phosphorus in South Florida Peat Marsh Wetlands*; U.S. Geological Survey Scientific Investigations Report 2008-5029. U.S. Geological Survey: Reston, VA, USA, 2008.
96. Paudel, R.; Min, J.; Jawitz, J.W. Management scenario evaluation for a large treatment wetland using a spatio-temporal phosphorus transport and cycling model. *Ecol. Eng.* **2010**, *36*, 1627–1638.
97. Jin, K.-R.; Ji, Z.-G. An integrated environment model for a constructed wetland—Hydrodynamics and transport processes. *Ecol. Eng.* **2015**, *84*, 416–426.
98. Ji, Z.-G.; Jin, K.-R. An integrated environmental model for a surface flow constructed wetland: Water quality processes. *Ecol. Eng.* **2016**, *86*, 247–261.
99. Jin, K.-R.; Ji, Z.-G.; James, R.T. Three-dimensional water quality and SAV modeling of a large shallow lake. *J. Great Lakes Res.* **2007**, *33*, 28–45.
100. Pietro, K.C.; Ivanoff, D. Comparison of long-term phosphorus removal performance of two large-scale constructed wetlands in South Florida, U.S.A. *Ecol. Eng.* **2015**, *79*, 143–157.
101. Cooke, J.G. Phosphorus removal processes in a wetland after a decade of receiving sewage effluent. *J. Environ. Qual.* **1992**, *21*, 733–739.

INDEX

A

Andosols, 24, 25, 29, 30, 31, 34, 35, 39

B

Bamboo splint, 1, 2, 9

Boreal Wetlands, 203, 204

C

Carbon Dynamics, 204

Clean Water Act, 12, 13, 18, 20

Coastal marsh, 182, 200

Coastal meadows, 182

Connectivity, 11, 12, 21

Constructed wetland, 2, 374

E

ecological footprint, 93, 94, 96, 97, 98, 99, 101, 102, 105

ecological security, 94, 96, 97, 99, 105, 106

embedded, 12, 13, 14, 15, 16, 17, 19, 70, 83, 85, 87, 88

Emissions, 45, 329, 338, 340, 341

Erosion Deposition Model, 70

F

first-order kinetics, 344, 345

G

Geographic isolation, 12

Gleyic, 24, 30, 31, 34, 35, 39

Global Warming, 204

H

Halophytes, 182

horizontal flow, 156, 178, 361

hydraulic conductivity, 26, 48, 49, 50, 51, 53, 60, 64, 67, 161, 166, 174, 178

Hydrology, 12, 20, 132, 133, 134, 135, 200, 244, 301, 347, 361, 405, 413

L

Latent heat fluxes, 112

M

Mire Microtopography, 204

Mulberry-Sided Fish Pond Ecosystem, 138

N

nitrate, 54, 139, 162, 169, 219, 220, 221, 222, 223, 225, 241, 243, 245, 290, 344, 361

Northern Great Plains, 70

numerical simulation, 55, 156

P

Palm silk, 2, 6

phosphorus, 7, 54, 62, 64, 67, 139, 145, 146, 158, 220, 412, 413, 414, 415

plant harvest, 220, 227

R

Revised Universal Soil Loss Equation, 69, 70, 71, 74

S

Salinity, 192, 305, 306, 307, 327

Sedimentation, 70, 88, 90, 325, 326, 384

Sphagnum Mosses, 203, 204, 215, 216

Stable surface flow wetland, 2

Streams, 12, 21, 275

Substrate, 2, 7, 47, 49

Surface energy, 112, 133

T

tracer test, 47, 48, 54

U

Upland, 12, 13, 19

W

Water balance, 112

Wetland draining, 24, 44

Wetland patch, 24